INTERNATIONAL CENTRE FOR MECHANICAL SCIENCES

COURSES AND LECTURES - No. 321

D1364166

PROGRESS IN COMPUTATIONAL ANALYSIS OF INELASTIC STRUCTURES

EDITED BY

E. STEIN
UNIVERSITY OF HANNOVER

SPRINGER - VERLAG WIEN - NEW YORK

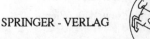

Le spese di stampa di questo volume sono in parte coperte da
contributi del Consiglio Nazionale delle Ricerche.

This volume contains 75 illustrations.

Printed in Italy

In order to make this volume available as economically and as rapidly as possible the authors' typescripts have been reproduced in their original forms. This method unfortunately has its typographical limitations but it is hoped that they in no way distract the reader.

ISBN 3-211-82429-4 Springer Verlag Wien-New York
ISBN 0-387-82429-4 Springer Verlag New York-Wien

PREFACE

The aim of the CISM summer course on "Progress in Computational Analysis of Inelastic Structures" was to teach and to discuss the latest results in computational plasticity including material instabilities and shake down problems. We were successful to get outstanding scientists as lecturers for their special topics.

Due to the proposals of CISM the contributions are published in the alphabetical order of the lecturers, namely P. Perzyna, A. Samuelsson, J.C. Simo, E. Stein and P. Wriggers. In the sequel the five contributions are introduced in the logical order as they were lectured.

At the beginning P. Wriggers gave a comprehensive outline of finite continuous deformation theory of solids and especially the consistent linearization of the weak form of equilibrium which is the starting point for finite element discretizations. For treating incompressible deformations, like in rubber-elasticity or in I_2*-plasticity, three-field-functionals due to Hu-Washizu are adequate. Furthermore iterative solvers for non-linear systems of algebraic equations are discussed, especially algorithms for detecting bifurcation and branch-switching.*

Next, J.C. Simo treated carefully the stable and efficient integration of the plastic flow rule under the condition of plastic incompressibility at finite strains as well as its error analysis. The exponential map algorithm yields the integration of a different first order system of equations and saves in compressibility whereas the radial return algorithm only fulfills the plastic incompressibility condition if the free energy function depends on logarithmic elastic strains. In order to ensure stability in the spatial domain, assumed displacement gradients are added to the isoparametric test functions. By these means advanced tasks of forming processes could be solved in a reliable and efficient way.

After having had a profound theoretical and numerical basis of inelastic deformations, P. Perzyna lectured on thermodynamics of thermo-viscoplastic deformations, and he gave consistent tangents for critical states of stress when material instabilities in the frame of Boltzmann-continua can occur due to jumps of the deformation rates. After those events a thermally driven, practically adiabatic shear band localization can occur. Critical hardening moduli for fully coupled thermodynamical deformations are given analytically and evaluated graphically.

A. Samuelsson then lectured on finite element algorithms for detecting shear band localization within inelastic deformation processes. A crucial issue is the finite element mesh dependence of critical load paths in the sequence of refined meshes. Special algorithms are presented which can fit the process using imperfections and regularizations. A couple of solved problems shows the efficiency and the limits of the method.

The fifth topic on computational plasticity-static shake down load-factors for systems with n-dimensional convex load spaces-was treated by E. Stein. A continuous microlayer model can describe arbitrary non-linear kinematic hardening materials. Alternating failure occurs if hardening does not increase the load factor. Special treatment is devoted to the solution of the inverse problems. Bertsekas-algorithms and an efficient reduced basis method are presented for progressive failure whereas pointwise uncoupled procedures can be used for alternating failure.

In total an integrated methodology for consistent theoretical and numerical models was presented by outstanding scientists in the field, being able to cover a wide range of applications in research and engineering technology.

The editor should like to thank all the lecturers for creating a stimulating atmosphere for the relatively large number of attendancies who were at most well prepared for fruitful discussions.

Computational plasticity is furthermore a fast developing topic, especially with couplings to other engineering problems as contact, forming, forging and ultimate load behaviour.

It is the pleasure of the authors and the editor to thank CISM for granting and supporting the course as well as permitting this publication. We furthermore thank all participants for lively and fruitful discussions.

We do hope that this volume will find a fruitful acceptance.

E. Stein

CONTENTS

CONSTITUTIVE EQUATIONS FOR THERMOINELASTICITY and INSTABILITY PHENOMENA IN THERMODYNAMIC FLOW PROCESSES

P. Perzyna
Polish Academy of Sciences, Warsaw, Poland

1 INTRODUCTION

In recent years it has been observed active research work in the field of the instability phenomena of plastic flow processes. Particularly the localization of plastic deformation along a shear band treated as a prelude to failure initation has been a matter of a great interest.

It has been shown that the onset of lacalization does depend critically on the assumed constitutive law.

RICE in his fundamental work on the localization of plastic deformation (cf. RICE [1976]) wrote: "The present study shows that conditions for localization relate closely to subtle and not well understood features of the constituitve description of plastic flow "...." While the constitutive modelling of these features needs to be improved in relation to the detailed mechanisms of deformation, so also is there need for a fuller assessment of the role of imperfections or initial non-uniformities in material properties in promoting localization. Indeed, the latter approach seems mandatory for rate-dependent plastic flow models and these, as well as range of thermomechanically coupled lacalization phenomena would seem to merit further study"

In the mean time different constititive features have been analyzed and their influence on the onset of localization have been investigated.

Particular attention has been focused on effects as follows:

(i) yield surface vertices (cf. RUDNICKI and RICE [1975] and NEEDLEMAN and RICE [1977]);

(ii) deviation from plastic "normality", i.e. deviation from an associated flow rule (cf. RUDNICKI and RICE [1975], NEEDLEMAN and RICE [1977], RICE and RUDNICKI [1980] and DUSZEK and PERZYNA [1988]);

(iii) the dilatational plastic flow due to nucleation and growth of microvoids (cf. RUDNICKI and RICE [1975], NEEDLEMAN and RICE [1977] and DUSZEK and PERZYNA [1988]);

(iv) strain induced anisotropy modelled as the kinamatic hardening rule (cf. MEAR and HUTCHINSON[1985], TVERGAARD [1987] and DUSZEK and PERZYNA [1988]);

(v) influence of the covariance terms (co-rotational terms) (cf. RUDNICKI and RICE [1975], RICE and RUDNICKI [1980] and LIPPMANN [1986]).

The investigation of the influence of thermo-mechanical coupling effects on the localization phenomenon needs greater attention.

In many technological processes such as plastic shaping and forming, low temperature processes, dynamic fragmentation and high velocity machining thermal effects may have determinated influence on the localization phenomenon.

Thermal effects may have more essetial influence on formation of shear bands in dynamic loading processes when the haet that is produced during plastic deformation is given insufficient time to be conducted away. Then the process considered is adiabatic and localization occurs more readily.

Experimental results which confirmed such conjecture have been recently reported by HARTLEY, DUFFY and HAWLEY [1987], MARCHAND and DUFFY [1988] and MARCHAND, CHO and DUFFY [1988].

Experimental investigations of HARTLEY, DUFFY and HAWLEY [1987] were performed under dynamic loading conditions for AISI 1018 cold rolled steel and AISI 1020 hot rolled steel. They observed that the formation of shear band for both steels is influenced very much by thermal effects. They showed that even small increase of local temperature (about 50–80°C) for 1018 CRS steel enhances local deformation and local heating, and as a result, causes initation of shear band formation. Similar results have been reported for a low alloy structural steel (HY-100) by MARCHAND and DUFFY [1988].

The main objective of the first part of the paper is a review of the investigations of the influence of thermo-mechanical couplings and thermal softening effects on adiabatic shear band localization criteria for finite rate independent deformation of an elastic-plastic body, cf. DUSZEK and PERZYNA [1991].

In chapter 2 a general internal state variable framework is presented. Particular attention is focused on the discussion of the spatial covariance constitutive structure.

In chapter 3 the constitutive equations for thermo-elastic-plasstic J_2 – flow theory are formulated within a framework of the rate type covariance structure with internal state variables. Two alternative descriptions of thermo-mechanical couplings for J_2 – flow thery are presented. In the first the Lie derivative is used to define objective rate of the Kirchhoff stress tensor, while in the second the Zaremba–Jaumann rate is utilized. Both constitutive structures formulated are invariant with respect to diffeomorphisms and are materially isomorphic.

Attention is focused on the coupling phenomena generated by the internal heat resulting from internal dissipation. By applying the Legendre transformation and basing on the careful analysis of the internal dissipation during plastic flow the identification procedure is developed. This method of identification permits to determine exact form of the evolution equation for the internal state variable vector which is responsible for dissipative nature of plastic flow phenomena.

Chapter 7 is devoted to the investigation of adiabatic process. A set of the coupled evolution equation for the Kirchhoff stress tensor and for temperature is investigated. The method has been developed which permits to obtain the fundamental rate equation for the Kirchhoff stress tensor. The matrix in this equation describes all thermo-mechanical couplings introduced. This important result allows to use the standard bifurcation method in examination of the adiabatic shear band localization criteria.

Main contribution to thermo-machanical couplings has been carefully discussed. Basing on this analysis the simplified evolution equation for the temperature is obtained.

The predictions, by applications of the localization criterion, are given in section 8.4.

The procedure has been developed which allows to discuss two separate new effects on the localization phenomenon along a shear band. One is thermo-mechanical coupling when there is no spatial covariance effects and other is the spatial covariance effects for assumed isothermal process.

For both seperate cases the criteria for adiabatic shear band localization are obtained in exact analytical form.

The influence of two important thermal effects, namely thermal expansion and thermal plastic softening on the criteria for localization of plastic deformation is investigated. Similar influence of the spatial covariance terms is also examined. Discussion of the results obtained is presented in section 8.5 .

It has been proved that thermal expansion implies that the inclination to instability for the axially symmetric compression is different from that for the axially symmetric tension. For tension the material is more sensitive to localization than for compression. This result is in agreement with experimental observations of the initiation of localized shear bands in maraging steel under condition of quasi-static plane strain, of ANAND and SPITZIG [1980]. Thermal plastic softening causes that material is more inclined to instability independently on the state of stress. It has been shown that even small increase of temperature for adiabatic process implies the plastic softening which becomes crucial for the initiation of the mechanism of localization. This result coincides with the conclusion drove from the experimental observations performed under dynamic loading conditions for steels reported by HARTLEY, DUFFY and HAWLEY [1987], MARCHAND and DUFFY [1988] and MARCHAND, CHO and DUFFY [1988].

Discussion of the influence of the covariance terms which arise from the difference between the Lie derivative and the material rate of the Kirchhoff strees tensor on the localization criteria is also presented. It has been shown that the influence of the covariance terms is important and may play a dominated role when the inception of localization phenomenon is expected to take place for small values of the rate of hardening modulus H, that is near to the maximum stress attained during the process. This result supports physically and experimentally justified conjecture.

The kinamatic hardening effect and softening effect generated by the micro-damage process are described by means of the internal state variable method in sections 3.6 and 3.7, respectively.

General constitutive equations for damaged solids are formulated in section 3.8 (cf. DUSZEK-PERZYNA, PERZYNA and STEIN [1989] and STEIN, DUSZEK-PERZYNA and PERZYNA [1989]).

Chapter 4 brings the discussion of constitutive properties of crystal. Kinematic of finite deformation is discussed. Main mechanisms on crystalline slip systems are investigated. This study gives the basis for derformination of the evolution equations for the internal state variables introduced. Hardening and softening effects are also described. The thermodynamic theory for elastic-viscoplastic crystal is developed within a framework of the rate type material structure with internal state variables. Thermomechanical couplings are investigated and the heat conduction equation coupled with the inelastic deformation process is obtained. Rate independent response of crystal is discussed as a limit case of the general rate dependent inelastic behaviour.

Chapter 5 is focused on the development of the rate dependent inelastic constitutive material structure for polycrystalline solids. The viscoplastic evolution law is postulated. The plastic potential is assumed different than the quasi-static yield criterion. Isotropic and kinematic hardening effects are described within a framework of the internal state variable method. The evolution law for the anisotropic hardening is postulated as the linear combination of the PRAGER and ZIEGLER rules. It has been found that making use of the simple geometrical relation permits to determine the coefficients in the evolution equation describing the kinamatic hardening. The procedure used is consistent with all constitutive requirements and leaves a room for the identification of matrial constants basing on available experimental results. Particular attention has been attached to the investigation of thermodynamic restrictions and thermo-mechanical couplings. Rate type constitutive equation is obtained. By performing a Legendre transformation the identification procedure for the material functions involved in the evolution equations is developed.

Rate independent thermo-plasticity theory of polycrystalline solids is studied in section 5.5 as a limint case of the general elastic-viscoplastic response. Alternative formulation of thermo-viscoplasticity is presented in chapter 6. Sections 7.4 and 7.5 are devoted to the analysis of adiabatic process for an elastic-viscoplastic model. The governing equations for elastic-viscoplastic solids in adiabatic process have been formulated. The results obtained allows to use in the investigation of the conditions for localization along shear bands the standard bifurcation method, cf. chapter 9.

2 INTERNAL STATE VARIABLE FRAMEWORK

2.1 Basic assumptions and definitions

Let assume that a continuum body is an open bounded set $\mathcal{B} \subset \mathbb{R}^3$, and let $\phi : \mathcal{B} \to \mathcal{S}$ be a C^1 configuration of $\mathcal{B}$ in $\mathcal{S}$. The tangent of ϕ is denoted $\mathbf{F}$ and is called the deformation gradient of ϕ .

Let $\{X^A\}$ and $\{x^a\}$ denote coordinate systems on $\mathcal{B}$ and $\mathcal{S}$, respectively. Then we refere to $\mathcal{B} \subset \mathbb{R}^3$ as the reference configuration of a continuum body with particles $\mathbf{X} \in \mathcal{B}$ and to $\mathcal{S} = \phi(\mathcal{B})$ as the current configuration with points $\mathbf{x} \in \mathcal{S}$. The matrix $\mathbf{F}(\mathbf{X}, t) = \frac{\partial \phi(\mathbf{X},t)}{\partial \mathbf{X}}$ with respect to the coordinate bases $\mathbf{E}_A(\mathbf{X})$ and $\mathbf{e}_a(\mathbf{x})$ is given by

$$F^a_A(\mathbf{X}, t) = \frac{\partial \phi^a}{\partial X^A}(\mathbf{X}, t), \tag{2.1}$$

where a mapping $\mathbf{x} = \phi(\mathbf{X}, t)$ represents a motion of a body $\mathcal{B}$.

In a neighbourhood of $\mathbf{X}$, i.e. in $\mathcal{N}(\mathbf{X})$ for every $\mathbf{X} \in \mathcal{B}$ we consider the local multiplicative decomposition

$$\mathbf{F} = \mathbf{F}^e \cdot \mathbf{F}^p, \tag{2.2}$$

where $(\mathbf{F}^e)^{-1}$ is the deformation gradient that release elastically the stress on the neighbourhood $\phi(\mathcal{N}(\mathbf{X}))$ in the current configuration.

Let define the total and elastic Finger deformation tensors

$$\mathbf{b} = \mathbf{F} \cdot \mathbf{F}^T, \quad \mathbf{b}^e = \mathbf{F}^e \cdot \mathbf{F}^{e^T}, \tag{2.3}$$

respectively, and the Eulerian strain tensors as follows

$$\mathbf{e} = \frac{1}{2}(\mathbf{g} - \mathbf{b}^{-1}), \quad \mathbf{e}^e = \frac{1}{2}(\mathbf{g} - \mathbf{b}^{e^{-1}}), \tag{2.4}$$

where $\mathbf{g}$ denotes the metric tensor in the current configuration.

By definition

$$\mathbf{e}^p = \mathbf{e} - \mathbf{e}^e = \frac{1}{2}(\mathbf{b}^{e^{-1}} - \mathbf{b}^{-1}) \tag{2.5}$$

we introduce the plastic Eulerian strain tensor.

In many fields of mechanics and particularly, in continuum mechanics very important rule plays the Lie derivative [1]. The Lie derivative of a spatial tensor field **t** with respect to the velocity field $\boldsymbol{v}$ can be defined as

$$L_{\boldsymbol{v}}\mathbf{t} = \phi_* \frac{\partial}{\partial t}(\phi^* \mathbf{t}) \tag{2.6}$$

where ϕ^* and ϕ_* denote the pull-back and push-forward operations, respectively.

We have the rates of defomation as follows

$$\begin{aligned} \mathbf{d} &= L_{\boldsymbol{v}}\mathbf{e} = \frac{1}{2}L_{\boldsymbol{v}}\mathbf{g} \\ d_{ab} &= \frac{1}{2}(L_{\boldsymbol{v}}\mathbf{g})_{ab} = \frac{1}{2}(g_{ac}v^c\mid_b + g_{cb}v^c\mid_a), \end{aligned} \tag{2.7}$$

where

$$v^a\mid_b = \frac{\partial v^a}{\partial x^b} + \gamma^a_{bc}v^c \tag{2.8}$$

and γ^a_{bc} denotes the Christoffel symbol for the general coordinate system $\{x^a\}$.

Similarly the rate of plastic deformation is given by

$$\mathbf{d}^p = L_{\boldsymbol{v}}\mathbf{e}^p = \frac{1}{2}L_{\boldsymbol{v}}(\mathbf{b}^{e^{-1}}) \tag{2.9}$$

and

$$\mathbf{d} = \mathbf{d}^e + \mathbf{d}^p. \tag{2.10}$$

For any scalar function we have

$$L_{\boldsymbol{v}}f = \dot{f} = \frac{\partial f}{\partial t} + \frac{\partial f}{\partial x^a}v^a. \tag{2.11}$$

Let $\boldsymbol{\tau}$ denote the Kirchhoff stress tensor related to the Cauchy stress tensor $\boldsymbol{\sigma}$ by

$$\boldsymbol{\tau} = J\boldsymbol{\sigma} = \frac{\rho_{Ref}}{\rho}\boldsymbol{\sigma} \tag{2.12}$$

where $\rho_{Ref}(\mathbf{X})$ and $\rho(\mathbf{x}, t)$ denote the mass density in the reference and current configuration, respectively, and the Jacobian J is the determinant of the linear transformation $\mathbf{F}(\mathbf{X}, t) = \frac{\partial}{\partial \mathbf{X}}\phi(\mathbf{X}, t)$.

It is noteworthy that any posible objective rate of stress tensor is a particular case of the Lie derative (cf. MARSDEN and HUGHES [1983]).

---

[1]For precise definition of the Lie derivative and its algebraic and dynamic interpretations please consult ABRAHAM, MARSDEN and RATIU [1988]. Aplication of the Lie derivative to theoretical mechanics may be found in ABRAHAM and MARSDEN [1978] and to continuum mechanics in MARSDEN and HUGHES [1983].

The Lie derivative of the Kirchhoff stress tensor $\boldsymbol{\tau}$ (provided we have only contravariant coordinate representation τ^{ab} in mind) gives

$$L_v\boldsymbol{\tau} = \phi_*\frac{\partial}{\partial t}(\phi^*\boldsymbol{\tau}) = \left\{\mathbf{F}\cdot\frac{\partial}{\partial t}[\mathbf{F}^{-1}\circ(\boldsymbol{\tau}\cdot\phi)\cdot\mathbf{F}^{-1^T}]\cdot\mathbf{F}^T\right\}\circ\phi^{-1} = \dot{\boldsymbol{\tau}}-(\mathbf{d}+\boldsymbol{\omega})\cdot\boldsymbol{\tau}-\boldsymbol{\tau}\cdot(\mathbf{d}+\boldsymbol{\omega})^T, \tag{2.13}$$

where $\circ$ denotes the composition of mappings and $\boldsymbol{\omega}$ the spin tensor is defined as follows

$$\omega_{ab} = \frac{1}{2}(g_{ac}v^c\,|_b - g_{cb}v^c\,|_a). \tag{2.14}$$

In coordinate system (2.13) reads

$$(L_{\boldsymbol{v}}\boldsymbol{\tau})^{ab} = F^a_A\frac{\partial}{\partial t}(F_c^{-1A}\tau^{cd}F_d^{-1B})F^b_B = \frac{\partial\tau^{ab}}{\partial t} + \frac{\partial\tau^{ab}}{\partial x^c}v^c - \tau^{cb}\frac{\partial v^a}{\partial x^c} - \tau^{ac}\frac{\partial v^b}{\partial x^c}. \tag{2.15}$$

Equation (2.15) defines the Oldroyd rate of the Kirchhoff stress tensor $\boldsymbol{\tau}$ (cf. OLDROYD [1950]).

The Zaremba–Jaumann stress rate or the co-rotated derivative of the Kirchhoff stress tensor $\boldsymbol{\tau}$ is defined as

$$\overset{\nabla}{\boldsymbol{\tau}} = \dot{\boldsymbol{\tau}} - \boldsymbol{\omega}\cdot\boldsymbol{\tau} + \boldsymbol{\tau}\cdot\boldsymbol{\omega}. \tag{2.16}$$

Comparison of (2.16) with (2.13) gives the result

$$L_{\boldsymbol{v}}\boldsymbol{\tau} = \overset{\nabla}{\boldsymbol{\tau}} - \mathbf{d}\cdot\boldsymbol{\tau} - \boldsymbol{\tau}\cdot\mathbf{d}. \tag{2.17}$$

The result (2.17) will be used in formulation of the alternative constitutive descriptions of thermo-elastic-plastic response of solids.

2.2 Objective constitutive structure

To describe the dissipation phenomena during a thermo-plastic flow process we have to introduce the internal state vector $\boldsymbol{\mu} \in V_n$, where V_n is n-dimensional vector space.

The intrinsic state of a particle $\mathbf{X} \in \mathcal{B}$ at time t is determined by a set of variables

$$s = (\mathbf{e}, \mathbf{F}, \vartheta; \boldsymbol{\mu}), \tag{2.18}$$

where $\mathbf{e}$ is the Eulerian strain tensor, $\mathbf{F}$ is the deformation gradient, ϑ denotes absolute temperature and $\boldsymbol{\mu}$ is the internal state vector.

It is postulated that there exists the free energy function

$$\psi = \widehat{\psi}(\mathbf{e}, \mathbf{F}, \vartheta; \boldsymbol{\mu}), \tag{2.19}$$

and the evolution equation for the internal state vector $\boldsymbol{\mu}$ is assumed in the form

$$L_{\boldsymbol{v}}\boldsymbol{\mu} = \widehat{\mathrm{m}}(s). \tag{2.20}$$

In the theory which we intend to develop the free energy function (2.19) and evolution equation (2.20) will play a fundamental rule.

To investigate a notion of objectivity let us consider a superposed rigid body motion given by a map (cf. TRUESDELL and NOLL [1965])

$$\mathbf{x}^{\#} = \mathbf{c}(t) + \mathbf{Q}(t) \cdot \mathbf{x} \tag{2.21}$$

or

$$\phi^{\#}(\mathbf{X}, t) = \mathbf{c}(t) + \mathbf{Q}(t) \cdot \phi(\mathbf{X}, t),$$

where $\mathbf{c}(t)$ is a vector fuction of time and $\mathbf{Q}(t)$ is a time dependent, proper orthogonal transformation.

The deformation gradient for a new motion is given by

$$\mathbf{F}^{\#} = \frac{\partial}{\partial \mathbf{X}} \phi^{\#} = \mathbf{Q}(t) \cdot \mathbf{F}. \tag{2.22}$$

A spatial tensor field is said to transform objectively under superposed rigid body motions if it transforms according to the standard rules of tensor analysis.

Similarly we can consider any superposed spatial diffeomorphism $\boldsymbol{\xi} : \mathbb{R}^3 \to \mathbb{R}^3$. This gives

$$\phi^{\#} = \boldsymbol{\xi} \circ \phi : \Gamma \to \mathbb{R}^3 \tag{2.23}$$

with deformation gradient

$$\mathbf{F}^{\#} = \frac{\partial \boldsymbol{\xi}}{\partial \mathbf{x}} \cdot \mathbf{F}, \tag{2.24}$$

where $T\boldsymbol{\xi} = \partial\boldsymbol{\xi}/\partial\mathbf{x}$ is the associated relative deformation gradient.

Generally, any spatial tensor field $\mathbf{t}$ is said to transform objectively under superposed diffeomorphism $\boldsymbol{\xi}$ if it transforms according to the rule

$$\mathbf{t}^{\#} = \boldsymbol{\xi}_* \mathbf{t}, \tag{2.25}$$

where $\boldsymbol{\xi}_*$ is the push-forward operation.

Let $\mathbf{t}$ be a given time dependent spatial tensor field on $\mathcal{S}$ and let $\mathbf{t}$ transform objectively. Let $\boldsymbol{v}^{\#}$ be the velocity field of $\phi^{\#}$. Then we have (cf. MARSDEN and HUGHES [1983])

$$L_{\boldsymbol{v}^{\#}} \mathbf{t}^{\#} = \boldsymbol{\xi}_*(L_{\boldsymbol{v}} \mathbf{t}). \tag{2.26}$$

It means that objective tensors have objective Lie derivative.

There are two levels of objectivity for the constitutive structures.

(i) Frame invariance. The constitutive structure should be invariant with respect to superposed rigid body motions (cf. TRUESDELL and NOLL [1965]).

(ii) Spatial covariance. The constitutive structure should be invariant with respect to diffeomorphisms (cf. MARSDEN and HUGHES [1983]).

The frame invariance leads to objectivity with respect to isometries, while the notion of spatial covariance furnishes the constitutive structure with objectivity property with respect to diffeomorphisms.

It is worthwhile to cinsider two important particular cases (cf. SIMO [1988]).

1. Let $\boldsymbol{\xi} : I\!R^3 \rightarrow I\!R^3$ be an isometry (rigid body motion), then the requirement of spatial covariance reduces to the frame invariance conditions.

2. By choosing $\boldsymbol{\xi} : I\!R^3 \rightarrow I\!R^3$ as the inverse of actual motion, i.e. $\boldsymbol{\xi} = \phi^{-1}$ we have

$$\boldsymbol{\xi}_* = \phi^*. \tag{2.27}$$

Then

$$\begin{aligned} \mathbf{e}^\# &= \phi^*\mathbf{e} = \mathbf{E}, \\ \mathbf{F}^\# &= \mathbf{F}^{-1}\cdot\mathbf{F} = \mathbf{1}, \\ \boldsymbol{\mu}^\# &= \phi^*\boldsymbol{\mu} = \mathbf{M}, \end{aligned} \tag{2.28}$$

where $\mathbf{E}$ is the Lagrangian strain tensor and $\mathbf{M}$ denotes the internal state vector in the material description. Finally (2.19) reduces to the free energy function

$$\psi = \hat{\psi}(\mathbf{E}, \Theta; \mathbf{M}) \tag{2.29}$$

for the material setting.

This result has very important consequence for the constitutive modeling.

In further considerations we shall take advantage of both notions of objectivity.

3 THERMODYNAMICS OF RATE INDEPENDENT PLASTICITY

3.1 Flow rule and isotropic hardening

Let us introduce the yield criterion in the form as follows

$$f - \kappa = 0, \tag{3.1}$$

where the yield function f is assumed as

$$f = \hat{f}(\boldsymbol{\tau}, \mathbf{g}) = J_2 = \frac{1}{2}\tau'^{ab}\tau'^{cd}g_{ac}g_{bd} \tag{3.2}$$

and the work-hardening-softening function κ is postulated in the form (cf. NEMES et al.[1989])

$$\kappa = \hat{\kappa}(\epsilon^p, \vartheta) = [\kappa_1 + (\kappa_0 - \kappa_1)e^{-h_1(\vartheta)\epsilon^p}]^2(1 - \omega\bar{\vartheta}), \tag{3.3}$$

where κ_0 is a material constant related to the initial yield stress, κ_1 is the saturation hardening stress, $h_1 = h_1(\vartheta)$ is a temperature dependent hardening function, ϵ^p denotes the equivalent plastic deformation

$$\epsilon^p = \int_0^t (\frac{2}{3}\mathbf{d}^p : \mathbf{d}^p)^{\frac{1}{2}} dt' \tag{3.4}$$

and

$$\bar{\vartheta} = \frac{\vartheta - \vartheta_0}{\vartheta_0}, \qquad \omega = \text{const.} \tag{3.5}$$

The function κ in the form (3.3) describes the saturation of the hardening of the material as the plastic deformation progresses.

Linear approximation of the function $e^{-h_1(\vartheta)\epsilon^p}$ gives

$$\kappa = \kappa_0^2[1 + h(\vartheta)\epsilon^p]^2(1 - \omega\bar{\vartheta}), \tag{3.6}$$

where

$$h(\vartheta) = \left(\frac{\kappa_1}{\kappa_0} - 1\right) h_1(\vartheta). \tag{3.7}$$

The flow rule is postulated in the form

$$\mathbf{d}^p = \Lambda \mathbf{P}, \tag{3.8}$$

where

$$\mathbf{P} = \frac{1}{2\sqrt{J_2}}\frac{\partial \hat{f}}{\partial \boldsymbol{\tau}} = \left(\frac{\boldsymbol{\tau}'}{2\sqrt{J_2}}\right)^{\flat} \tag{3.9}$$

and the symbol $\flat$ denotes the index lowering operator.

Fulfilment of the consistency condition $\dot{f} - \dot{\kappa} = 0$ gives

$$\Lambda = \langle \frac{1}{H}(\mathbf{P} : \dot{\boldsymbol{\tau}} + \pi\dot{\vartheta})\rangle, \tag{3.10}$$

where the symbol $\langle (x) \rangle$ defines the ramp function

$$\langle (x) \rangle = \frac{x + |x|}{2}, \tag{3.11}$$

and is used to express the loading/unloading criterion, the isotropic hardening modulus H is determined by

$$H = \frac{1}{2\sqrt{3J_2}}\frac{\partial \hat{\kappa}}{\partial \epsilon^p} = \frac{h_1(\vartheta)(\kappa_1 - \kappa_0)}{\sqrt{3J_2}}\left[\kappa_1 + (\kappa_0 - \kappa_1)e^{-h_1(\vartheta)\epsilon^p}\right](1 - \omega\bar{\vartheta})e^{-h_1(\vartheta)\epsilon^p}, \tag{3.12}$$

$$\begin{aligned} \pi &= -\frac{1}{2\sqrt{J_2}}\frac{\partial \hat{\kappa}}{\partial \vartheta} = \frac{1}{2\sqrt{J_2}}\left[\kappa_1 + (\kappa_0 - \kappa_1)e^{-h_1(\vartheta)\epsilon^p}\right]\left\{\frac{\omega}{\vartheta_0}\left[\kappa_1 + (\kappa_0 - \kappa_1)e^{-h_1(\vartheta)\epsilon^p}\right]\right. \\ &\left. + 2(\kappa_0 - \kappa_1)\frac{dh_1(\vartheta)}{d\vartheta}\epsilon^p(1 - \omega\bar{\vartheta})e^{-h_1(\vartheta)\epsilon^p}\right\}. \end{aligned} \tag{3.13}$$

Finally the flow rule takes the form as follows

$$\mathbf{d}^p = \langle \frac{1}{H}(\mathbf{P} : \dot{\boldsymbol{\tau}} + \pi\dot{\vartheta})\rangle \mathbf{P}. \tag{3.14}$$

It is noteworthy that our consideration is valid for any material funktion $\kappa = \widehat{\kappa}(\epsilon^p, \vartheta)$. More, we can also postulate evolution equation for κ, e.g.,

$$\dot{\kappa} = \widehat{h}_1(s)(\mathbf{d}^p : \mathbf{d}^p)^{\frac{1}{2}} + \widehat{h}_2(s)\dot{\vartheta} \tag{3.15}$$

then

$$H = \widehat{h_1}(s)\frac{1}{2\sqrt{2J_2}} \tag{3.16}$$

and

$$\pi = -\frac{1}{2\sqrt{J_2}}\widehat{h}_2(s). \tag{3.17}$$

3.2 Evolution equation for the internal state vector

It is reasonable to present here a brief discussion of main features of a rate indepe-dent plastic model of a material.

The first important property is connected with permanent deformations. This is the result of different paths assumed for loading and unloading processes. Unloading process starting from achieved elastic-plastic state follows a path in the stress space different from that of loading process.

The second feature of plastic model is its time independency. So, the constitutive equations as well as the evolution equations for an elastic-plastic material have to be invariant under the time scale changes.

These both features, namely the occurrence of permanent deformations and time independent behaviour of a material are characteristic for inviscid plastic models.

The internal state vector $\boldsymbol{\mu}$ is introduced to describe dissipation effects occurring during thermo-plastic flow process. The evolution equation for the internal state vector $\boldsymbol{\mu}$ postulated in the form (2.20) has to satisfy these two main properties of a plastic model.

To fulfil this requirement we have to assume

$$L_{\boldsymbol{v}}\boldsymbol{\mu} = \mathrm{m}(s)\langle\frac{1}{H}(\mathbf{P} : \dot{\boldsymbol{\tau}} + \pi\dot{\vartheta})\rangle, \tag{3.18}$$

where the material function $\mathrm{m}(s)$ remains to be determined.

3.3 Thermodynamic restrictions

Consider balance principles as follows:

(i) Conservation of mass. Let assume that $\phi(\mathbf{X}, t)$ is a C^1 regular motion. A function $\rho(\mathbf{x}, t)$ is said to obey conservation of mass if

$$\dot{\rho} + \rho \operatorname{div}\boldsymbol{v} = 0 \tag{3.19}$$

or

$$\rho(\mathbf{x}, t)J(\mathbf{X}, t) = \rho_{Ref}(\mathbf{X}). \tag{3.20}$$

(ii) Balance of momentum. Assume that conservation of mass and balance of momentum hold. If there is no external body force field, then

$$\rho \dot{\boldsymbol{v}} = \mathrm{div}\boldsymbol{\sigma}. \tag{3.21}$$

(iii) Balance of moment of momentum. Let conservation of mass and balance of momentum hold. Then balance of moment of momentum holds if and olny if $\boldsymbol{\tau}$ is symmetric.

(iv) Balance of energy. Assume the following balance principles hold: conservation of mass, balance of momentum, balance of moment of momentum and balance of energy. If there is no external heat supply then

$$\rho(\dot{\psi} + \dot{\vartheta}\eta + \eta\dot{\vartheta}) + \mathrm{div}\mathbf{q} = \frac{\rho}{\rho_{Ref}}\boldsymbol{\tau} : \mathbf{d}, \tag{3.22}$$

where η denotes the specific (per unit mass) entropy and $\mathbf{q}$ is the heat vector field.

(v) Entropy production inequality. Assume consertvation of mass, balance of momentum, moment of momentum, energy and the entropy production inequality hold. Then the reduced dissipation inequality is satisfied:

$$\frac{1}{\rho_{Ref}}\boldsymbol{\tau} : \mathbf{d} - (\eta\dot{\vartheta} + \dot{\psi}) - \frac{1}{\rho\vartheta}\mathbf{q} \cdot \mathrm{grad}\vartheta \geq 0. \tag{3.23}$$

Let introduce <u>the axiom of entropy production</u>: For any regular motion of a body $\mathcal{B}$ the constitutive functions are assumed to satify the reduced dissipation inequality (3.23).

Then the constitutive assumption (2.19) and the evolution equation for the internal state vector $\boldsymbol{\mu}$ (3.18) together with the reduced dissipation inequality (2.23) lead to the results as follows

$$\boldsymbol{\tau} = \rho_{Ref}\frac{\partial\hat{\psi}}{\partial\mathbf{e}}, \qquad \eta = -\frac{\partial\hat{\psi}}{\partial\vartheta},$$
$$-\frac{\partial\hat{\psi}}{\partial\boldsymbol{\mu}} \cdot L_{\boldsymbol{v}}\boldsymbol{\mu} - \frac{1}{\rho\vartheta}\mathbf{q} \cdot \mathrm{grad}\vartheta \geq 0. \tag{3.24}$$

Let define the rate of internal dissipation by

$$\vartheta\hat{i} = -\frac{\partial\hat{\psi}}{\partial\boldsymbol{\mu}} \cdot L_{\boldsymbol{v}}\boldsymbol{\mu} = -\frac{\partial\hat{\psi}}{\partial\boldsymbol{\mu}} \cdot \mathbf{m}(s)\langle\frac{1}{H}(\mathbf{P} : \dot{\boldsymbol{\tau}} + \pi\dot{\vartheta})\rangle. \tag{3.25}$$

Equation (3.25) expresses very important feature of thermo-plastic response of a material, namely that the rate of internal dissipation occurs only during loading process.

3.4 Rate type constitutive relation (for J_2 – flow theory)

Operating on the stress relation $(3.24)_1$ with the Lie derivative, keeping the history constant (the internal state vector constant), we obtain

$$L_{\upsilon}\boldsymbol{\tau} = \mathcal{L}^e \cdot \mathbf{d}^e - \mathcal{L}^{th}\dot{\vartheta} \tag{3.26}$$

where

$$\mathcal{L}^e = \rho_{Ref}\frac{\partial^2\widehat{\psi}}{\partial \mathbf{e}^2}, \quad \mathcal{L}^{th} = -\rho_{Ref}\frac{\partial^2\widehat{\psi}}{\partial \mathbf{e}\partial\vartheta} \tag{3.27}$$

denote the elastic moduli and the thermal stress coefficients, respectively.

Let generalize the relation (3.26) for an elasto-plastic flow process. Then taking into account (2.10) we can write

$$L_{\upsilon}\boldsymbol{\tau} = \mathcal{L}^e \cdot (\mathbf{d} - \mathbf{d}^p) - \mathcal{L}^{th}\dot{\vartheta}. \tag{3.28}$$

Notice that in view of Eqs (2.13) and (2.16) the flow rule (3.14) can be written accordingly in two equivalent forms:

$$\mathbf{d}^p = \langle\frac{1}{H}[\mathbf{P} : (L_{\upsilon}\boldsymbol{\tau} + \mathbf{d}\cdot\boldsymbol{\tau} + \boldsymbol{\tau}\cdot\mathbf{d}) + \pi\dot{\vartheta}]\rangle\mathbf{P} \tag{3.29}$$

or

$$\mathbf{d}^p = \langle\frac{1}{H}(\mathbf{P} : \overset{\nabla}{\boldsymbol{\tau}} + \pi\dot{\vartheta})\rangle\mathbf{P}. \tag{3.30}$$

Substituting (3.29) into (3.28) yields the evolution equation for the Kirchhoff stress tensor $\boldsymbol{\tau}$ in the form as follows

$$L_{\upsilon}\boldsymbol{\tau} = \mathcal{L}\cdot\mathbf{d} - \mathbf{z}\dot{\vartheta}, \tag{3.31}$$

where

$$\begin{aligned}
\mathcal{L} &= \left[\mathbf{I} - \frac{\frac{1}{H}\mathcal{L}^e\cdot\mathbf{PP}}{1+\frac{1}{H}(\mathcal{L}^e\cdot\mathbf{P}):\mathbf{P}}\right]\cdot\left[\mathcal{L}^e - \frac{1}{H}\mathcal{L}^e\cdot\mathbf{P}(\mathbf{P}\cdot\boldsymbol{\tau} + \boldsymbol{\tau}\cdot\mathbf{P})\right],\\
\mathbf{z} &= \left[\mathbf{I} - \frac{\frac{1}{H}\mathcal{L}^e\cdot\mathbf{PP}}{1+\frac{1}{H}(\mathcal{L}^e\cdot\mathbf{P}):\mathbf{P}}\right]\cdot\left[\mathcal{L}^{th} + \frac{1}{H}\pi\mathcal{L}^e\cdot\mathbf{P}\right].
\end{aligned} \tag{3.32}$$

Substituting the flow rule (3.30) and the relation (2.17) into Eq.(3.28) gives the alternative form of the rate type constitutive equation

$$\overset{\nabla}{\boldsymbol{\tau}} = \widehat{\mathcal{L}}\cdot\mathbf{d} - \mathbf{z}\dot{\vartheta}, \tag{3.33}$$

where

$$\widehat{\mathcal{L}} = \left[\mathbf{I} - \frac{\frac{1}{H}\mathcal{L}^e\cdot\mathbf{PP}}{1+\frac{1}{H}(\mathcal{L}^e\cdot\mathbf{P}):\mathbf{P}}\right]\cdot[\mathcal{L}^e + \mathbf{g}\boldsymbol{\tau} + \boldsymbol{\tau}\mathbf{g}] \tag{3.34}$$

and $\mathbf{z}$ is determined by $\text{Eq.}(3.32)_2$.

It is noteworthy that the rate type formulations (3.31) and (3.33) are materially isomorphic or equivalent, that is each of them describes the same material.

There is new an extensive technical literature dealing with an elastic-plastic model of solids in which there is no distinction made between the matrix $\mathcal{L}^e + \mathbf{g}\boldsymbol{\tau} + \boldsymbol{\tau}\mathbf{g}$ and $\mathcal{L}^e$. In other words for practical purposes in such a model the term $g^{ac}\tau^{bd} + g^{ad}\tau^{bc}$ compares to the matrix $(\mathcal{L}^e)^{abcd}$ is neglected.

The simplified theory obtained in this way is objective olny with respect to superposed rigid body motions.

Such a theory does not meet the much stronger condition of spatial covariance.

3.5 Thermo-mechanical couplings (for J$_2$ – flow theory)

Substituting $\dot{\psi}$ into (3.22) and taking into account the results (3.24) gives

$$\rho\vartheta\dot{\eta} = -\mathrm{div}\mathbf{q} + \rho\vartheta\hat{\imath}. \tag{3.35}$$

Operating on the entropy relation $(3.24)_2$ with the Lie derivative and substituting the result together with the assumed Fourier constitutive law for the heat flux in the form

$$\mathbf{q} = -k\mathrm{grad}\vartheta, \tag{3.36}$$

where k is the conductivity coefficient, into (3.35) we obtain the heat conduction equation as follows

$$\rho c_p\dot{\vartheta} = \mathrm{div}(k\mathrm{grad}\vartheta) + \vartheta\frac{\rho}{\rho_{Ref}}\frac{\partial\boldsymbol{\tau}}{\partial\vartheta} : \mathbf{d} + \rho\chi\langle\frac{1}{H}(\mathbf{P} : \dot{\boldsymbol{\tau}} + \pi\dot{\vartheta})\rangle, \tag{3.37}$$

where

$$\chi = -\left(\frac{\partial\widehat{\psi}}{\partial\boldsymbol{\mu}} - \vartheta\frac{\partial^2\widehat{\psi}}{\partial\vartheta\partial\boldsymbol{\mu}}\right)\cdot\mathbf{m}(s), \tag{3.38}$$

and c_p denotes the specific heat and is determined by

$$c_p = -\vartheta\frac{\partial^2\widehat{\psi}}{\partial\vartheta^2}. \tag{3.39}$$

The main problem remains to be solved is connected with the determination of the constitutive function $\mathbf{m}(s)$. This function plays the crucial role in the description of the rate of internal dissipation (3.25) as well as of the main contribution to thermo-mechanical coupling phenomena (cf. Eqs(3.37) and (3.38)).

To find the answer to this problem let us perform a Legendre transformation [2]. Define the complementary free energy by

$$\varphi = \widehat{\varphi}(\boldsymbol{\tau}, \mathbf{F}, \vartheta; \boldsymbol{\mu}) = \frac{1}{\rho_{Ref}}\boldsymbol{\tau} : \mathbf{e} - \widehat{\psi}(\mathbf{e}, \mathbf{F}, \vartheta; \boldsymbol{\mu}). \tag{3.40}$$

[2]For the application of a Legendre transformation to analysis of elasto-plastic properties of material see HILL and RICE [1973].

Differentiating (3.40) in $\boldsymbol{\tau}$, we have

$$\mathbf{e} = \rho_{Ref} \frac{\partial \widehat{\varphi}}{\partial \boldsymbol{\tau}}. \tag{3.41}$$

Operating on the last result with the Lie derivative gives

$$\mathbf{d} = \mathbf{c} \cdot L_{\boldsymbol{\upsilon}} \boldsymbol{\tau} + \mathbf{r} \dot{\vartheta} + \rho_{Ref} \frac{\partial^2 \widehat{\varphi}}{\partial \boldsymbol{\tau} \partial \boldsymbol{\mu}} \cdot L_{\boldsymbol{\upsilon}} \boldsymbol{\mu}, \tag{3.42}$$

where

$$\mathbf{c} = \rho_{Ref} \frac{\partial^2 \widehat{\varphi}}{\partial \boldsymbol{\tau}^2}, \quad \mathbf{r} = \rho_{Ref} \frac{\partial^2 \widehat{\varphi}}{\partial \boldsymbol{\tau} \partial \vartheta} \tag{3.43}$$

define the material complince tensors.

Identifying the last term in Eq.(3.42) with the rate of plastic deformation yields the identity [3]

$$\rho_{Ref} \frac{\partial^2 \widehat{\varphi}}{\partial \boldsymbol{\tau} \partial \boldsymbol{\mu}} \cdot \mathbf{m}(s) \langle \frac{1}{H} (\mathbf{P} : \dot{\boldsymbol{\tau}} + \pi \dot{\vartheta}) \rangle = \langle \frac{1}{H} (\mathbf{P} : \dot{\boldsymbol{\tau}} + \pi \dot{\vartheta}) \rangle \mathbf{P}. \tag{3.44}$$

So we get the result

$$\mathbf{m}(s) = \frac{1}{\rho_{Ref}} \left[\frac{\partial^2 \widehat{\varphi}}{\partial \boldsymbol{\tau} \partial \boldsymbol{\mu}} \right]^{-1} : \mathbf{P}. \tag{3.45}$$

Answer to the problem of determination of the material function $\mathbf{m}(s)$ can be found in different way, namely by using the stress relation (3.24) and operating on it directly with the Lie derivative.

Making use of the definition (3.38) and taking advantage of the result (3.45) we get

$$\chi = -\frac{1}{\rho_{Ref}} \left(\frac{\partial \widehat{\psi}}{\partial \boldsymbol{\mu}} - \vartheta \frac{\partial^2 \widehat{\psi}}{\partial \vartheta \partial \boldsymbol{\mu}} \right) \cdot \left[\frac{\partial^2 \widehat{\varphi}}{\partial \vartheta \partial \boldsymbol{\mu}} \right]^{-1} : \mathbf{P}. \tag{3.46}$$

Substituting the result (3.46) into the heat conduction equation (3.37) we obtain

$$\rho c_p \dot{\vartheta} = \text{div}\,(k \text{ grad } \vartheta) + \vartheta \frac{\rho}{\rho_{Ref}} \frac{\partial \boldsymbol{\tau}}{\partial \vartheta} : \mathbf{d} + \zeta \boldsymbol{\tau} : \mathbf{d}^p + \frac{\rho}{\rho_{Ref}} \vartheta \frac{\partial^2 \widehat{\psi}}{\partial \vartheta \partial \boldsymbol{\mu}} \cdot \left[\frac{\partial^2 \widehat{\varphi}}{\partial \vartheta \partial \boldsymbol{\mu}} \right]^{-1} : \mathbf{d}^p, \tag{3.47}$$

where new denotation

$$\zeta \boldsymbol{\tau} = -\frac{\rho}{\rho_{Ref}} \frac{\partial \widehat{\psi}}{\partial \boldsymbol{\mu}} \cdot \left[\frac{\partial^2 \widehat{\varphi}}{\partial \vartheta \partial \boldsymbol{\mu}} \right]^{-1} \tag{3.48}$$

is introduced.

To determine ζ we have the identity

$$\zeta \rho_{Ref} \frac{\partial \widehat{\psi}}{\partial \mathbf{e}} = -\frac{\rho}{\rho_{Ref}} \frac{\partial \widehat{\psi}}{\partial \boldsymbol{\mu}} \cdot \left[\frac{\partial^2 \widehat{\varphi}}{\partial \vartheta \partial \boldsymbol{\mu}} \right]^{-1}. \tag{3.49}$$

[3] Similar identification of the rate of plastic deformation was first performed by RICE [1971] (cf. also HILL and RICE [1973]).

3.6 Induced anisotropy. Kinematic hardening

Let focus now attention on the intrintic state $s = (\mathbf{e}, \mathbf{F}, \vartheta; \boldsymbol{\mu})$ and assume the interpretation of the internal state variable $\boldsymbol{\mu}$ as follows

$$\boldsymbol{\mu} = (\boldsymbol{\zeta}, \boldsymbol{\alpha}, \xi) \tag{3.50}$$

where $\boldsymbol{\zeta} \in V_{n-7}$ the new internal state vector is introduced to describe the dissipation effects generated by plastic flow phenomena only, $\boldsymbol{\alpha}$ is the residual stress (the back stress) and aims to describe the strain induced anisotropy effects (kinematic hardening) and ξ is the porosity or the volume fraction parameter brought in to take account of micro-damage effects.

Let assume that the evolution equation for the anisotropic internal state variable $\boldsymbol{\alpha}$ has a general, nonlinear form

$$L_{\boldsymbol{v}}\boldsymbol{\alpha} = \bar{\mathbf{A}}(\boldsymbol{\tau}, \boldsymbol{\alpha}, \vartheta, \xi, \mathbf{d}^p), \tag{3.51}$$

where $\bar{\mathbf{A}}$ is an isotropic symmetric tensor function.

In practical aplications it would be useful to simplify the evolution law (3.51) by replacement of $\boldsymbol{\tau}$ and $\boldsymbol{\alpha}$ by $\tilde{\boldsymbol{\tau}} = \boldsymbol{\tau} - \boldsymbol{\alpha}$, so we have

$$L_{\boldsymbol{v}}\boldsymbol{\alpha} = \mathbf{A}(\tilde{\boldsymbol{\tau}}, \vartheta, \xi, \mathbf{d}^p). \tag{3.52}$$

Making use of the tensorial representation of the function $\mathbf{A}$ (cf. TRUESDELL and NOLL [1965]) and taking into account that there is no change of $\boldsymbol{\alpha}$ when $\tilde{\boldsymbol{\tau}} = 0$ and $\mathbf{d}^p = 0$ the evolution law (3.52) can be written in the form

$$\begin{aligned} L_{\boldsymbol{v}}\boldsymbol{\alpha} &= a_1\mathbf{d}^p + a_2\tilde{\boldsymbol{\tau}} + a_3\mathbf{d}^{p^2} + a_4\tilde{\boldsymbol{\tau}}^2 + a_5(\mathbf{d}^p \cdot \tilde{\boldsymbol{\tau}} + \tilde{\boldsymbol{\tau}} \cdot \mathbf{d}^p) \\ &+a_6(\mathbf{d}^{p^2} \cdot \tilde{\boldsymbol{\tau}} + \tilde{\boldsymbol{\tau}} \cdot \mathbf{d}^{p^2}) + a_7(\mathbf{d}^p \cdot \tilde{\boldsymbol{\tau}}^2 + \tilde{\boldsymbol{\tau}}^2 \cdot \mathbf{d}^p) + a_8(\mathbf{d}^{p^2} \cdot \tilde{\boldsymbol{\tau}}^2 + \tilde{\boldsymbol{\tau}}^2 \cdot \mathbf{d}^{p^2}), \end{aligned} \tag{3.53}$$

where $a_1, ..., a_8$ are functions of ten basic invariants of $\tilde{\boldsymbol{\tau}}$ and $\mathbf{d}^p$, temperature ϑ and porosity ξ.

A linear approximation of the general evolution law (3.53) leads to the result (cf. DUSZEK AND PERZYNA [1991])

$$L_{\boldsymbol{v}}\boldsymbol{\alpha} = a_1\mathbf{d}^p + a_2\tilde{\boldsymbol{\tau}}. \tag{3.54}$$

This evolution law represents the linear combination of the PRAGER and ZIEGLER kinematic hardening rules (cf. PRAGER [1955] and ZIEGLER [1959]).

3.7 Micro-damage process. Softening effects

To describe the intrinsic micro-damage process let us introduce the density of microcracks (the number of microcracks per unit volume) N and the average size of a microcrack R. These two internal state variables are sugested by experimental observations as well as physical analysis of facture mechanisms, cf. PERZYNA [1986].

The intrinsic micro-damage process consist of nucleation, growth and coalescence of microvoids. Recent experimental observation results have shown that coalescence mechanism can be treated as nucleation and growth process on a smaller scale, cf. SHOCKEY, SEAMAN and CURRAN [1985]. This conjecture simplifies very much the description of the intrinsic micro-damage by taking account only of the nucleation mechanism and the growth process. The density of microvoids N is responsible for the description of the nucleation phenomenon, and the average size of a microvoid R aims at handling the growth mechanism.

Let assume that the nucleation mechanism occurs mainly at second phase particles, by decohesion of the particle-matrix interface and by the particle cracking. This assumption is based on the experimental observations for metals. The debonding process at MnS-matrix interface in steel is shown in Fig.1. Manganese sulfide particles have a low bonding strength with the matrix. The cracking phenomenon of the second phase particles in nickel is shown in Fig.2. In this figure some second phase particles that are fractured can be seen.

Thus we can write

$$\dot{N} = \bar{k}_1(\epsilon^p, N, \vartheta)\tilde{\boldsymbol{\tau}} : \mathbf{d}^p + \bar{k}_2(\tilde{J}_1, N, \vartheta)\dot{\tilde{J}}_1, \tag{3.55}$$

where the first term describes debonding of second-phase particles from the matrix as the plastic work progressively increases, and the second term is responsible for the cracking of the second-phase particles as the mean stress increases. The debonding material function $\bar{k}_1$ depends on the equivalent plastic deformation ϵ^p, the actual density of microvoids N and temperature ϑ. While the cracking material function $\bar{k}_2$ depends on the mean stress $\tilde{J}_1$, the actual density of microvoids N as well as temperature ϑ.

The growth process is assumed to be controled only by the plastic flow phenomenon. Therefore we can postulate

$$\dot{R} = \bar{k}_3(\epsilon^p, R, \vartheta)\mathbf{d}^p : \mathbf{g}, \tag{3.56}$$

where $\bar{k}_3$ denotes the growth material function which depends on the average size of a microvoid R, the equivalent plastic deformation ϵ^p and temperature ϑ. This assumption is also justified by the experimental observation results, cf. Fig.3.

A very important feature of this description is to take advantage of the porosity parameter ξ introduced as the state variable in (3.50). To do this it is sufficient to postulate that the microcracks have spherical shapes, so that

$$\xi = 4\pi R^3 N/3 \tag{3.57}$$

and

$$\dot{\xi} = \xi\left(\frac{\dot{N}}{N} + 3\frac{\dot{R}}{R}\right). \tag{3.58}$$

The last result together with (3.55) and (3.56) gives the evolution equation for the porosity parameter ξ in the form (cf. GURSON [1977], NEEDLEMAN and RICE [1978])

$$\dot{\xi} = k_1\tilde{\tau} : \mathbf{d}^p + k_2\dot{\tilde{J}}_1 + k_3\mathbf{d}^p : \mathbf{g}, \tag{3.59}$$

Figure 1: Debonding at manganese sulfide particle–matrix interface in steel (after MEYERS and AIMONE [1983])

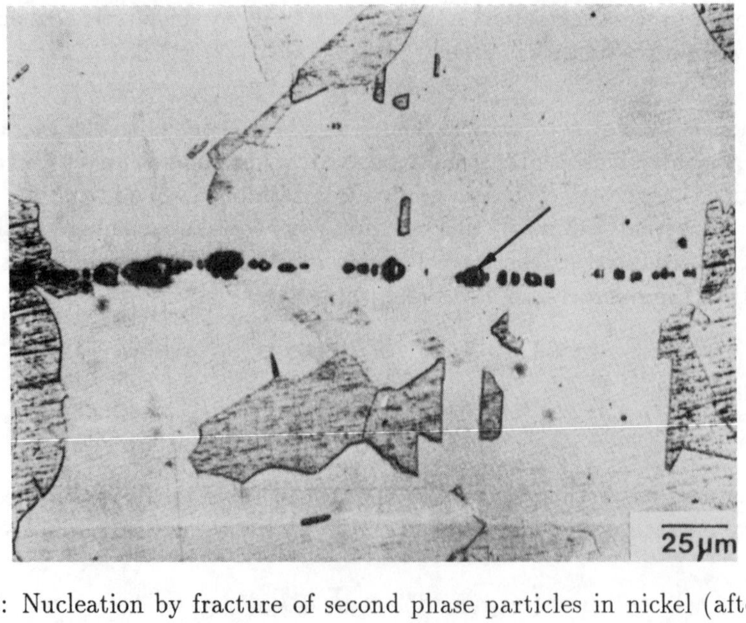

Figure 2: Nucleation by fracture of second phase particles in nickel (after MEYERS and AIMONE [1983])

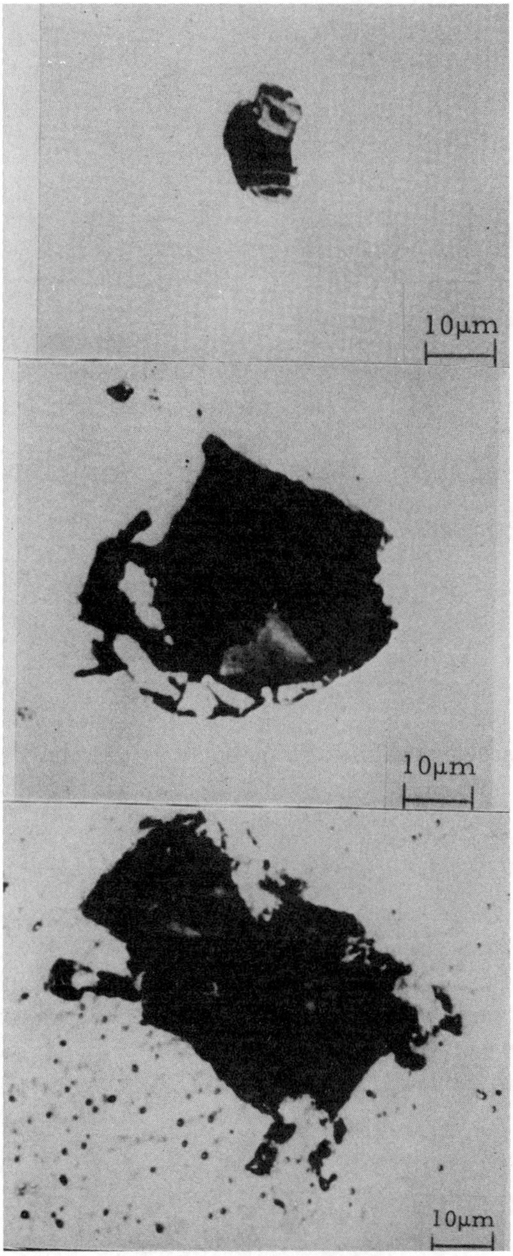

Figure 3: Voids in the centers of tensile specimens of 18 Ni maraging steel strained various amounts plastically (after COX and LOW [1974])

provided it is postulated that

$$\begin{aligned} \frac{\xi}{N}\bar{k}_1(\epsilon^p, N, \vartheta) &= k_1(\epsilon^p, \xi, \vartheta), \\ \frac{\xi}{N}\bar{k}_2(\tilde{J}_1, N, \vartheta) &= k_2(\tilde{J}_1, \xi, \vartheta), \\ \frac{\xi}{R}\bar{k}_3(\epsilon^p, R, \vartheta) &= k_3(\epsilon^p, \xi, \vartheta), \end{aligned} \tag{3.60}$$

where k_1, k_2 and k_3 denote new material functions.

3.8 General constitutive equations

To take account of the induced anisotropy effects as well as of the intrinsic micro-damage process let introduce the yield criterion in the form as follows

$$\tilde{\varphi} = f - \kappa = 0, \tag{3.61}$$

where the yield function f is now assumed as [4]

$$\begin{aligned} f &= \tilde{f}(\tilde{\boldsymbol{\tau}}, \mathbf{g}, \vartheta, \xi) = \tilde{J}_2 + [n_1(\vartheta) + n_2(\vartheta)\xi]\tilde{J}_1, \\ \tilde{J}_2 &= \frac{1}{2}\tau'^{ab}\tau'^{cd}g_{ac}g_{db}, \qquad \tilde{J}_1 = \tilde{\tau}'^{ab}g_{ab}, \qquad \tilde{\boldsymbol{\tau}} = \boldsymbol{\tau} - \boldsymbol{\alpha}, \end{aligned} \tag{3.62}$$

and the isotropic hardening-softening function κ is postulated as (cf. NEMES et al. [1990])

$$\kappa = \tilde{\kappa}(\epsilon^p, \xi, \vartheta) = \left[\kappa_1 + (\kappa_0 - \kappa_1)e^{-h_1(\vartheta)\epsilon^p}\right]^2 \left[1 - \frac{\xi}{\xi^F(\vartheta)}\right](1 - w\bar{\vartheta}), \tag{3.63}$$

with the additional material function $\xi^F(\vartheta)$.

It is noteworthy to add that for a proposed model of a porous material the isotropic hardening-softening material function (3.63) satisfies the following fracture criterion, when $\xi \to \xi^F$ then $\kappa \to 0$.

Let assume again the flow rule in the form (3.8), i.e.

$$\mathbf{d}^p = \Lambda \mathbf{P} \tag{3.64}$$

where

$$\mathbf{P} = \frac{1}{2\sqrt{\tilde{J}_2}} \left(\frac{\partial \tilde{\varphi}}{\partial \boldsymbol{\tau}}\Big|_{\xi=\mathrm{const}}\right)^b. \tag{3.65}$$

[4]The yield function in the form (3.62) has been first suggested by SHIMA and OYANE [1976] for porous solids.

From the consistency condition

$$\dot{f} - \dot{\kappa} = 0 \tag{3.66}$$

and the geometrical relation [5] (cf. DUSZEK and PERZYNA [1991])

$$(L_{\boldsymbol{v}}\boldsymbol{\alpha} - r\mathbf{d}^p) : \mathbf{Q} = 0, \tag{3.67}$$

where r is new material coefficient, and

$$\mathbf{Q} = \frac{1}{2\sqrt{\tilde{J}_2}} \left[\frac{\partial \tilde{\varphi}}{\partial \boldsymbol{\tau}} \Big|_{\xi=\mathrm{const}} + \frac{\partial \tilde{\varphi}}{\partial \xi} \frac{\partial \xi}{\partial \boldsymbol{\tau}} \right]^b,$$

we can determine the coeffitiens Λ, a_1 and a_2. This leads to the results as follows

$$\begin{aligned}
\mathbf{d}^p &= \langle \frac{1}{H} \{ \mathbf{Q} : [\dot{\boldsymbol{\tau}} - (\mathbf{d} \cdot \boldsymbol{\alpha} + \boldsymbol{\alpha} \cdot \mathbf{d})] + \pi \dot{\vartheta} \} \rangle \mathbf{P}, \\
L_{\boldsymbol{v}}\boldsymbol{\alpha} &= \left\langle \frac{1}{H} \{ \mathbf{Q} : [\dot{\boldsymbol{\tau}} - (\mathbf{d} \cdot \boldsymbol{\alpha} + \boldsymbol{\alpha} \cdot \mathbf{d})] + \pi \dot{\vartheta} \} \right\rangle \left(r_1 \mathbf{P} + r_2 \frac{\mathbf{P} : \mathbf{Q}}{\tilde{\boldsymbol{\tau}} : \mathbf{Q}} \tilde{\boldsymbol{\tau}} \right), \\
\dot{\xi} &= \left\langle \frac{1}{H} \{ \mathbf{Q} : [\dot{\boldsymbol{\tau}} - (\mathbf{d} \cdot \boldsymbol{\alpha} + \boldsymbol{\alpha} \cdot \mathbf{d})] + \pi \dot{\vartheta} \} \right\rangle (k_1 \tilde{\boldsymbol{\tau}} : \mathbf{P} + k_3 \mathbf{P} : \mathbf{g}) + k_2 \dot{\tilde{\boldsymbol{\tau}}} : \mathbf{g},
\end{aligned} \tag{3.68}$$

where

$$\begin{aligned}
P_{ab} &= \frac{1}{2\sqrt{\tilde{J}_2}} \tilde{\tau}'^{cd} g_{ca} g_{db} + A g_{ab}, \\
Q_{ab} &= \frac{1}{2\sqrt{\tilde{J}_2}} \tilde{\tau}'^{cd} g_{ca} g_{db} + B g_{ab},
\end{aligned} \tag{3.69}$$

$$\begin{aligned}
A &= \frac{1}{\sqrt{\tilde{J}_2}} (n_1 + n_2 \xi) \tilde{\tau}^{ab} g_{ab}, \\
B &= A + \frac{k_2}{2\sqrt{\tilde{J}_2}} \left\{ n_2 (\tilde{\tau}^{ab} g_{ab})^2 + [\kappa_1 + (\kappa_0 - \kappa_1) e^{-h_1 \epsilon^P}]^2 \frac{1 - w\bar{\vartheta}}{\xi^F} \right\}, \quad H = H^* + H^{**}, \\
H^* &= -\frac{1}{2\sqrt{\tilde{J}_2}} \left\{ n_2 \tilde{J}_1^2 + [\kappa_1 + (\kappa_0 - \kappa_1) e^{-h_1 \epsilon^P}]^2 \frac{1 - w\bar{\vartheta}}{\xi^F} \right\} [k_1 (\sqrt{\tilde{J}_2} \\
&\quad + A \tilde{J}_1) + 3 A k_3] + \frac{h_1 (\kappa_1 - \kappa_0)}{\sqrt{3\tilde{J}_2}} [\kappa_1 + (\kappa_0 - \kappa_1) e^{-h_1 \epsilon^P}] (1 \\
&\quad - \left. \frac{\xi}{\xi^F} \right) (1 - w\bar{\vartheta}) (1 + 6A^2)^{\frac{1}{2}} e^{-h_1 \epsilon^P}, \\
H^{**} &= r \mathbf{P} : \mathbf{Q}, \qquad r = r_1 + r_2, \qquad \pi = \frac{1}{2\sqrt{\tilde{J}_2}} \frac{\partial \tilde{\varphi}}{\partial \vartheta}.
\end{aligned} \tag{3.70}$$

[5] For interpretation of the geometrical relation (3.67) see Fig.4.

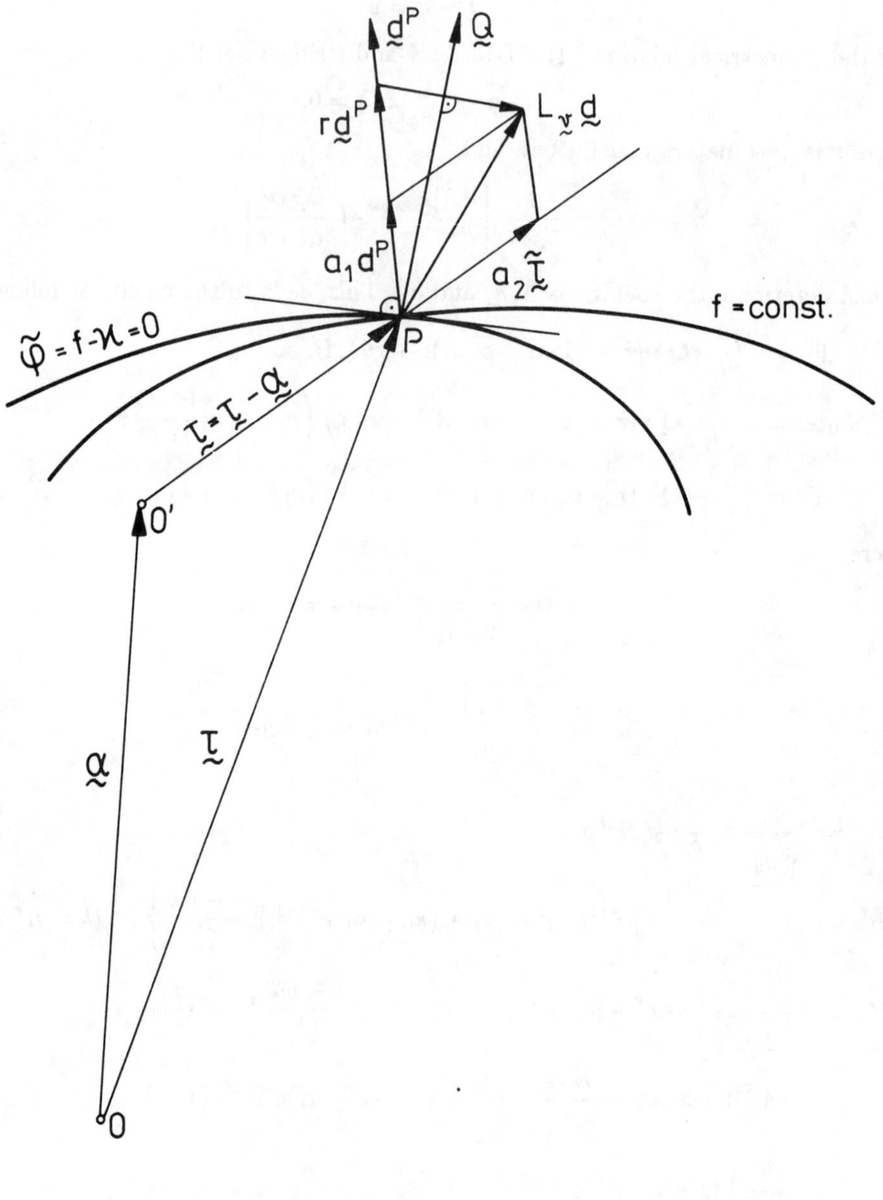

Figure 4: Interpretation of the geometrical relation (3.67)

To take account of loading criterion the bracket $\langle\ \ \rangle$ which defines the ramp function (cf. (3.11)) is introduced.

The loading process takes place when

$$\langle \frac{1}{H}\{\mathbf{Q}:[\dot{\boldsymbol{\tau}}-(\mathbf{d}\cdot\boldsymbol{\alpha}+\boldsymbol{\alpha}\cdot\mathbf{d})]+\pi\dot{\vartheta}\}\rangle \neq 0. \tag{3.71}$$

For the internal state vector $\boldsymbol{\zeta}$ we have the evolution equation as follows

$$L_{\boldsymbol{v}}\boldsymbol{\zeta}=\mathbf{Z}(s)\langle \frac{1}{H}\{\mathbf{Q}:[\dot{\boldsymbol{\tau}}-(\mathbf{d}\cdot\boldsymbol{\alpha}+\boldsymbol{\alpha}\cdot\mathbf{d})]+\pi\dot{\vartheta}\}\rangle, \tag{3.72}$$

where the material function $\mathbf{Z}(s)$ can be determined by using the same procedure as it was shown in section 3.5 (cf. the determination of the material function $\mathbf{m}(s)$).

All dissipative phenomena except the nucleation mechanism generated by cracking of second phase particles are directly combined with plastic flow process. The cracking mechanism can take place independently of plastic flow. It means that it can occur even in elastic range (i.e. when $\langle\frac{1}{H}\{\mathbf{Q}:[\dot{\boldsymbol{\tau}}-(\mathbf{d}\cdot\boldsymbol{\alpha}+\boldsymbol{\alpha}\cdot\mathbf{d})]+\pi\dot{\vartheta}\}\rangle=0$) provided the threshold velue for $\widetilde{J}_1$ is exceeded (cf. DUSZEK and PERZYNA [1988]).

To take advantage of the axiom of entropy production we have to postulate that k_2 vanishes when $\widetilde{J}_1 \leq \widetilde{J}_1^N$, where $\widetilde{J}_1^N$ denotes the threshold value for a material.

For damaged solid body the mass density $\rho(\mathbf{x},t)$ (in spatial coordinates) is given by the relation

$$\rho=\rho_M(1-\xi)+\rho_v\xi, \tag{3.73}$$

where ρ_M is the mass density of the matrix material and ρ_v the mass density of voids.

If

$$\rho_v << \rho_M, \tag{3.74}$$

then

$$\rho=\rho_M(1-\xi). \tag{3.75}$$

A function $\rho(\mathbf{x},t)$ is said to obey conservation of mass if

$$\dot{\rho}_M(1-\xi)+\rho_M(1-\xi)\mathrm{div}\boldsymbol{v}-\rho_M\dot{\xi}=0 \tag{3.76}$$

or

$$\rho_M(1-\xi)J(\mathbf{X},t)=\beta_M(\mathbf{X})(1-\xi_0)=\rho_{Ref}. \tag{3.77}$$

For damaged solid body the balance equations (3.19) and (3.20) have to be replaced by (3.76) and (3.77), respectively.

The fulfilment of the axiom of entropy production leads to the results (3.24), where the rate of internal dissipation is now given by

$$\vartheta\hat{i}=-\frac{\partial\hat{\psi}}{\partial\boldsymbol{\mu}}\cdot L_{\boldsymbol{v}}\boldsymbol{\mu}=-\Bigg[\frac{\partial\hat{\psi}}{\partial\boldsymbol{\zeta}}\cdot\mathbf{Z}(s)+\frac{\partial\hat{\psi}}{\partial\boldsymbol{\alpha}}:(r_1\mathbf{P}+r_2\frac{\mathbf{P}:\mathbf{Q}}{\tilde{\boldsymbol{\tau}}:\mathbf{Q}}\tilde{\boldsymbol{\tau}}) \tag{3.78}$$

$$+\frac{\partial\hat{\psi}}{\partial\xi}(k_1\tilde{\boldsymbol{\tau}}:\mathbf{P}+k_3\mathbf{P}:\mathbf{g})\Bigg]\langle\frac{1}{H}\{\mathbf{Q}:[\dot{\boldsymbol{\tau}}-(\mathbf{d}\cdot\boldsymbol{\alpha}+\boldsymbol{\alpha}\cdot\mathbf{d})]+\pi\dot{\vartheta}\}\rangle-\frac{\partial\hat{\psi}}{\partial\xi}k_2\dot{\tilde{\boldsymbol{\tau}}}:\mathbf{g}.$$

The rate type constitutive equation for the Kirchhoff stress tensor $\boldsymbol{\tau}$ and the heat conduction equation take now the form as follows

$$
\begin{aligned}
L_{\boldsymbol{v}}\boldsymbol{\tau} &= \mathcal{L} : \mathbf{d} - \mathbf{z}\dot{\vartheta}, \qquad (3.79)\\
\rho c_p \dot{\vartheta} &= \operatorname{div}(k \operatorname{grad}\vartheta) + \vartheta \frac{\rho}{\rho_{Ref}} \frac{\partial \boldsymbol{\tau}}{\partial \vartheta} : \mathbf{d} \\
&\quad + \rho \bar{\chi}_1 \langle \frac{1}{H} \{\mathbf{Q} : [\dot{\boldsymbol{\tau}} - (\mathbf{d}\cdot\boldsymbol{\alpha} + \boldsymbol{\alpha}\cdot\mathbf{d})] + \pi\dot{\vartheta}\} \rangle \\
&\quad + \rho \bar{\chi}_2 [L_{\boldsymbol{v}}\boldsymbol{\tau} : \mathbf{g} + (\mathbf{g}\cdot\tilde{\boldsymbol{\tau}} + \tilde{\boldsymbol{\tau}}\cdot\mathbf{g}) : \mathbf{d}],
\end{aligned}
$$

where

$$
\begin{aligned}
\mathcal{L} &= \left[I - \frac{\frac{1}{H}\mathcal{L}^e : \mathbf{PQ}}{1 + \frac{1}{H}(\mathcal{L}^e : \mathbf{P}) : \mathbf{Q}} \right] \cdot \left[\mathcal{L}^e - \frac{1}{H}\mathcal{L}^e : \mathbf{P}(\mathbf{Q}\cdot\tilde{\boldsymbol{\tau}} + \tilde{\boldsymbol{\tau}}\cdot\mathbf{Q}) \right], \qquad (3.80)\\
\mathbf{z} &= \left[I - \frac{\frac{1}{H}\mathcal{L}^e : \mathbf{PQ}}{1 + \frac{1}{H}(\mathcal{L}^e : \mathbf{P}) : \mathbf{Q}} \right] \cdot \left[\mathcal{L}^{th} + \frac{1}{H}\pi\mathcal{L}^e : \mathbf{P} \right],
\end{aligned}
$$

$$
\begin{aligned}
\bar{\chi}_1 &= -\left\{ \left(\frac{\partial\hat{\psi}}{\partial\zeta} - \vartheta\frac{\partial^2\hat{\psi}}{\partial\vartheta\partial\zeta} \right) \cdot \mathbf{Z}(s) + \left(\frac{\partial\hat{\psi}}{\partial\xi} - \vartheta\frac{\partial^2\hat{\psi}}{\partial\vartheta\partial\xi} \right) (k_1\tilde{\boldsymbol{\tau}} : \mathbf{P} + k_3\mathbf{g} : \mathbf{P}) \right.\\
&\quad \left. + \left[\left(\frac{\partial\hat{\psi}}{\partial\boldsymbol{\alpha}} - \vartheta\frac{\partial^2\hat{\psi}}{\partial\vartheta\partial\boldsymbol{\alpha}} \right) - k_2\mathbf{g} \left(\frac{\partial\hat{\psi}}{\partial\xi} - \vartheta\frac{\partial^2\hat{\psi}}{\partial\vartheta\partial\xi} \right) \right] : \left(r_1\mathbf{P} + r_2\frac{\mathbf{P}:\mathbf{Q}}{\tilde{\boldsymbol{\tau}}:\mathbf{Q}}\tilde{\boldsymbol{\tau}} \right) \right\}, \qquad (3.81)\\
\bar{\chi}_2 &= -\left(\frac{\partial\hat{\psi}}{\partial\xi} - \vartheta\frac{\partial^2\hat{\psi}}{\partial\vartheta\partial\xi} \right) k_2.
\end{aligned}
$$

Let consider the case for which we can assume $k_2 = 0$, i.e. the cracking mechanism of second phase particles is neglected. This assumption leads to very important results:

(i) $\mathbf{P} = \mathbf{Q}$, then the normality rule does apply, i.e.

$$
\mathbf{d}^p = \langle \frac{1}{H} \{\mathbf{P} : [\dot{\boldsymbol{\tau}} - (\mathbf{d}\cdot\boldsymbol{\alpha} + \boldsymbol{\alpha}\cdot\mathbf{d})] + \pi\dot{\vartheta}\} \rangle \mathbf{P}. \qquad (3.82)
$$

(ii) The evolution equation for the internal state vector $\boldsymbol{\mu}$ takes now the form as follows

$$
L_{\boldsymbol{v}}\boldsymbol{\mu} = \mathbf{m}(s) \langle \frac{1}{H} \{\mathbf{P} : [\dot{\boldsymbol{\tau}} - (\mathbf{d}\cdot\boldsymbol{\alpha} + \boldsymbol{\alpha}\cdot\mathbf{d})] + \pi\dot{\vartheta}\} \rangle, \qquad (3.83)
$$

where

$$
\mathbf{m}(s) = \begin{cases} \mathbf{Z}(s), \\ r_1\mathbf{P} + r_2\frac{\mathbf{P}:\mathbf{P}}{\tilde{\boldsymbol{\tau}}:\mathbf{P}}\tilde{\boldsymbol{\tau}}, \\ k_1\tilde{\boldsymbol{\tau}} : \mathbf{P} + k_3\mathbf{P} : \mathbf{g}. \end{cases} \qquad (3.84)
$$

(iii) The heat conduction equation has now the form

$$\rho c_p \dot{\vartheta} = \mathrm{div}(k \,\mathrm{grad}\, \vartheta) + \vartheta \frac{\rho}{\rho_{Ref}} \frac{\partial \boldsymbol{\tau}}{\partial \vartheta} : \mathbf{d} + \rho \bar{\chi} \langle \frac{1}{H} \{ \mathbf{P} : [\dot{\boldsymbol{\tau}} - (\mathbf{d} \cdot \boldsymbol{\alpha} + \boldsymbol{\alpha} \cdot \mathbf{d})] + \pi \dot{\vartheta} \} \rangle, \quad (3.85)$$

where

$$\bar{\chi} = \bar{\chi}_1 \quad \text{and} \quad \bar{\chi}_2 = 0. \quad (3.86)$$

4 CONSTITUTIVE STRUCTURE FOR CRYSTAL

4.1 Kinematics of finite deformations of crystal

It is understood that $\mathbf{F}^e$ is the lattice contributions to $\mathbf{F}$, and is associated with stretching and rotation of the lattice, $\mathbf{F}^p$ describes the deformation solely due to plastic shearing on crystallographic slip systems, cf. Fig.5.

A particular slip system α is specified by the slip vectors $\mathbf{s}_0^{(\alpha)}, \mathbf{m}_0^{(\alpha)}$, where $\mathbf{s}_0^{(\alpha)}$ gives the slip direction and $\mathbf{m}_0^{(\alpha)}$ is the slip plane normal. The vectors $\mathbf{s}_0^{(\alpha)}$ and $\mathbf{m}_0^{(\alpha)}$ in the undeformed lattice are taken to be orthonormal. As the crystal derforms the vectors $\mathbf{s}^{(\alpha)}$ and $\mathbf{m}^{(\alpha)}$ are stretched and rotated according to $\mathbf{F}^e$. In the deformed lattice we have

$$\mathbf{s}^{(\alpha)} = \mathbf{F}^e \cdot \mathbf{s}_0^{(\alpha)}, \qquad \mathbf{m}^{(\alpha)} = \mathbf{m}_0^{(\alpha)} \cdot \mathbf{F}^{e^{-1}}. \quad (4.1)$$

Let us define the Eulerian velocity gradient in the current state of the crystal by

$$\mathbf{l} = \dot{\mathbf{F}} \cdot \mathbf{F}^{-1} = \dot{\mathbf{F}}^e \cdot \mathbf{F}^{e^{-1}} + \mathbf{F}^e \cdot \dot{\mathbf{F}}^p \cdot \mathbf{F}^{p^{-1}} \cdot \mathbf{F}^{e^{-1}}, \quad (4.2)$$

and postulate for the plastic part

$$\mathbf{l}^p = \dot{\mathbf{F}} \cdot \mathbf{F}^{-1} - \dot{\mathbf{F}}^e \cdot \mathbf{F}^{e^{-1}} = \mathbf{F}^e \cdot \dot{\mathbf{F}}^e \cdot \mathbf{F}^{p^{-1}} \cdot \mathbf{F}^{e^{-1}} = \sum_{\alpha=1}^{n} \mathbf{s}^{(\alpha)} \mathbf{m}^{(\alpha)} \dot{\gamma}^{(\alpha)}, \quad (4.3)$$

where $\dot{\gamma}^{(\alpha)}$ is the rate of shearing on slip system α.

For each slip system one can define the symmetric and antisymmetric tensors

$$\mathbf{N}^{(\alpha)} = \frac{1}{2}[\mathbf{s}^{(\alpha)}\mathbf{m}^{(\alpha)} + \mathbf{m}^{(\alpha)}\mathbf{s}^{(\alpha)}], \qquad \mathbf{W}^{(\alpha)} = \frac{1}{2}[\mathbf{s}^{(\alpha)}\mathbf{m}^{(\alpha)} - \mathbf{m}^{(\alpha)}\mathbf{s}^{(\alpha)}]. \quad (4.4)$$

Then instead of (4.2) we can write

$$\mathbf{l} = \mathbf{l}^e + \mathbf{l}^p = \mathbf{d} + \boldsymbol{\omega} = \mathbf{d}^e + \boldsymbol{\omega}^e + \mathbf{d}^p + \boldsymbol{\omega}^p, \quad (4.5)$$

where $\mathbf{d}$ is the symmetric rate of the stretching tensor and $\boldsymbol{\omega}$ is the anti-symmetric spin rate. The elastic rates of stretching and spin $\mathbf{d}^e$ and $\boldsymbol{\omega}^e$ are the symmetric and anti-symmetric parts of $\dot{\mathbf{F}}^e \cdot \mathbf{F}^{e^{-1}}$, respectively. The plastic parts of the rate of stretching and spin are determined by the relations

$$\mathbf{d}^p = \sum_{n=1}^{n} \dot{\gamma}^{(\alpha)} \mathbf{N}^{(\alpha)}, \qquad \boldsymbol{\omega}^p = \sum_{n=1}^{n} \dot{\gamma}^{(\alpha)} \mathbf{W}^{(\alpha)}. \quad (4.6)$$

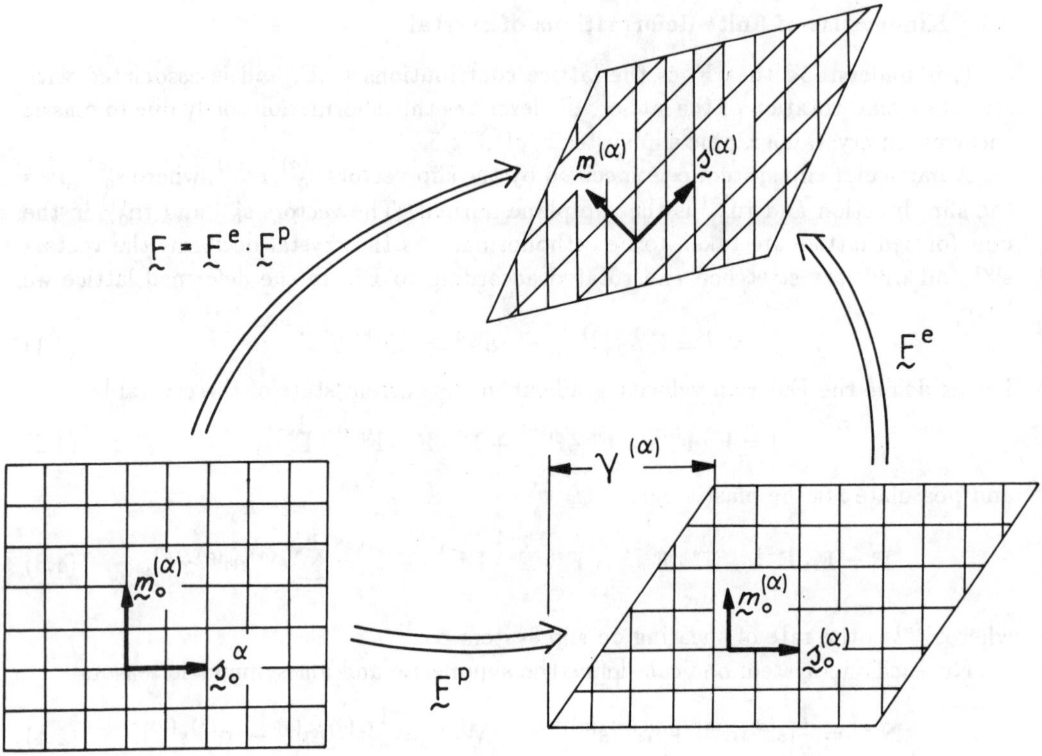

Figure 5: Decomposition of the deformation gradient for single crystals

4.2 Fundamental definition

Let $\boldsymbol{\tau}$ denotes the Kirchhoff stress tensor, and let take the rate of stress working per unit reference volume

$$\boldsymbol{\tau} : \mathbf{d} = \boldsymbol{\tau} : \mathbf{d}^e + \boldsymbol{\tau} : \mathbf{d}^p = \boldsymbol{\tau} : \mathbf{d}^e + \sum_{\alpha=1}^{n} \tau^{(\alpha)} \dot{\gamma}^{(\alpha)}, \tag{4.7}$$

where

$$\tau^{(\alpha)} = \boldsymbol{\tau} : \mathbf{N}^{(\alpha)} \tag{4.8}$$

is the Schmid resolved shear stress on slip system α .

4.3 Temperature and rate dependent kinetic laws

The rate and temperature dependence of the flow stress of metal crystals can be explained by different physical mechanisms of dislocation motion. The microscopic processes combine in various ways to give several groups of dissipative mechanisms, each of which can be limited in its range of temperature and strain rate changes.

It is reasonable to discuss some examples of the dissipative mechanisms in detail.

(i) Diclocation creep mechanism. A pure empirical power relation between strain rate and stress is often used to characterize steady creep of metals and other materials at temperature above one-third of the melting point. At this range of temperature there is sufficient mobility of vacancies to allow dislocation to climb as well as glide. Deformation is possible at a lower stress than would be needed for glide alone.

The connection between the shear rate and resolved shear stress on the α system for high temperature can be described by the semi-empirical equation (cf. ASHBY and FROST [1975], HUTCHINSON [1976] and ASARO and NEEDLEMAN [1985], PEIRCE, ASARO and NEEDLEMAN [1983], PAN and RICE [1983])

$$\dot{\gamma}^{(\alpha)} = \eta_c^{(\alpha)} \left[\frac{\tau^{(\alpha)}}{g_c^{(\alpha)}} \right]^{\frac{1}{m}} \operatorname{sgn} \tau^{(\alpha)} \tag{4.9}$$

where $\eta_c^{(\alpha)}$ is a convenient reference creep rate for the α system, and in general it is temperature dependent coefficient. The reference stress $g_c^{(\alpha)}$ for the α system is a strong function of temperature and shear strain. The exponent m also depends on temperature although somewhat less strongly.

ASHBY and FROST [1975] have broadly surveyed polycrystal and crystal data for many metals to determine the range of temperature and stress over which steady creep can be reasonably approximated by a power law stress dependence, cf. Fig.6.

(ii) Viscoplastic flow mechanism. It has been proposed that the rate dependent plastic flow is descibed by the relation (cf. BINGHAM [1922], HOHENEMSER and PRAGER [1932], MALVERN [1951] and PERZYNA [1963])

$$\dot{\gamma}^{(\alpha)} = \eta_v^{(\alpha)} \langle \Phi \left[\frac{\tau^{(\alpha)}}{g_v^{(\alpha)}} - 1 \right] \rangle \operatorname{sgn} \tau^{(\alpha)}, \tag{4.10}$$

where

$$\langle\Phi\rangle = \begin{cases} 0 & \text{if} \quad \tau^{(\alpha)} \le g_v^{(\alpha)}, \\ \Phi & \text{if} \quad \tau^{(\alpha)} > g_v^{(\alpha)}, \end{cases} \tag{4.11}$$

$\eta_v^{(\alpha)}$ has dimension of rate of shearing and is assumed as a material viscosity coefficient for the slip system α,$g_v^{(\alpha)}$ has interpretation of the critical shear stress in the system and function Φ is determined basing on the available experimental results for particular material. For nonisothermal processes $\eta_v^{(\alpha)}$ and $g_v^{(\alpha)}$ are assumed to depend on temperature.

Experimental justifications of the assumed postulates of the model and the discussion of its range of applicability have been given by LINDHOLM [1964, 1968] and CAMPBELL [1973], cf. Fig.7.

(iii) Interaction of the thermally activated and phonon damping mechanisms (cf. KUMAR and KUMBLE [1969], TEODOSIU and SIDOROFF [1976] and PERZYNA [1977]). If a dislocation is moving trough the rows of barriers, then its velocity can be determined by the expression

$$v = AL^{-1}/(t_S + t_B), \tag{4.12}$$

where AL^{-1} is the average distance of dislocation movement after each thermal activation, t_S is the time duration a dislocation spent at the obstacle and t_B is the time duration of travelling between the barriers.

The rate of shearing in the α slip system is given by the relationship

$$\dot{\gamma}^{(\alpha)} = \eta_T^{(\alpha)} \langle \exp\{\varphi[(\tau^{(\alpha)} - g_T^{(\alpha)})Lb]/k\vartheta\} + ABL^{-1}\nu/(\tau^{(\alpha)} - g_D^{(\alpha)})b\rangle^{-1}, \tag{4.13}$$

where

$$\eta_T^{(\alpha)} = AL^{-1}\nu b\rho_M, \quad g_T^{(\alpha)} = \tau_\mu^{(\alpha)}, \quad \eta_D^{(\alpha)} = \eta_T^{(\alpha)} \frac{b}{ABL^{-1}\nu} = \frac{\rho_M b^2}{B}, \tag{4.14}$$

ν is the frequency of vibration of the dislocation, φ is the activation energy (Gibbs free energy), k is the Boltzmann constant, ϑ is actual absolute temperature, ρ_M is the mobile dislocation density, b denotes the Burger's vector, $\tau_\mu^{(\alpha)}$ is the athermal stress, $g_D^{(\alpha)}$ is interpreted as the stress needed to overcome the forest dislocation barrier to the dislocation motion and B is called the dislocation drag coefficient.

If the time duration t_B taken by the dislocation to travel between the barriers in a viscous phonon medium is negligible when compared with the time duration t_S dislocation spent at the obstacle, then

$$v = \frac{AL^{-1}}{t_S} \tag{4.15}$$

and we can focus our attention on the analysis of the thermally activated process (cf. SEEGER [1955], EVANS and RAWLINGS [1969] and KOCKS, ARGON and ASHBY [1975]).

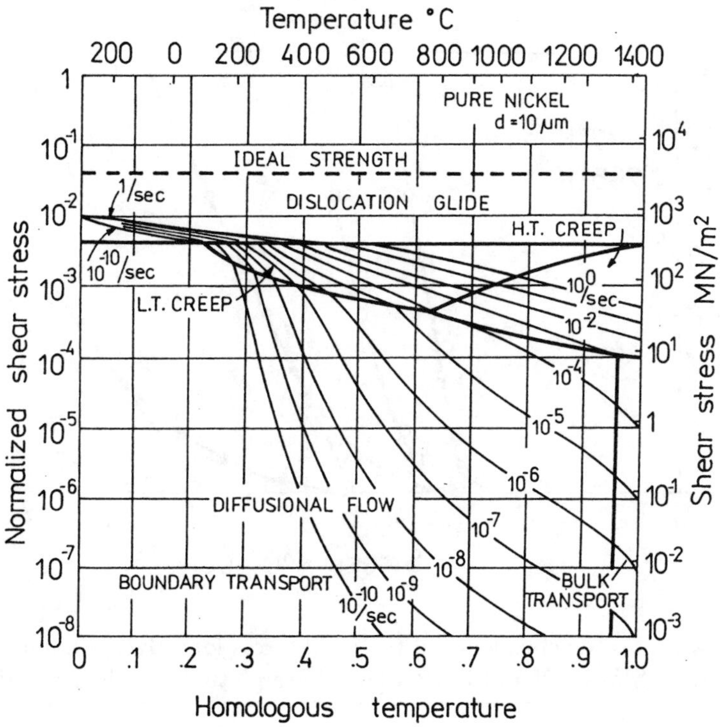

Figure 6: The deformation mechanism map for nickel with a grain size of 10 μm (after ASHBY and FROST [1975])

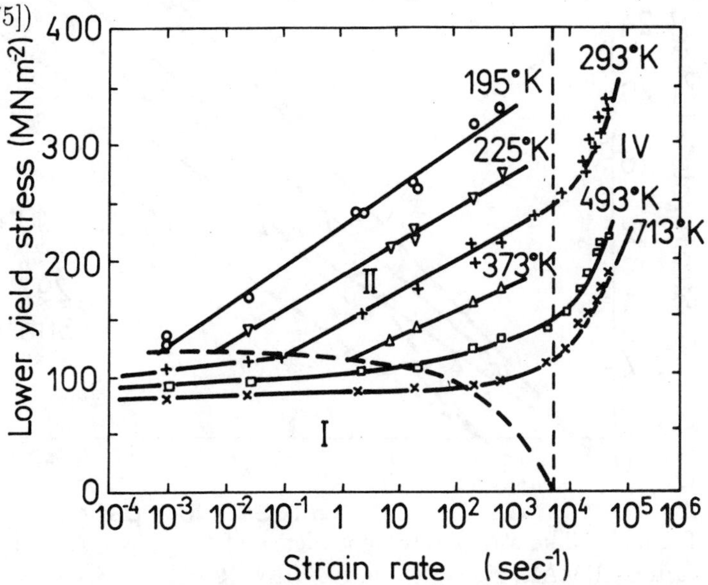

Figure 7: Variation of lower yield stress with strain rate for mild steel at constant temperature (after CAMPBELLand FERGUSON [1970])

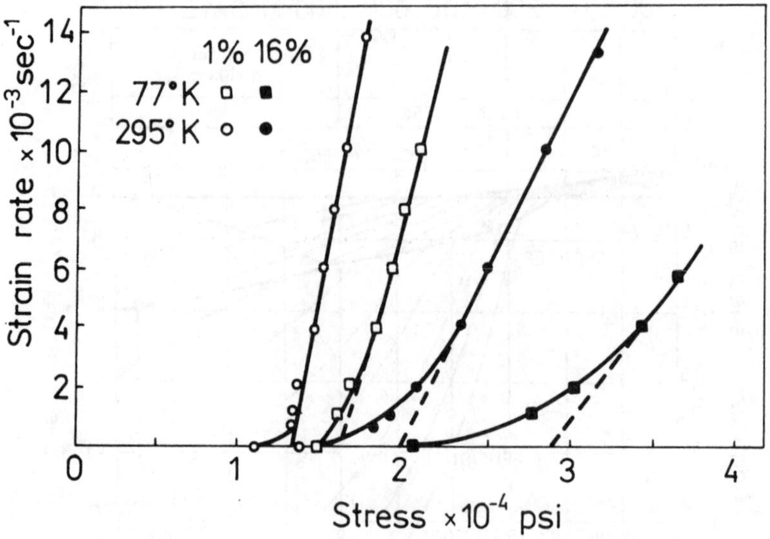

Figure 8: The strain rate dependence of the yield stress for polycrystalline aluminium. Data of HAUSER, SIMMON and DORN [1961] are plotted on a linear scale

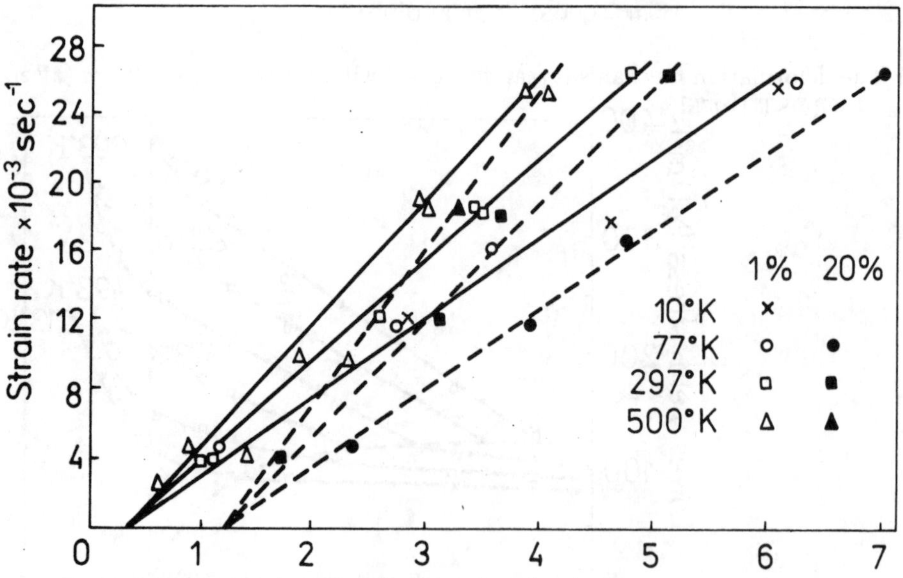

Figure 9: The strain rate dependence of the flow stress of aluminium single crystals (after KUMAR, HAUSER and DORN [1968])

When the ratio t_B/t_S increases then the dislocation velocity (4.12) can be approximated by

$$v = \frac{AL^{-1}}{t_B} \tag{4.16}$$

for the phonon damping mechanism (cf. NABARRO [1967]).

Experimental justification of the interaction of the thermally activated and phonon damping mechanisms has been given by KUMAR and KUMBLE [1969] and SHIOIRI, SATOH and NISHIMURA [1979]. Particular importance have experimental investigations performed by CAMPBELL and FERGUSON [1970], HAUSER, SIMMONS and DORN [1961] and KUMAR, HAUSER and DORN [1968], cf. Figs.8 and 9.

4.4 Hardening and softening effects

To describe hardening effects the internal state variables $g^{(\alpha)}$ have been introduced (cf. PERZYNA [1988]). The fundamental hardening law is assumed in the form as follows (cf. ASARO [1983], ASARO and NEEDLEMAN [1985] and PEIRCE, ASARO and NEEDLEMAN [1982,1983])

$$\dot{g}^{(\alpha)} = \sum_{\beta=1}^{n} h_{\alpha\beta}(\vartheta, \gamma)\dot{\gamma}^{(\beta)} \quad \text{if} \quad \tau^{(\alpha)} > g^{(\alpha)}, \tag{4.17}$$

where $h_{\alpha\beta}$ is the temperature and shear strain dependent instantaneous plane hardening rate and γ is defined as

$$\gamma = \sum_{\alpha=1}^{n} \gamma^{(\alpha)}, \tag{4.18}$$

$h_{\alpha\alpha}$ is the selfhardening rate on system α and $h_{\alpha\beta}(\alpha \neq \beta)$ is the latent-hardening rate of system α caused by slip on system β .

If the work hardening law is postulated in the form

$$g^{(\alpha)} = g^{(\alpha)}(\gamma, \vartheta), \tag{4.19}$$

then $g^{(\alpha)}$ $(\alpha = 1, ...n)$ do not play a role of the internal state variables.

For practical applications a great importance has the thermal-softening effect. Of course, to describe thermal softening effect we have to postulate a particular form for the relationship $h_{\alpha\beta} = h_{\alpha\beta}(\gamma, \vartheta)$ or $g^{(\alpha)} = g^{(\alpha)}(\gamma, \vartheta)$. Serval empirical relationships have been proposed basing on experimental results. One example of the hardening-softening relationship will be given in Sec. 4.8.

4.5 Thermodynamic restrictions

Let introduce the two fundamental postulates: (i) Existence of the free energy function. It is assume that free energy function is given by

$$\psi = \widehat{\psi}(\mathbf{e}, \mathbf{F}, \vartheta; \gamma^{(\alpha)}, g^{(\alpha)}). \tag{4.20}$$

(ii) The axiom of entropy production. For any regular motion of crystal (called $\mathcal{B}$) the constitutive functions are assumed to satisfy the reduced dissipation inequality (3.23).

Then the constitutive assumption (4.20) and the evolution equations for $\gamma^{(\alpha)}$ and $g^{(\alpha)}$ discussed in Sections 4.3 and 4.4 together with the reduced dissipation inequality (3.23) lead to the results as follows

$$\begin{aligned} \boldsymbol{\tau} &= \rho_{Ref}\frac{\partial\widehat{\psi}}{\partial\mathbf{e}}, \qquad \eta = -\frac{\partial\widehat{\psi}}{\partial\vartheta}, \\ \vartheta\widehat{i} &- \frac{1}{\rho\vartheta}\mathbf{q}\cdot\mathrm{grad}\vartheta \geq 0, \end{aligned} \tag{4.21}$$

where

$$\vartheta\widehat{i} = -\sum_{\alpha=1}^{n}\left\{\left[\frac{\partial\widehat{\psi}}{\partial\gamma^{(\alpha)}} + \sum_{\beta=1}^{n}\frac{\partial\widehat{\psi}}{\partial g^{(\beta)}}h_{\beta\alpha}(\vartheta,\gamma)\right]\dot{\gamma}^{(\alpha)}\right\} \tag{4.22}$$

denotes the rate of the internal dissipation.

Introducing the denotation

$$\tau^{(\alpha)} = -\rho\left[\frac{\partial\widehat{\psi}}{\partial\gamma^{(\alpha)}} + \sum_{\beta=1}^{n}\frac{\partial\widehat{\psi}}{\partial g^{(\beta)}}h_{\beta\alpha}(\gamma,\vartheta)\right] \tag{4.23}$$

we have

$$\vartheta\widehat{i} = \sum_{\alpha=1}^{n}\frac{1}{\rho}\tau^{(\alpha)}\dot{\gamma}^{(\alpha)}. \tag{4.24}$$

This implies additional restriction in the form of the identity

$$\rho_{Ref}\frac{\partial\widehat{\psi}}{\partial\mathbf{e}} : \mathbf{N}^{(\alpha)} = -\rho\left[\frac{\partial\widehat{\psi}}{\partial\gamma^{(\alpha)}} + \sum_{\beta=1}^{n}\frac{\partial\widehat{\psi}}{\partial g^{(\beta)}}h_{\beta\alpha}(\vartheta,\gamma)\right]. \tag{4.25}$$

4.6 Rate type constitutive relation

This consideration follows the formulation of the rate type constitutive relation presented by HILL and RICE [1972,1973], cf. also ASARO [1982, 1983].

The basic assumption is that the response of the material is always of Green type, i.e., that the free energy function exists, and is given in the form (4.20).

Let assume that the thermodynamic restrictions give the results presented in the form (4.21). We say that a set of the internal state variables $(\gamma^{(\alpha)}, g^{(\alpha)})$, of course together with the evolution equations and the initial values for $\gamma^{(\alpha)}$ and $g^{(\alpha)}$, describes the prior history of inelastic deformation of the crystal.

Let take the stress relation $(4.21)_1$ and let consider purely elastic deformation - temperature process. Such a process is characterized by fixed prior history, i.e. keeping the internal state variables $\gamma^{(\alpha)}$ and $g^{(\alpha)}$ constant. For such a process let operate on the stress relation $(4.21)_1$ with the Lie derivative. The result of this operation is as follows

$$\overset{el}{(L_{\boldsymbol{v}}\boldsymbol{\tau})} = \mathcal{L}^e \cdot \mathbf{d}^e - \mathbf{z}\dot{\vartheta}$$

where

$$\mathcal{L}^e = \rho_{Ref}\frac{\partial^2\widehat{\psi}}{\partial \mathbf{e}^2}, \qquad \mathbf{z} = -\rho_{Ref}\frac{\partial^2\widehat{\psi}}{\partial \mathbf{e}\partial\vartheta}, \tag{4.26}$$

$$\overset{el}{(L_{\boldsymbol{v}}\boldsymbol{\tau})}{}^{ab} = \dot{\tau}^{ab} - \tau^{ac}(d^e_{dc} + \omega^e_{dc})g^{db} - \tau^{cd}(d^e_{dc} + \omega^e_{dc})g^{ad}. \tag{4.27}$$

On the other hand for general deformation - temperature process (when the internal state variables $\gamma^{(\alpha)}$ and $g^{(\alpha)}$ vary) we have

$$(L_{\boldsymbol{v}}\boldsymbol{\tau})^{ab} = \dot{\tau}^{ab} - \tau^{ac}(d_{dc} + \omega_{dc})g^{db} - \tau^{cb}(d_{dc} + \omega_{dc})g^{ad}. \tag{4.28}$$

Diference between these two rates is

$$\overset{el}{(L_{\boldsymbol{v}}\boldsymbol{\tau})}{}^{ab} - (L_{\boldsymbol{v}}\boldsymbol{\tau})^{ab} = \tau^{ac}(d^p_{dc} + \omega^p_{dc})g^{db} + \tau^{cb}(d^p_{dc} + \omega^p_{dc})g^{ad}. \tag{4.29}$$

Denoting by

$$\mathbf{b}^{(\alpha)} = (\mathbf{N}^{(\alpha)} + \mathbf{W}^{(\alpha)})\cdot\boldsymbol{\tau} + \boldsymbol{\tau}\cdot(\mathbf{N}^{(\alpha)} - \mathbf{W}^{(\alpha)}) \tag{4.30}$$

we finally have the resulting constitutive law of the rate type in the form [6]

$$L_{\boldsymbol{v}}\boldsymbol{\tau} = \mathcal{L}^e \cdot \mathbf{d} - \sum_{\alpha=1}^{n}[\mathcal{L}^e\cdot\mathbf{N}^{(\alpha)} + \mathbf{b}^{(\alpha)}]\dot{\gamma}^{(\alpha)} - \mathbf{z}\dot{\vartheta}. \tag{4.31}$$

The result obtained in the form (4.31) has similar shape as the rate equation formulated by HILL and RICE [1972] (cf. also recent paper by ASARO [1983]). Main difference is in the definition of the Kirchhoff stress rate. HILL and RICE [1972] used the Zaremba–Jaumann rate, while in the present paper the Lie derivative of the Kirchhoff stress tensor was used. That is why the term proportional to $\mathbf{b}^{(\alpha)}$, which represents the stiffening of the lattice due to microscopic phenomena associated with continued slipping depend in our case on the rate of plastic spin as well as on the rate of plastic deformation, while in the HILL and RICE paper [1972] similar term, denoted by $\boldsymbol{\beta}^{(\alpha)}$, depends on the rate of plastic spin only.

4.7 Thermo-mechanical couplings

Substituting $\dot{\psi}$ into (3.22) and taking advantage of the results (4.21) gives

$$\rho\vartheta\dot{\eta} = -\text{div}\mathbf{q} + \rho\vartheta\widehat{\imath}. \tag{4.32}$$

[6] By performing a Legendre transformation it is possible to obtain the inverse form of the rate type constitutive relation (4.31).

Operating on the entropy relation $(4.21)_2$ with the Lie derivative and substituting the result together with the Fourier constitutive law for the heat vector field into (4.32) we obtain the heat conduction equation in the form as follows

$$\begin{aligned}\rho c_p \dot{\vartheta} &= \operatorname{div}(k \operatorname{grad}\vartheta) + \vartheta \frac{\rho}{\rho_{Ref}} \frac{\partial \boldsymbol{\tau}}{\partial \vartheta} : \mathbf{d} + \rho \vartheta \widehat{i} \\ &+ \rho \vartheta \left[\sum_{\alpha=1}^{n} \left(\frac{\partial^2 \widehat{\psi}}{\partial \vartheta \partial \gamma^{(\alpha)}} + \sum_{\beta=1}^{n} \frac{\partial^2 \widehat{\psi}}{\partial \vartheta \partial g^{(\beta)}} h_{\beta\alpha} \right) \dot{\gamma}^{(\alpha)} \right],\end{aligned} \tag{4.33}$$

where

$$c_p = -\vartheta \frac{\partial^2 \widehat{\psi}}{\partial \vartheta^2} \tag{4.34}$$

denotes the specific heat.

On the right hand side of Eq.(4.33) the term $\rho\vartheta\widehat{i} = \sum_{\alpha=1}^{n} \tau^{(\alpha)} \dot{\gamma}^{\alpha}$ represents the internal heating generated by the rate of internal dissippation. This term describes the main contribution to the thermo–mechanical coupling phenomena.

Beside this main term there appear two additional terms responsible for the cross coupling effects, namely the term prortional to $\frac{\partial \boldsymbol{\tau}}{\partial \vartheta} : \mathbf{d}$ generated by the temperature dependence of the stress relation, and the term proportional to $\sum_{\alpha=1}^{n} \left[\frac{\partial^2 \widehat{\psi}}{\partial \vartheta \partial \gamma^{(\alpha)}} + \right.$ $\left. + \sum_{\beta=1}^{n} \frac{\partial^2 \widehat{\psi}}{\partial \vartheta \partial g^{(\beta)}} h_{\beta\alpha}(\vartheta, \gamma) \right] \dot{\gamma}^{(\alpha)}$ implied by the temperature dependence of the generalized forces conjugate to the internal state variables $\gamma^{(\alpha)}$ and $g^{(\alpha)}$. The first of these two additional terms has not dissipative character, while the second is very dissipative in its nature.

4.8 Rate independent response

Let us consider the viscoelastic flow model described by the evolution equations (4.10) and (4.17). From (4.10) we have

$$\tau^{(\alpha)} = g_v^{(\alpha)} \left\{ 1 + \Phi^{-1} \left(\frac{\dot{\gamma}^{(\alpha)}}{\eta_v^{(\alpha)}} \right) \right\}. \tag{4.35}$$

Let the viscosity coeficients $\eta_v^{(\alpha)}$ tend to infinity, i.e. $\eta_v^{(\alpha)} \to \infty$, then from (4.35)

$$\tau^{(\alpha)} = g_v^{(\alpha)} \tag{4.36}$$

and (4.10) gives

$$\dot{\gamma}^{(\alpha)} = \Lambda^{(\alpha)}. \tag{4.37}$$

The conficient $\Lambda^{(\alpha)}$ is undetermined in the range of the viscoplastic response. The rate of shearings $\dot{\gamma}^{(\alpha)}$ can be determined in the range of the plastic response from the hardening law (4.17) together with (4.36). We have the result as follows

$$\dot{\gamma}^{(\alpha)} = \sum_{\beta=1}^{n} h^{-1}_{\alpha\beta}(\gamma\vartheta)\dot{\tau}^{(\beta)}. \tag{4.38}$$

The rate type constitutive relation takes the form

$$L_{\boldsymbol{v}}\boldsymbol{\tau} = \mathcal{L}^e \cdot \mathbf{d} - \mathbf{z}\dot{\vartheta} - \sum_{\alpha=1}^{n}[\mathcal{L}^e \cdot \mathbf{N}^{(\alpha)} + \mathbf{b}^{(\alpha)}] \sum_{\beta=1}^{n} h^{-1}_{\alpha\beta}(\gamma,\vartheta)\dot{\tau}^{(\beta)}. \tag{4.39}$$

The heat conduction equation has the form

$$\begin{aligned} \rho c_p \dot{\vartheta} &= \text{div } (k \text{ grad } \vartheta) + \vartheta \frac{\rho}{\rho_{Ref}} \frac{\partial \boldsymbol{\tau}}{\partial \vartheta} : \mathbf{d} + \sum_{\alpha=1}^{n}\sum_{\beta=1}^{n} \tau^{(\alpha)} h^{-1}_{\alpha\beta}(\gamma,\vartheta)\dot{\tau}^{(\beta)} \\ &+ \rho\vartheta \left[\sum_{\alpha=1}^{n}\sum_{\beta=1}^{n} \left(\frac{\partial^2 \widehat{\psi}}{\partial\vartheta\partial\gamma^{(\alpha)}} h^{-1}_{\alpha\beta}(\gamma,\vartheta) + \frac{\partial^2 \widehat{\psi}}{\partial\vartheta\partial g^{(\beta)}} \right) \dot{\tau}^{(\beta)} \right]. \end{aligned} \tag{4.40}$$

We shall have different situation when the hardening-softening relationship will be given by the expression

$$\tau^{(\alpha)} = g^{(\alpha)}(\gamma,\vartheta), \tag{4.41}$$

e.g.

$$g^{(\alpha)}(\gamma,\vartheta) = [\kappa_1^{(\alpha)} + (\kappa_0^{(\alpha)} - \kappa_1^{(\alpha)})e^{-h_1(\vartheta)\gamma}] \left[1 - \omega_0 (\frac{\vartheta - \vartheta_0}{\vartheta_0})^m\right]. \tag{4.42}$$

Then, of course $g^{(\alpha)}$ $(\alpha = 1, ..., n)$ do not play a role of the internal state variables. For rates we have relationship

$$\dot{\tau}^{(\alpha)} = \sum_{\beta=1}^{n} \frac{\partial g^{(\alpha)}}{\partial \gamma^{(\beta)}} \dot{\gamma}^{(\beta)} + \frac{\partial g^{(\alpha)}}{\partial \vartheta}\dot{\vartheta} \tag{4.43}$$

or

$$\dot{\gamma}^{(\alpha)} = \sum_{\beta=1}^{n} h^{-1}_{\alpha\beta}\dot{\tau}^{(\beta)} + h^{(\alpha)}\dot{\vartheta} \tag{4.44}$$

where

$$h_{\alpha\beta} = \frac{\partial g^{(\alpha)}}{\partial \gamma^{(\beta)}} \quad \text{and} \quad -\frac{\partial g^{(\alpha)}}{\partial \vartheta} \left(\sum_{\beta=1}^{n} \frac{\partial g^{(\alpha)}}{\partial \gamma^{(\beta)}} \right)^{-1} = h^{(\alpha)}. \tag{4.45}$$

For adiabatic process we obtain a set of two coupled evolution equations

$$\begin{aligned} L_{\boldsymbol{v}}\boldsymbol{\tau} &= \mathcal{L}^e \cdot \mathbf{d} - \mathbf{z}\dot{\vartheta} - \sum_{\alpha=1}^{n} \left[\mathcal{L}^e \cdot \mathbf{N}^{(\alpha)} + \mathbf{b}^{(\alpha)}\right] \left(\sum_{\beta=1}^{n} h^{-1}_{\alpha\beta}\dot{\tau}^{(\beta)} + h^{(\alpha)}\dot{\vartheta} \right) \\ \rho c_p \dot{\vartheta} &= \vartheta \frac{\rho}{\rho_{Ref}} \frac{\partial \boldsymbol{\tau}}{\partial \vartheta} : \mathbf{d} + \sum_{\alpha=1}^{n}\sum_{\beta=1}^{n} \tau^{(\alpha)} \left(h^{-1}_{\alpha\beta}\dot{\tau}^{(\beta)} + h^{(\alpha)}\dot{\vartheta} \right) \\ &+ \rho\vartheta \sum_{\alpha=1}^{n}\sum_{\beta=1}^{n} \frac{\partial^2 \widehat{\psi}}{\partial\vartheta\partial\gamma^{(\alpha)}} \left(h^{-1}_{\alpha\beta}\dot{\tau}^{(\beta)} + h^{(\alpha)}\dot{\vartheta} \right). \end{aligned} \tag{4.46}$$

This set of equations has fundamental importance in the investigation of the conditions for localization of plastic deformation along shear bands for monocrystals.

5 THERMODYNAMICS OF RATE DEPENDENT PLASTICITY (VISCOPLASTICITY)

5.1 Definitions and assumptions

An exhanstive investigation of the thermodynamic, large deformation and rate dependent theory of plasticity for polycrystalline solids with a finite set of the internal state variables constitutes the subject of the present chapter.

The main question arises how the propositions presented in the previous chapter for crystals can be extended to develope thermodynamic theory for large deformation, temperature and rate dependent of polycrystalline solids.

In the description of crystal we introduced a set of internal state variables which consists of the shearings $\gamma^{(\alpha)}$ and the critical shear stresses $g^{(\alpha)}$. The shearings $\gamma^{(\alpha)}$ described the plastic deformation and plastic spin, while the critical shear stresses $g^{(\alpha)}$ described the hardening - softening effects.

To describe all these effects in macroscopic - phenomenological theory of polycrystalline solids we have to introduce again such notions as isotropic and kinematic hardening effects. We shall also take advantage of micro-damage process to model softening of a material.

It is postulated that to describe the constitutive properties of polycrystalline solids we define the intrinsic state again by s (cf. Eq.(2.18)) and the internal state vector $\boldsymbol{\mu}$ by Eq.(3.50).

It is noteworthy that such internal state variable characterization would seem suitable but it is not unique. One could introduce the infinite and complete set of internal state variables which would describe any detail of the dislocation rearrangement in polycrystals. Since it is not likely that all the particulars on a microscopic scale are important for the macroscopic elasto – viscoplastic behaviour of the polycrystalline aggregate, but rather certain averages only are macroscopically effective and important for phenomenological description, one is inclined to introduce a finite set of the internal state variables.

It is assumed that the free energy function exists and has the form (2.19) with the internal vector $\boldsymbol{\mu}$ given by (3.50).

Let

$$\tilde{\varphi} = f - \kappa = 0 \tag{5.1}$$

denotes the quasi-static yield criterion for rate independent plastic response, where f and κ are determined by (3.62) and (3.63), respectively.

In the rate dependent range of plastic flow process the viscoplastic evolution law for crystalline damaged solids is postulated in the form

$$\mathbf{d}^P = \frac{\lambda}{\beta} \langle \Phi(f-\kappa) \rangle \mathbf{P}, \qquad \mathbf{P} = \frac{1}{2\sqrt{\tilde{J}_2}} \frac{\partial f}{\partial \boldsymbol{\tau}}, \tag{5.2}$$

where λ denotes the viscosity coefficient, β is the control function, Φ represents the viscoplastics overstress function, the symbol $\langle \Phi(f-\kappa) \rangle$ is understood according to the definiction (cf. the viscoplastic kinetic law (4.10)).

$$\langle \Phi(f-\kappa) \rangle = \begin{cases} 0 & \text{if} \quad f-\kappa \leq 0, \\ \Phi(f-\kappa) & \text{if} \quad f-\kappa > 0. \end{cases} \tag{5.3}$$

The viscoplastic overstress function Φ can be determined basing on available experimental results for dynamic loading processes.

5.2 Kinematic hardeming and mirco-damage process

To determine the evolution equation for the internal state vector $\boldsymbol{\mu} = (\zeta, \boldsymbol{\alpha}, \xi)$ let first postulate the evolution equation for the back stress $\boldsymbol{\alpha}$ as the linear combination of the PRAGER and ZIEGLER kinematic hardening rules (3.54).

Assuming again the geometrical relation (3.67) we have

$$a_2 = (r - a_1)\frac{\lambda}{\beta} \langle \Phi(f-\kappa) \rangle \frac{\mathbf{P} : \mathbf{Q}}{\tilde{\boldsymbol{\tau}} : \mathbf{Q}}. \tag{5.4}$$

Denoting

$$a_1 = r_1, \qquad r_2 = r - a_1 \tag{5.5}$$

we obtain the evolution law for the back stress $\boldsymbol{\alpha}$ in the form

$$L_{\boldsymbol{v}}\boldsymbol{\alpha} = \left(r_1 \mathbf{P} + r_2 \frac{\mathbf{P} : \mathbf{Q}}{\tilde{\boldsymbol{\tau}} : \mathbf{Q}} \tilde{\boldsymbol{\tau}} \right) \frac{\lambda}{\beta} \langle \Phi(f-\kappa) \rangle . \tag{5.6}$$

Let postulate the evolution for the porosity parameter ξ in the form (3.59), with additional assumption

$$k_2 = 0. \tag{5.7}$$

Finally we have the evolution equation

$$L_{\boldsymbol{v}}\boldsymbol{\mu} = \mathbf{m}(s)\frac{\lambda}{\beta} \langle \Phi(f-\kappa) \rangle , \tag{5.8}$$

where the material function $\mathbf{m}(s)$ is determined by (3.84).

5.3 Rate type constitutive structure and thermo-mechanical conplings

The axiom of entropy production for the constitutive assumption (5.2) and the evolution equations (5.6) and (5.8) leads to the results (3.24), where the rate of internal dissipation is determined by

$$\vartheta \hat{i} = -\frac{\partial \hat{\psi}}{\partial \boldsymbol{\mu}} \cdot \mathbf{m}(s) \frac{\lambda}{\beta} \langle \Phi(f - \kappa) \rangle . \tag{5.9}$$

Operating on the stress relation (3.24) with the Lie derivetive, keeping the history constant (i.e. the internal state vector $\boldsymbol{\mu} = \text{const}$), we obtain

$$L_{\boldsymbol{v}} \boldsymbol{\tau} = \mathcal{L}^e : \mathbf{d} - \mathcal{L}^{th} \dot{\vartheta} - \frac{\lambda}{\beta} \langle \Phi(f - \kappa) \rangle \, \mathcal{L}^e : \mathbf{P}. \tag{5.10}$$

Operating on the entropy relation $(3.24)_2$ with Lie derivative and substituting the result together with the Fourier constitutive law for the heat flux (3.36) into (3.35) we obtain the heat conduction equation in the form

$$\rho c_p \dot{\vartheta} = \text{div}\, (k \text{ grad } \vartheta) + \frac{\rho}{\rho_{Ref}} \vartheta \frac{\partial \boldsymbol{\tau}}{\partial \vartheta} : \mathbf{d} + \rho \chi \frac{\lambda}{\beta} \langle \Phi(f - \kappa) \rangle \tag{5.11}$$

where

$$\chi = -\left(\frac{\partial \hat{\psi}}{\partial \boldsymbol{\mu}} - \vartheta \frac{\partial^2 \hat{\psi}}{\partial \vartheta \partial \boldsymbol{\mu}} \right) \cdot \mathbf{m}(s). \tag{5.12}$$

5.4 Rate independent plasticity as a limit case

From (5.2) we have

$$f - \kappa = \Phi^{-1} \left[\frac{\beta}{\lambda} \dot{\epsilon}^P \left(\frac{2}{3} \mathbf{P} : \mathbf{P} \right)^{-\frac{1}{2}} \right] . \tag{5.13}$$

Let $\beta \to 0$, then (5.13), (5.2) and (5.8) give the results

$$f - \kappa = 0, \qquad \mathbf{d}^P = \Lambda \mathbf{P}, \qquad L_{\boldsymbol{v}} \boldsymbol{\mu} = \Lambda \mathbf{m}(s), \tag{5.14}$$

respectively.
The coefficient Λ has to be detrmined from the consistency condition

$$\dot{f} - \dot{\kappa} = 0 \tag{5.15}$$

which gives

$$\Lambda = \left\langle \frac{1}{H} \{ \mathbf{P} : [\dot{\boldsymbol{\tau}} - (\mathbf{d} \cdot \boldsymbol{\alpha} + \boldsymbol{\alpha} \cdot \mathbf{d})] + \pi \dot{\vartheta} \} \right\rangle . \tag{5.16}$$

The hardening moduls $H = H^* + H^{**}$ consists of the isotropic hardening modulus H^* and the kinematic hardening modulus H^{**}, cf. Eqs(3.70).

6 ALTERNATIVE FORMULATION OF THERMOVISCOPLASTICITY

6.1 Basic postulates

In recent years the particular model of viscoplasticity theory of a porous material has been used in several studies of locilization and ductile fracture, cf. PAN , SAJE and NEEDLEMAN [1983], NEEDLEMAN and TVERGAARD [1984] and TVERGAARD and NEEDLEMAN [1986]. In this model rate sensitivity effect is incorporated by the assumption that the yield stress of matrix material does depend on the strain rate. In other words instead of the overstress conception of viscoplasticity an idea of the yield stress *vs* strain rate dependence of matrix material is used.

Thus, this theory of viscoplasticity is developed within the framework of plasticity provided the yield stress of matrix material is strain rate dependent function.

To do this let assume the isotropic hardeming parameter κ as the material function

$$\kappa = \hat{\kappa}(\epsilon^P, \xi, \vartheta, \sigma_M), \tag{6.1}$$

where σ_M represents the yield stress of the matrix material.

As an example we can consider

$$\kappa = \sigma_M^2 \left[\iota + (1-\iota)e^{-h(\vartheta)\epsilon^P}\right]^2 \left[1 - \frac{\xi}{\xi^F(\vartheta)}\right](1 - w\bar{\vartheta}), \tag{6.2}$$

where $\iota = \sigma_s/\sigma_M$ denotes the ratio between the saturated stress and the yield stress of the matrix material.

Let assume all postulates introduced for inviscid plasticity in Section 3.8 with $k_2 = 0$ and let replace the isotropic hardening softening function (3.63) by (6.2).

Let postulate that in the matrix material the microcopic effective plastic strain rate $\dot{\epsilon}_M^P$ is represented by the temperature and rate dependent kinetic law as follows:

(i) Dislocation creep mechanism (cf. with. Eq.(4.9))

$$\dot{\epsilon}_M^P = \dot{\epsilon}_0 \left[\frac{\sigma_M}{g(\epsilon_M^P)}\right]^{\frac{1}{m}}, \tag{6.3}$$

where m is the strain rate hardening exponent, $\dot{\epsilon}_0$ is a reference strain rate and ϵ_M^P is the current value of the effective plastic strain representing the actual microscopic strain state in the matrix material. The function $g(\epsilon_M^P)$ represents the effective tensile flow stress in the matrix material in a tensile test carried out at a strain rate such that $\dot{\epsilon}_M^P = \dot{\epsilon}_0$, and σ_M denotes the yield stress of the matrix material.

(ii) Viscoplastic flow mechanism (cf. Eq.(4.10))

$$\dot{\epsilon}_M^P = \dot{\epsilon}_0 \left\langle \Phi \left[\frac{\sigma_M}{g(\epsilon_M^P)} - 1\right]\right\rangle, \tag{6.4}$$

where Φ is the empirical function (the viscoplastic overstress function) and can be determined basing on the avilable experimental results.

(iii) Thermally activated mechanism (cf. Eq.(4.13.))

$$\dot{\epsilon}^P_M = \dot{\epsilon}_0 \left\langle -\exp\left[\frac{\sigma_M}{g(\epsilon^P_M)} - 1\right]\right\rangle, \tag{6.5}$$

where

$$\dot{\epsilon}_0 = A\,L^{-1}\nu b \rho_M \tag{6.6}$$

and $g(\epsilon^P_M)$ is interpreted as the athermal stress.

Additionally we assume the equivalent plastic work expression

$$\tilde{\sigma} : \mathbf{d}^P = (1-\xi)\sigma_M \dot{\epsilon}^P_M. \tag{6.7}$$

6.2 Constitutive equations and loading criterion

The further precedure is the same as in the case of the theory of plasticity of porous solids developed in Section 3.8. Finally we obtain the flow rule and the evolution equation for the internal state vector $\boldsymbol{\mu}$ in the form as follows

$$\begin{aligned} \mathbf{d}^P &= \left\langle \frac{1}{H}\left\{\mathbf{P} : [\dot{\boldsymbol{\tau}} - (\mathbf{d}\cdot\boldsymbol{\alpha} + \boldsymbol{\alpha}\cdot\mathbf{d})] + \pi\dot{\vartheta} - k\dot{\sigma}_M\right\}\right\rangle \mathbf{P}, \\ L_{\boldsymbol{v}}\boldsymbol{\mu} &= \mathbf{m}(s)\left\langle \frac{1}{H}\left\{\mathbf{P} : [\dot{\boldsymbol{\tau}} - (\mathbf{d}\cdot\boldsymbol{\alpha} + \boldsymbol{\alpha}\cdot\mathbf{d})] + \pi\dot{\vartheta} - k\dot{\sigma}_M\right\}\right\rangle, \end{aligned} \tag{6.8}$$

where

$$k = \frac{1}{2\sqrt{\tilde{J}_2}}\frac{\partial\hat{\kappa}}{\partial\sigma_M}. \tag{6.9}$$

The loading process takes place when

$$\left\langle \frac{1}{H}\left\{\mathbf{P} : [\dot{\boldsymbol{\tau}} - (\mathbf{d}\cdot\boldsymbol{\alpha} + \boldsymbol{\alpha}\cdot\mathbf{d})] + \pi\dot{\vartheta} - k\dot{\sigma}_M\right\}\right\rangle \neq 0. \tag{6.10}$$

6.3 Rate type relation and thermo-mechanical couplings

Let proceed again as in Section 3.8. Then we obtain the rate type constitutive equation for the Kirchhoff stress tensor $\boldsymbol{\tau}$ and the heat conduction equation in the following form

$$\begin{aligned} L_{\boldsymbol{v}}\boldsymbol{\tau} &= \mathcal{L} : \mathbf{d} - \mathbf{z}\dot{\vartheta} + \mathcal{K}\dot{\sigma}_M, \\ \rho c_p \dot{\vartheta} &= \operatorname{div}(k \operatorname{grad} \vartheta) + \vartheta \frac{\rho}{\rho_{Ref}}\frac{\partial\boldsymbol{\tau}}{\partial\vartheta} : \mathbf{d} \\ &\quad + \rho\bar{\chi}\left\langle \frac{1}{H}\left\{\mathbf{P} : [\dot{\boldsymbol{\tau}} - (\mathbf{d}\cdot\boldsymbol{\alpha} + \boldsymbol{\alpha}\cdot\mathbf{d})] + \pi\dot{\vartheta} - k\dot{\sigma}_M\right\}\right\rangle, \end{aligned} \tag{6.11}$$

where we have introduced the new denotation

$$\mathcal{K} = \left[\mathbf{I} - \frac{\frac{1}{H}\mathcal{L}^e : \mathbf{PP}}{1 + \frac{1}{H}(\mathcal{L}^e : \mathbf{P}) : \mathbf{P}}\right] \cdot k\mathcal{L}^e : \mathbf{P}. \tag{6.12}$$

7 ADIABATIC PROCESS

7.1 Fundamental equation for adiabatic proces (for J_2 – flow theory)

The thermodynamic process is assumed to be adiabatic, i.e.,

$$\mathbf{q} = 0. \tag{7.1}$$

This assumption is satisfied for the thermo-plastic flow process before localization takes place, then the distribution of plastic deformation as well as rate of plastic deformation is homogeneous.

The term div(k grad ϑ) in the heat conduction equation (3.37) vanishes. Then to describe thermo-mechanical constitutive properties of a material we have two coupled evolution equations, namely for the Kirchoff stress tensor (3.31) and for temperature

$$c_p\dot{\vartheta} = \vartheta \frac{1}{\rho_{Ref}} \frac{\partial \boldsymbol{\tau}}{\partial \vartheta} : \mathbf{d} + \chi \left\langle \frac{1}{H}(\mathbf{P} : \dot{\boldsymbol{\tau}} + \pi\dot{\vartheta}) \right\rangle. \tag{7.2}$$

Equation (7.2) can be written in the form

$$\dot{\vartheta} = \frac{\vartheta H}{\rho_{Ref}(c_p H - \chi\pi)} \frac{\partial \boldsymbol{\tau}}{\partial \vartheta} : \mathbf{d} + \frac{\chi}{c_p H - \chi\pi} \mathbf{P} : \dot{\boldsymbol{\tau}}. \tag{7.3}$$

Taking advantage of Eq.(2.13) in (7.3) and substituting the result into Eq.(3.31) gives

$$L_{\boldsymbol{\upsilon}}\boldsymbol{\tau} = \ \mathbb{L} : \mathbf{d}, \tag{7.4}$$

where

$$\mathbb{L} = \left[\mathbf{I} + \frac{\chi}{c_p H - \chi\pi}\mathbf{z}\mathbf{P}\right]^{-1} \cdot \left\{\mathcal{L} - \frac{1}{c_p H - \chi\pi}\left[\frac{\vartheta H}{\rho_{Ref}}\mathbf{z}\frac{\partial \boldsymbol{\tau}}{\partial \vartheta} + 2\chi\mathbf{z}(\mathbf{P}\cdot\boldsymbol{\tau} + \boldsymbol{\tau}\cdot\mathbf{P})\right]\right\}. \tag{7.5}$$

The result (7.4) is of great importance to constitutive modelling of thermo-mechanical coupling phenomena in adiabatic processes and to the investigation of the conditions for the localization of plastic deformation.

It is noteworthy that the fundamental matrix $\mathbb{L}$ (cf. Eq.(7.5)) in the evolution equation (7.4) desribes all introduced thermo-mechanical coupling effects.

Proceed similarly but replacing the Lie derivative equation for the Kirchhoff stress tensor (3.31) by the Zaremba–Jaumann rate equation (3.33) we get

$$\overset{\nabla}{\boldsymbol{\tau}} = \widehat{\mathbb{L}} : \mathbf{d}, \tag{7.6}$$

where

$$\widehat{\mathbb{L}} = \left[I + \frac{\chi}{c_p H - \chi\pi}\mathbf{z}\mathbf{P}\right]^{-1} \cdot \left[\hat{\mathcal{L}} - \frac{H\vartheta}{\rho_{Ref}(c_p H - \chi\pi)}\mathbf{z}\frac{\partial \boldsymbol{\tau}}{\partial \vartheta}\right]. \tag{7.7}$$

It is worthwhile to point out once more that evolution equation (7.6) is materially isomorphic with (7.4).

7.2 Main contribution to thermo-mechanical coupling (for $\mathbf{J}_2$ – flow theory)

Let us consider the heat condution equation (3.47). For adiabatic process this equation takes form

$$\begin{aligned}\rho c_p \dot{\vartheta} &= \vartheta \frac{\rho}{\rho_{Ref}} \frac{\partial \boldsymbol{\tau}}{\partial \vartheta} : \mathbf{d} + \zeta \boldsymbol{\tau} : \mathbf{d}^P \\ &+ \frac{\rho}{\rho_{Ref}} \vartheta \frac{\partial^2 \widehat{\psi}}{\partial \vartheta \partial \boldsymbol{\mu}} \cdot \left[\frac{\partial^2 \widehat{\varphi}}{\partial \boldsymbol{\tau} \partial \boldsymbol{\mu}} \right]^{-1} : \mathbf{d}^P. \end{aligned} \tag{7.8}$$

The second term on the right hand side represents main contribution to the thermo-mechanical coupling phenomena. It generates the internal heating caused by the rate of internal dissipation during the adiabatic process considered.

The first and third terms represent the cross coupling effects, the first is caused by the dependence of the stress tensor on temperatute while the third is induced by the same dependence of the generalized force conjugates to the internal state vector $\boldsymbol{\mu}$.

The cross coupling effects influence the evolution of temperature (cf. Eq.(7.8)) through the second order terms when compare with the internal dissipation term. Their contribution to internal heating during the adiabatic process considered is small.

This suggests that these two terms can be neglected in some considerations like the investigation of the conditions of the localization of plastic deformation along shear band.

So, it is reasonable to consider the evolution equation for temperature in the form

$$\rho c_p \dot{\vartheta} = \zeta \boldsymbol{\tau} : \mathbf{d}^P, \tag{7.9}$$

where ζ is determined by (3.48).

Let us consider a set of two coupled evolution equations, for the stress tensor (3.31) and for temperature (7.9). This set is reduced again to the fundamental equation of the form

$$L_{\boldsymbol{v}} \boldsymbol{\tau} = \widetilde{\mathbb{L}} : \mathbf{d}, \tag{7.10}$$

where

$$\widetilde{\mathbb{L}} = \left[\mathbf{I} + \frac{\zeta \sqrt{J_2}}{H \rho c_p - \zeta \pi \sqrt{J_2}} \mathbf{z} \mathbf{P} \right]^{-1} \cdot \left\{ \mathcal{L} - \frac{2 \zeta \sqrt{J_2}}{H \rho c_p - \zeta \pi \sqrt{J_2}} \mathbf{z} [\mathbf{P} \cdot \boldsymbol{\tau} + \boldsymbol{\tau} \cdot \mathbf{P}] \right\}. \tag{7.11}$$

Let us consider now a similar set of two coupled evolution equations for the stress tensor (3.33) and for temperature (7.9). This gives

$$\overset{\nabla}{\boldsymbol{\tau}} = \widehat{\widetilde{\mathbb{L}}} : \mathbf{d} \tag{7.12}$$

where

$$\widehat{\widetilde{\mathbb{L}}} = \left[\mathbf{I} + \frac{\zeta \sqrt{J_2}}{H \rho c_p - \zeta \pi \sqrt{J_2}} \mathbf{z} \mathbf{P} \right]^{-1} \cdot \widehat{\mathcal{L}}. \tag{7.13}$$

7.3 Rate independent response of polycrystals (damaged solids)

For adiabatic plastic flow process the evolution equation for temperature can be written in the form (cf. Eq.(3.85))

$$\dot{\vartheta} = \mathbf{M} : L_{\upsilon}\boldsymbol{\tau} + \mathbf{N} : \mathbf{d}, \tag{7.14}$$

where

$$\begin{aligned} \mathbf{M} &= \frac{\bar{\chi}\mathbf{P}}{Hc_p - \bar{\chi}\pi}, \\ \mathbf{N} &= \left[\frac{\vartheta H}{\rho_{Ref}}\frac{\partial \boldsymbol{\tau}}{\partial \vartheta} + \bar{\chi}(\mathbf{P}\cdot\tilde{\boldsymbol{\tau}} + \tilde{\boldsymbol{\tau}}\cdot\mathbf{P})\right](Hc_p - \bar{\chi}\pi)^{-1}. \end{aligned} \tag{7.15}$$

Substituting the evolution equation for temperature into the evolution equation for the Kirchhoff stress tensor $(3.79)_1$ gives

$$L_{\upsilon}\boldsymbol{\tau} = \mathbb{L} : \mathbf{d} \tag{7.16}$$

where

$$\mathbb{L} = \left(\mathbf{I} - \frac{\mathbf{zM}}{1 + \mathbf{z} : \mathbf{M}}\right) \cdot (\mathcal{L} - \mathbf{zN}). \tag{7.17}$$

The last result allows to use in the investigation of criteria for localization along shear band the standard bifurcation method.

It is noteworthy that the fundamental matrix $\mathbb{L}$ describes all introduced thermo-mechanical coupling effects.

The matrix $\mathbb{L}$ can be expressed in the form

$$\mathbb{L} = \bar{\mathcal{L}} + \mathcal{T} + \mathcal{C}, \tag{7.18}$$

where

$$\begin{aligned} \bar{\mathcal{L}} &= \mathcal{L}^e - \frac{\frac{1}{H}\mathcal{L}^e\cdot\mathbf{PP}\cdot\mathcal{L}^e}{1 + \frac{1}{H}(\mathcal{L}^e\cdot\mathbf{P}):\mathbf{P}}, \\ \mathcal{T} &= -\frac{(\mathbf{zM})\cdot\mathcal{L}^e}{1+\mathbf{z}:\mathbf{M}} + \frac{\mathbf{zM}\cdot\left(\frac{1}{H}\mathcal{L}^e\cdot\mathbf{PP}\cdot\mathcal{L}^e\right)}{(1+\mathbf{z}:\mathbf{M})\left[1 + \frac{1}{H}(\mathcal{L}^e\cdot\mathbf{P}):\mathbf{P}\right]}, \\ \mathcal{C} &= -\frac{1}{H}\mathcal{L}^e\cdot\mathbf{P}(\mathbf{P}\cdot\tilde{\boldsymbol{\tau}} + \tilde{\boldsymbol{\tau}}\cdot\mathbf{P}) - \mathbf{zN} + \frac{\mathbf{zM}\cdot\mathbf{zN}}{1+\mathbf{z}:\mathbf{M}} \\ &\quad + \frac{\frac{1}{H^2}(\mathcal{L}^e\cdot\mathbf{PP})\cdot(\mathcal{L}^e\cdot\mathbf{P})(\mathbf{P}\cdot\tilde{\boldsymbol{\tau}} + \tilde{\boldsymbol{\tau}}\cdot\mathbf{P})}{1 + \frac{1}{H}(\mathcal{L}^e\cdot\mathbf{P}):\mathbf{P}}. \end{aligned} \tag{7.19}$$

The matrix $\bar{\mathcal{L}}$ describes the elastic-plastic proporties of damaged solid, $\mathcal{T}$ takes account of the couplings between the elastic-plastic damaged solid properties and thermal effects, while $\mathcal{C}$ arises as a result of the influence for covariance terms (i.e. terms which are generated by the difference between the Lie derivative and the material rate of the Kirchhoff stress tensor) on constitutive proporties of the material.

7.4 Elastic-viscoplastic response

In technological applications the thermodynamic process is frequently considered under adiabatic conditions. Particular importance for technological purposes have adiabatic processes in which one can observe the localization of plastic derofmation along shear bands. The phenomenon of localization is directly related to the adiabatic conditions of the process considered. It is generally accepted that shear bands are narrow zones of highly non-homogeneous deformation that take shape by a thermo-mechanical instability process and accompany the thermo-mechanical coupling effects and the thermal softening of yielding. It has been experimentally proved (cf. HARTLEY, DUFFY and HAWLEY [1987] and MARCHAND and DUFFY [1988]) that shear bands nucleate due to the presense of a local inhomogeneity or defects causing enhanced local deformation and local heating.

The heat conduction equation (5.11) for adiabatic process takes the form

$$\dot{\vartheta} = \frac{1}{c_p \rho_{Ref}} \vartheta \frac{\partial \boldsymbol{\tau}}{\partial \vartheta} : \mathbf{d} + \frac{1}{c_p} \chi \frac{\lambda}{\beta} \langle \Phi(\varphi) \rangle. \tag{7.20}$$

Substituting (7.20) into (5.10) gives

$$L_{\boldsymbol{v}} \boldsymbol{\tau} = \mathcal{E} : \mathbf{d} + \mathcal{F} \tag{7.21}$$

where

$$\begin{aligned} \mathcal{E} &= \left[\mathcal{L}^e - \frac{1}{c_p \rho_{Ref}} \vartheta (\mathcal{L}^{th} \frac{\partial \boldsymbol{\tau}}{\partial \vartheta}) \right], \\ \mathcal{F} &= -\left[\mathcal{L}^e \cdot \mathbf{P} + \mathcal{L}^{th} \frac{\chi}{c_p} \right] \frac{\lambda}{\beta} \langle \Phi(\varphi) \rangle. \end{aligned} \tag{7.22}$$

7.5 Rate dependent response of polycrystals (alternative description)

For adiabatic process from $(6.11)_2$ we have the evolution equation for temperature

$$\dot{\vartheta} = \mathbf{M} : L_{\boldsymbol{v}} \boldsymbol{\tau} + \mathbf{N} : \mathbf{d} + R \dot{\sigma}_M \tag{7.23}$$

where

$$\begin{aligned} \mathbf{M} &= \frac{\bar{\chi} \mathbf{P}}{H c_p - \bar{\chi} \pi}, \\ \mathbf{N} &= \left[\frac{\vartheta H}{\rho_{Ref}} \frac{\partial \boldsymbol{\tau}}{\partial \vartheta} + \bar{\chi} (\mathbf{P} \cdot \tilde{\boldsymbol{\tau}} + \tilde{\boldsymbol{\tau}} \cdot \mathbf{P}) \right] (H c_p - \bar{\chi} \pi)^{-1}, \\ R &= -\frac{\bar{\chi} \frac{\partial \widehat{\kappa}}{\partial \sigma_M}}{2 \sqrt{\widetilde{J}_2} H (H c_p - \bar{\chi} \pi)}. \end{aligned} \tag{7.24}$$

Substituting (7.23) into $(6.11)_1$ yields the fundamental rate equation in the form

$$L_{\upsilon}\tau = \mathbb{L} : \mathbf{d} + \bar{\mathbb{K}}\dot{\sigma}_M. \tag{7.25}$$

where

$$\begin{aligned} \mathbb{L} &= \left(\mathbf{I} - \frac{\mathbf{z}M}{1+\mathbf{z}:\mathbf{M}}\right)\cdot(\mathcal{L} - \mathbf{z}\mathbf{N}), \\ \bar{\mathbb{K}} &= \left(\mathbf{I} - \frac{\mathbf{z}M}{1+\mathbf{z}:\mathbf{M}}\right) : (\mathcal{K} - \mathbf{z}R). \end{aligned} \tag{7.26}$$

Since for viscoplastic flow mechanism (cf. Eq.(6.4)) we have

$$\dot{\sigma}_M = \frac{g}{\dot{\epsilon}_0}\frac{\partial\Phi^{-1}\left(\frac{\dot{\epsilon}^P_M}{\dot{\epsilon}_0}\right)}{\partial\left(\frac{\dot{\epsilon}^P_M}{\dot{\epsilon}_0}\right)}\ddot{\epsilon}^P_M, \tag{7.27}$$

provided $g = \text{const}$, then the fundamental rate equation (7.25) can be written in the form as follows

$$L_{\upsilon}\tau = \mathbb{L} : \mathbf{d} + \mathbb{K}\ddot{\epsilon}^P_M, \tag{7.28}$$

where the denotation

$$\mathbb{K} = \left(\mathbf{I} - \frac{\mathbf{z}M}{1+\mathbf{z}:\mathbf{M}}\right) : (\mathcal{K} - \mathbf{z}R)\frac{g}{\dot{\epsilon}_0}\frac{\partial\Phi^{-1}\left(\frac{\dot{\epsilon}^P_M}{\dot{\epsilon}_0}\right)}{\partial\left(\frac{\dot{\epsilon}^P_M}{\dot{\epsilon}_0}\right)} \tag{7.29}$$

is introduced.

7.6 Rate independent response of monocrystals. Adiabatic process for single slip

Let consider only single slip adiabatic process. Then we have (cf. Eq.(4.41))

$$\tau = g(\gamma, \vartheta). \tag{7.30}$$

Differentiation (7.30) with respect to time gives

$$\dot{\gamma} = \frac{1}{h}\dot{\tau} + \pi\dot{\vartheta}, \tag{7.31}$$

where the denotations

$$h = \frac{\partial g}{\partial \gamma}, \qquad \pi = -\frac{1}{h}\frac{\partial g}{\partial \vartheta} \tag{7.32}$$

are introduced.

For single slip we can write

$$\tau = \tau^{ab} s_a m_b. \tag{7.33}$$

Operating on this relation with the Lie derivative $(\overset{el}{L_{\boldsymbol{v}}})$, i.e. taking the rate of change seen by an observer who streches and rotates with the crystal lattice, we obtain

$$\dot{\tau} = (\overset{el}{L_{\boldsymbol{v}}}\boldsymbol{\tau})_{ab} s_a m_b + \tau^{ab}(\overset{el}{L_{\boldsymbol{v}}}\mathbf{s})_a m_b + \tau^{ab} s_a (\overset{el}{L_{\boldsymbol{v}}}\mathbf{m})_b. \tag{7.34}$$

To detrmine the Schmid stress rate let take the Lie derivative $(\overset{el}{L_{\boldsymbol{v}}})$ of the slip vectors $\mathbf{s}$ and $\mathbf{m}$. Thus we have

$$\begin{aligned} (\overset{el}{L_{\boldsymbol{v}}}\mathbf{s})_a &= s_b(d^e_{ca} + \omega^e_{ca})g^{bc} + \dot{s}_a, \\ (\overset{el}{L_{\boldsymbol{v}}}\mathbf{m})_b &= m_c(d^e_{db} + \omega^e_{db})g^{cd} + \dot{m}_b. \end{aligned} \tag{7.35}$$

Substituting (7.35) into (7.34) gives

$$\dot{\tau} = (\overset{el}{L_{\boldsymbol{v}}}\boldsymbol{\tau}) : \mathbf{N} + 2\boldsymbol{\tau}\cdot(\mathbf{N}+\mathbf{W})\cdot\mathbf{g}\cdot\mathbf{d}^e. \tag{7.36}$$

Taking advantage of the rate relation for the elastic range

$$(\overset{el}{L_{\boldsymbol{v}}}\boldsymbol{\tau}) = \mathcal{L}^e : \mathbf{d}^e - \theta\mathbf{z}\dot{\vartheta} \tag{7.37}$$

and substituting (7.36) and (7.37) into (7.31) we obtain

$$\dot{\gamma} = \frac{1}{h}[\mathcal{L}^e : \mathbf{N} + 2\boldsymbol{\tau}\cdot(\mathbf{N}+\mathbf{W})\cdot\mathbf{g}] : \mathbf{d}^e - \frac{1}{h}\theta\mathbf{z} : \mathbf{N}\dot{\vartheta} + \pi\dot{\vartheta}. \tag{7.38}$$

Let recall the rate evolution equation for the Kirchhoff stress (4.31) and take advantage of the heat conduction equation for adiabetic process ($\mathbf{q} = 0$), when the second order terms are neglected, we have

$$\begin{aligned} L_{\boldsymbol{v}}\boldsymbol{\tau} &= \mathcal{L}^e : \mathbf{d} - \theta\mathbf{z}\dot{\vartheta} - [\mathcal{L}^e : \mathbf{N} + \mathbf{b}]\dot{\gamma}, \\ \rho c_p\dot{\vartheta} &= \chi\tau\dot{\gamma}. \end{aligned} \tag{7.39}$$

Using

$$\mathbf{d}^e = \mathbf{d} - \mathbf{N}\dot{\gamma} \tag{7.40}$$

and eliminating $\dot{\gamma}$ and $\dot{\vartheta}$ from (7.38) and (7.39) we obtain the fundamental rate equation for the Kirchhoff stress tensor $\boldsymbol{\tau}$ in the form as follows

$$L_{\boldsymbol{v}}\boldsymbol{\tau} = \mathbb{L} : \mathbf{d}, \tag{7.41}$$

where

$$\mathbb{L} = \left[\mathcal{L}^{e} - \frac{(\mathbf{Q} + \Theta\frac{\mathbf{z}}{G}\boldsymbol{\tau} : \mathbf{N})\mathbf{Q}}{h + \mathbf{Q} : \mathbf{N} + (\Theta\frac{\mathbf{z}:\mathbf{N}}{G} - \Pi)\boldsymbol{\tau} : \mathbf{N}}\right] \tag{7.42}$$

with the denotations as follows

$$\begin{aligned} \mathbf{Q} &= \mathcal{L}^{e} : \mathbf{N} + \mathbf{b}, \\ \Theta &= \frac{\chi\theta}{\rho c_p} G, \\ \Pi &= \frac{\chi}{\rho c_p} h\pi = -\frac{\chi}{\rho c_p}\frac{\partial g}{\partial \vartheta}. \end{aligned} \tag{7.43}$$

If the Zaremba–Jaumann rete is used then we arive at the rate equation

$$\overset{\nabla}{\boldsymbol{\tau}} = \mathbb{L}^{\#} : \mathbf{d}, \tag{7.44}$$

where

$$\mathbb{L}^{\#} = \left[\mathcal{L}^{e} - \frac{(\mathbf{Q}^{\#} + \Theta\frac{\mathbf{z}}{G}\boldsymbol{\tau} : \mathbf{N})\mathbf{Q}^{\#}}{h + \mathbf{Q}^{\#} : \mathbf{N} + (\Theta\frac{\mathbf{z}:\mathbf{N}}{G} - \Pi)\boldsymbol{\tau} : \mathbf{N}}\right] \tag{7.45}$$

and

$$\begin{aligned} \mathbf{Q}^{\#} &= \mathcal{L}^{e} : \mathbf{N} + \boldsymbol{\beta}, \\ \boldsymbol{\beta} &= \mathbf{W}\cdot\boldsymbol{\tau} - \boldsymbol{\tau}\cdot\mathbf{W}. \end{aligned} \tag{7.46}$$

Assuming isothermal process, i.e. $\vartheta = \text{const}$, then the fundamental rate equation (7.44) reduces to

$$\overset{\nabla}{\boldsymbol{\tau}} = \mathbb{L}^{o} : \mathbf{d} \tag{7.47}$$

where

$$\mathbb{L}^{o} = \left[\mathcal{L}^{e} - \frac{\mathbf{Q}^{\#}\mathbf{Q}^{\#}}{h + \mathbf{Q}^{\#} : \mathbf{N}}\right]. \tag{7.48}$$

The result (7.47) is the same as has been obtained by ASARO and RICE [1977] (cf. Eq.(2.24) in their paper) provided in their considertion the tensor $\boldsymbol{\alpha}$ of non-Schmid effects is neglected.

8 THERMODYNAMIC RATE INDEPENDENT PLASTIC FLOW PROCESS FOR POLYCRYSTALLINE SOLIDS

8.1 General formulation

By Ψ we denote a set containing all values of the solution functions, and let $\mathcal{A}(I\!R^{+}, \Psi)$ denote a set of functions φ defined on the set of non-negative real number $I\!R^{+}$ with values in Ψ, i.e.

$$\varphi : I\!R^{+} \to \Psi. \tag{8.1}$$

For each $\tau \in \mathcal{T} \subset \mathbb{R}^+$ and $\varphi \in \mathcal{A}(\mathbb{R}^+, \Psi)$ we introduce a notion φ_τ for translate of φ defined on $\mathbb{R}^+$ by

$$\varphi_\tau(t) = \varphi(\tau + t), \qquad t \in \mathbb{R}^+. \tag{8.2}$$

For our purposes we can assume $\mathcal{T}$ as an interval $[0, d_P]$, where d_P denotes the duration of a process.

Definition 1 (cf. KNOPS and WILKES [1973]). A dynamical system is a set $\mathcal{A}(\mathbb{R}^+, \Psi)$ of functions defined on $\mathbb{R}^+$ taking values in Ψ such that

(i) $\varphi_\tau \in \mathcal{A}(\mathbb{R}^+, \Psi)$ whenever $\varphi \in \mathcal{A}(\mathbb{R}^+, \Psi)$ and $\tau \in \mathcal{T} \subset \mathbb{R}^+$;

(ii) $\lim_{t\to 0} \varphi_\tau(t) = \varphi(\tau), \quad \varphi \in \mathcal{A}(\mathbb{R}^+, \Psi) \quad \tau \in \mathcal{T}$.

We can say that function φ is the solution and the motion is the translate φ_τ.

Similarly to KNOPS and WILKES [1973] we define the trajectory as the set of all pairs $\{t, \varphi_\tau(t)\}$ for $t \in \mathbb{R}^+$, i.e. as a graph of the motion, cf. Fig.10, and the orbit as the projection of the trajectory onto Ψ, that is the set of values $pr_\Psi\,(\varphi_\tau(t))$.

We may treat our thermodynamic plastic flow process as a dynamical system corresponding to a triple $(\mathbb{R}^+, \mathcal{T}, \Psi)$, provided the following identification is assumed

$$\varphi \equiv (\boldsymbol{\phi}, \boldsymbol{v}, \vartheta, \boldsymbol{\mu}), \tag{8.3}$$

where $\boldsymbol{\phi}$ is a motion, $\boldsymbol{v}$ denotes the spatial velocity of the motion, ϑ is temperature and $\boldsymbol{\mu} \in V_n$ represents the internal state vector.

Let $Y(\tau)$ be the subset of Ψ defined as follows

$$Y(\tau) = \{\varphi(\tau) : \varphi \in \mathcal{A}(\mathbb{R}^+, \Psi)\}. \tag{8.4}$$

This may be regarded as the set of initial values of the motion φ_τ i.e. $Y(\tau) = \Psi_\tau$.

It will be convenient to introduce a nonlinear operator $\mathbb{T}_{(\cdot)}$ such that

$$\varphi_\tau = \mathbb{T}_{(\cdot)}\varphi(\tau), \tag{8.5}$$

i.e. φ_τ definied by Eq.(8.5) is an element of $\mathcal{A}(\mathbb{R}^+, \Psi)$ for any $\varphi(\tau) \in \Psi_\tau \equiv Y(\tau)$ and $\mathbb{T}_{(\cdot)}$ is a mapping

$$\mathbb{T}_{(\cdot)} : \Psi \to \mathcal{A}(\mathbb{R}^+, \Psi). \tag{8.6}$$

For particular process the nonlinear operator $\mathbb{T}_{(\cdot)}$ is defined by the initial–boundary–value problem considered.

For a thermodynamic plastic flow process the mapping $\mathbb{T}_{(\cdot)}$ is defined by the initial–boundary–value problem as follows :

Find $\boldsymbol{\phi}, \boldsymbol{v}, \vartheta$ and $\boldsymbol{\mu}$ as function of $\mathbf{x}$ and t such that

(i) the field equations

$$\begin{aligned}
\rho\dot{\boldsymbol{v}} &= \operatorname{div}\boldsymbol{\sigma} \\
L_{\boldsymbol{v}}\boldsymbol{\tau} &= \mathcal{L}\cdot\mathbf{d} - \mathrm{z}\dot{\vartheta}, \\
L_{\boldsymbol{v}}\boldsymbol{\mu} &= \mathrm{m}(s)\langle\frac{1}{H}\{\mathbf{P} : [\dot{\boldsymbol{\tau}} - (\mathbf{d}\cdot\boldsymbol{\alpha} + \boldsymbol{\alpha}\cdot\mathbf{d})] + \pi\dot{\vartheta}\}\rangle,
\end{aligned} \tag{8.7}$$

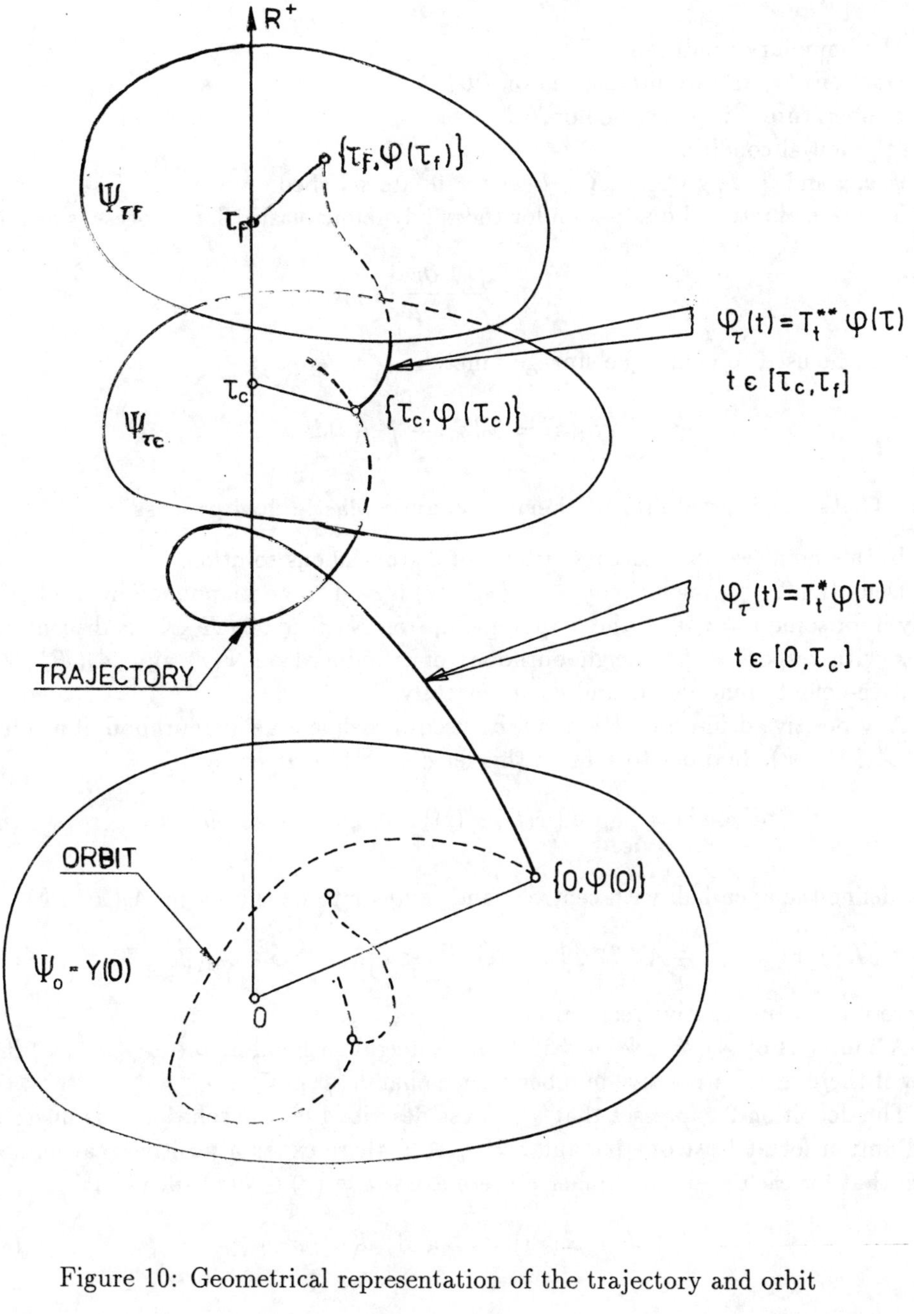

Figure 10: Geometrical representation of the trajectory and orbit

$$\rho c_p \dot{\vartheta} = \operatorname{div}(k \operatorname{grad} \vartheta) + \vartheta \frac{\rho}{\rho_{Ref}} \frac{\partial \boldsymbol{\tau}}{\partial \vartheta} : \mathbf{d}$$
$$+ \rho \bar{\chi} \langle \frac{1}{H} \{ \mathbf{P} : [\dot{\tau} - (\mathbf{d} \cdot \boldsymbol{\alpha} + \boldsymbol{\alpha} \cdot \mathbf{d})] + \pi \dot{\vartheta} \} \rangle;$$

(ii) the boundary conditions
traction $(\boldsymbol{\tau} \cdot \mathbf{n})^i$ are presciribed on $\partial \mathcal{B}$,
temperature ϑ is prescribed on $\partial \mathcal{B}$
(iii) the initial conditions
$\phi, \boldsymbol{v}, \boldsymbol{\mu}$ and ϑ are given at $X \in \mathcal{B}$ at $t = 0$; are satisfied.

The rate of internal dissipation for thermodynamic plastic flow process is as follows

$$\hat{i}(s) = -\frac{1}{\vartheta} \frac{\partial \hat{\psi}}{\partial \boldsymbol{\mu}} \cdot L_{\boldsymbol{v}} \boldsymbol{\mu} \tag{8.8}$$

and can be used to define the storage function

$$S_t(s) = S_0(s_0) + \int_0^t \hat{i}(s) dt. \tag{8.9}$$

8.2 Criteria of instability of thermodynamic plastic flow process

In this point we shall discuss criteria of discontinuous solution.

Definition 2. A solution $\varphi \in \mathcal{A}_d(\mathbb{R}^+, \Psi)$ is said to be Liapounov unstable if and only if for some $\tau = \tau_c \in \mathcal{T}$ the mapping $\mathbb{T}_{(.)}$ from $\Psi(\tau)$ to $\mathcal{A}_d(\mathbb{R}^+, \Psi)$ is discontinuous at φ with respect to the neighbourhoods of φ induced on $\Psi(\tau)$ and $\mathcal{A}_d(\mathbb{R}^+, \Psi)$ by positive-definite function d_τ and d, respectively.

Any positive-definite function can be used to define a neighbourhood of a solution $\varphi \in \mathcal{A}_d(\mathbb{R}^+, \Psi)$. In order to achieve this let $d(\varphi, \varphi^1)$ be given by

$$d(\varphi, \varphi^1) = \sup_{t \in \mathbb{R}^+} d\left(\varphi(t), \varphi^1(t)\right), \qquad \varphi, \varphi^1 \in \mathcal{A}_d(\mathbb{R}^+, \Psi), \tag{8.10}$$

and define the open ball with centre φ and radius r to be the set in $\mathcal{A}_d(\mathbb{R}^+, \Psi)$

$$K(\varphi, r) = \left\{ \varphi^1 \in \mathcal{A}_d(\mathbb{R}^+, \Psi) : d(\varphi, \varphi^1) < r \right\}, \qquad \varphi \in \mathcal{A}_d(\mathbb{R}^+, \Psi), \tag{8.11}$$

where r is a non–negative real number.

A subset A of $\mathcal{A}_d(\mathbb{R}^+, \Psi)$ is said to be a neighbourhood of $\varphi \in \mathcal{A}_d(\mathbb{R}^+, \Psi)$ if and only if there exists a positive number r such that $K(\varphi, r) \subseteq A$.

The definition 2 expresses that a process described by the solution φ is unstable if and only if for at least one instant $\tau = \tau_c \in \mathcal{T}$ there exists a positive real number ε such that for each positive number δ there exists a $\varphi^1(\tau) \in \Psi(\tau)$ such that

$$d_\tau\left(\varphi(\tau), \varphi^1(\tau)\right) < \delta \quad \text{and} \quad d(\varphi_\tau, {\varphi^1}_\tau) \geq \varepsilon \tag{8.12}$$

where $\varphi^1 = \mathbb{T}_{(\cdot)}\varphi^1(\tau)$, and

$$d(\varphi_\tau, {\varphi^1}_\tau) = \sup_{t\in \mathbb{R}^+} d\left(\varphi_\tau(t), {\varphi^1}_\tau(t)\right) \tag{8.13}$$

with $\varphi, \varphi^1 \in \mathcal{A}_d(\mathbb{R}^+, \Psi)$.

Theorem 1. A solution $\varphi \in \mathcal{A}_d(\mathbb{R}^+, \Psi)$ is unstable if and only if there exist positive definite functions (Liapounov functions)

$$\mathrm{V}_{\tau,t}, \qquad \tau \in \mathcal{T} \text{ defined on } \Psi \times \Psi \tag{8.14}$$

such that

(i) the mapping $\mathbb{T}_{(\cdot)}$ from $\Psi(\tau)$ to $\mathcal{A}_{V_\tau}(\mathbb{R}^+, \Psi)$ at φ for some $\dot{\tau} = \tau_c \in \mathcal{T}$ (cf. Fig.11);

(ii) the identity mapping I is continuous from $\mathcal{A}_d(\mathbb{R}^+, \Psi)$ to $\mathcal{A}_{V_\tau}(\mathbb{R}^+, \Psi)$ (cf. Fig.11), where

$$\mathrm{V}_\tau(\varphi_\tau, {\varphi^1}_\tau) = \sup_{t\in \mathbb{R}^+} V_{\tau,t}\left(\varphi_\tau(t), {\varphi^1}_\tau(t)\right). \tag{8.15}$$

We may introduce the Liapounov function as follows

$$V_{\tau,t}\left(\varphi_\tau(t), \varphi^\#\right) = \int_{\mathcal{B}} \left\{\left[\hat{\psi}\left(\varphi_\tau(t)\right) - \hat{\psi}(\varphi^\#)\right] + \frac{1}{2}\rho \boldsymbol{v}^2\right\} dV, \tag{8.16}$$

where $\varphi^\# = (\phi^\#, 0, \vartheta^\#, \boldsymbol{\mu}^\#)$ is an equalibrium solution, i.e.

$$\boldsymbol{v}^\# = 0, \qquad L_{\boldsymbol{v}}\boldsymbol{\mu}^\# = 0 \tag{8.17}$$

(it means that temperature $\vartheta = \vartheta^\#$ and strain $\mathbf{e} = \mathbf{e}^\#$ are constant in $\mathcal{B}$).

Let us consider the solutioin

$$\varphi_\tau(t) = \mathbb{T}_{(t)}\varphi(\tau) \in \mathcal{A}(\mathbb{R}^+, \Psi) \tag{8.18}$$

of the thermodynamic plastic flow process. For some $\tau = \tau_c \in \mathcal{T}$ the solution φ is not unique.

Starting from the point $\{\tau_c, \varphi_\tau(\tau_c)\}$ on the trajectory we can expect more than one solution (cf. Fig.10).

So we have at $\{\tau_c, \varphi(\tau_c)\}$ branching of the solution. In other words the mapping $\mathbb{T}_{(\cdot)}$ is a nonsingle–valued function (is a multi-function).

8.3 Quasi-static and adiabatic approximation

In many practical situations the thermodynamic plastic flow process can be treated as quasi-static and adiabatic.

Then the mapping $\mathbb{T}_{(\cdot)}$ is defined by the following initial-boundary value problem:
Find $\varphi = (\phi, \boldsymbol{v}, \vartheta, \boldsymbol{\mu})$ as function of $\mathbf{x}$ and t such that

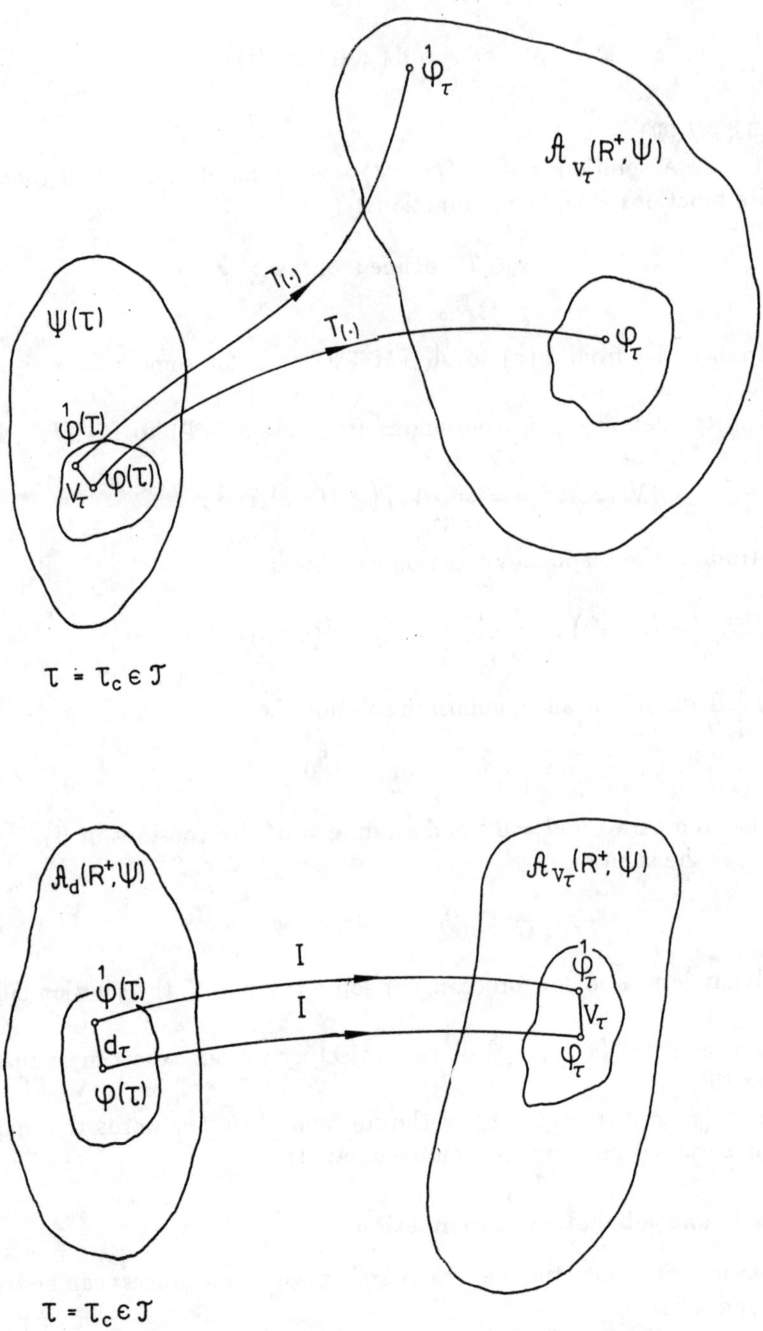

Figure 11: The discontinuous mapping $\mathbb{T}_{(\cdot)}$ and the continuous identity mapping I

(i) the field equations

$$
\begin{aligned}
\text{div}\boldsymbol{\sigma} &= 0, \\
L_{\boldsymbol{v}}\boldsymbol{\tau} &= \mathcal{L}\cdot\mathbf{d} - \mathbf{z}\dot{\vartheta}, \\
L_{\boldsymbol{v}}\boldsymbol{\mu} &= \mathrm{m}(s)\langle\frac{1}{H}\{\mathbf{P}:[\dot{\boldsymbol{\tau}} - (\mathbf{d}\cdot\boldsymbol{\alpha}+\boldsymbol{\alpha}\cdot\mathbf{d})] + \pi\dot{\vartheta}\}\rangle, \\
\rho c_p\dot{\vartheta} &= \vartheta\frac{\rho}{\rho_{Ref}}\frac{\partial\boldsymbol{\tau}}{\partial\vartheta} : +\rho\chi\left\langle\frac{1}{H}\{\mathbf{P}:[\dot{\boldsymbol{\tau}} - (\mathbf{d}\cdot\boldsymbol{\alpha}+\boldsymbol{\alpha}\cdot\mathbf{d})] + \pi\dot{\vartheta}\}\right\rangle;
\end{aligned}
\tag{8.19}
$$

(ii) traction $(\boldsymbol{\tau}\cdot\mathbf{n})^i$ are prescribed on $\partial\mathcal{B}$
temperature ϑ is prescribed on $\partial\mathcal{B}$;
(iii) $\phi, \boldsymbol{v}, \boldsymbol{\mu}$ and ϑ are given at $X\in\mathcal{B}$ at $t=0$; are satisfied.

8.4 Criteria for shear band localization. Application of the standard bifurcation method

Basing on the results obtained in the previous Chapter the field equations of (8.19)) can be reduced to

$$
\begin{aligned}
\text{div}\boldsymbol{\sigma} &= 0 \\
L_{\boldsymbol{v}}\boldsymbol{\tau} &= \mathbb{L}:\mathbf{d}
\end{aligned}
\tag{8.20}
$$

with prescribed trations $(\boldsymbol{\tau}\cdot\mathbf{n})^i$ on $\partial\mathcal{B}$.

Let $\{X^I\}$ and $\{x^i\}$ denote Cartesian coordinate systems on $\mathcal{B}$ and $\mathcal{S}$, respectively.

Let assume that the boundary conditions are such that a body $\mathcal{B}$ is sustained uniform stress $\boldsymbol{\tau}^{\circ}$ and uniform temperature ϑ°.

The response to a homogeneous velocity gradient field $\left(\frac{\partial\boldsymbol{v}}{\partial\mathbf{x}}\right)^{\circ}$ is the homogeneous stress rate $\dot{\boldsymbol{\tau}}^{\circ}$ wchich satisfies the quasi-static field equations (8.20).

Conditions are sought for which the state of the solid body $\mathcal{B}$ allows the field equations to be satisfied for an alternate field

$$
\frac{\partial\boldsymbol{v}}{\partial\mathbf{x}} = \left(\frac{\partial\boldsymbol{v}}{\partial\mathbf{x}}\right)^{\circ} + \Delta\left(\frac{\partial\boldsymbol{v}}{\partial\mathbf{x}}\right) \tag{8.21}
$$

in which the jump $\Delta(\partial\boldsymbol{v}/\partial\mathbf{x})$ is a function only of distance across a planar band and vanishes outside the band. If the velocity is to be continuous at bifurcation, the compatibility condition

$$
\Delta\left(\frac{\partial\boldsymbol{v}}{\partial\mathbf{x}}\right) = \mathbf{k}\mathbf{n} \tag{8.22}
$$

must be satisfield, where $\mathbf{n}$ denotes the unit normal to the plane of the band and the magnitude of jump $\mathbf{k}$ is function of distance across the band $(\mathbf{n}\cdot\mathbf{x})$ only, and is zero outside.

Equilibrium must also be satisfied at the inception of bifurcation. This is expressed in rate form as (provided choosing the reference state to coincide, instantaneously, with the current state)

$$\frac{\partial}{\partial x_i}\left(\frac{\partial \sigma^{ij}}{\partial t}\right) = \frac{\partial}{\partial x_i}\left(\dot{\tau}^{ij} - \tau^{ij} d_{kk} - v^k \frac{\partial \tau^{ij}}{\partial x^k}\right) = 0. \tag{8.23}$$

Because bifurcation is from the homogeneous state $\boldsymbol{\tau} = \boldsymbol{\tau}^{\circ}$ the rate equilibrium condition reduces at the instant considered to

$$\frac{\partial \dot{\tau}^{ij}}{\partial x^i} - \tau^{ij} \frac{\partial d^{kk}}{\partial x^i} = 0. \tag{8.24}$$

Equations (8.22) and (8.24) require that

$$\mathbf{n} \cdot \Delta \dot{\boldsymbol{\tau}} - (\mathbf{n} \cdot \boldsymbol{\tau})(\mathbf{k} \cdot \mathbf{n}) = 0, \tag{8.25}$$

where $\Delta \dot{\boldsymbol{\tau}} = \dot{\boldsymbol{\tau}} - \dot{\boldsymbol{\tau}}^{\circ}$.

Taking advantage of the rate type constitutive relation $(8.20)_2$ the homogeneous field outside the band has to satisfy

$$L_{\boldsymbol{v}} \boldsymbol{\tau}^{\circ} = \mathbb{L}^{\circ} : \mathbf{d}^{\circ}, \tag{8.26}$$

and inside the band the corresponding equation

$$L_{\boldsymbol{v}} \boldsymbol{\tau} = \mathbb{L} : \mathbf{d}. \tag{8.27}$$

The compatibility condition (8.22) can be expressed in terms of $\mathbf{d}$ by

$$\Delta \mathbf{d} = \mathbf{d} - \mathbf{d}^{\circ} = \frac{1}{2}(\mathbf{kn} + \mathbf{nk}). \tag{8.28}$$

Assuming that the constitutive response remains continuous at the inception of localization, i.e.

$$\Delta \mathbb{L} = \mathbb{L} - \mathbb{L}^{\circ} = 0, \tag{8.29}$$

then Eqs(8.25)-(8.28) yield

$$\left(n_i \, \mathbb{L}^{ijkl} n_l + n_i \tau^{il} n_l \delta^{jk}\right) k_k = 0. \tag{8.30}$$

The last term in the bracket arises due to the difference between $L_{\boldsymbol{v}} \boldsymbol{\tau}$ and $\dot{\boldsymbol{\tau}}$.

The onset of the localization occurs at the first instant in the deformation history for which a nontrivial solution of Eq.(8.30) exists.

Thus, the necessary condition for a localized shear band to be form is

$$\det\left[n_i \, \mathbb{L}^{ijkl} n_l + n_i \tau^{il} n_l \delta^{jk}\right] = 0. \tag{8.31}$$

For simplicity let introduce rectangular Cartesian coordinates $\{x^i\}$ in such a way that $\mathbf{n}$ is in the x_2-direction. Then Eq.(8.31) takes the form as follows

$$\det\left[\ \mathbb{L}^{2jk2} + \tau^{22}\delta^{jk}\right] = 0. \tag{8.32}$$

To make possible analytical investigation of criteria for localization we introduce some simplifications.

(i) The evolution equation for temperature (7.14) is replaced by (cf. the discussion presented in section 7.2)

$$c_p\dot{\vartheta} = \chi\tilde{\boldsymbol{\tau}} : \mathbf{d}^P, \tag{8.33}$$

where

$$\begin{aligned} \chi \;=\; & -\left[\frac{\partial\hat{\psi}}{\partial\zeta}\cdot\mathbf{Z}(s) + \frac{\partial\hat{\psi}}{\partial\boldsymbol{\alpha}} : \left(r_1\mathbf{P} + r_2\frac{\mathbf{P}:\mathbf{P}}{\tilde{\boldsymbol{\tau}}:\mathbf{P}}\tilde{\boldsymbol{\tau}}\right)\right.\\ & \left. + \frac{\partial\hat{\psi}}{\partial\xi}\left(k_1\tilde{\boldsymbol{\tau}} : \mathbf{P} + k_3\mathbf{P} : \mathbf{g}\right)\right]\frac{1}{\tilde{\boldsymbol{\tau}}:\mathbf{P}}. \end{aligned} \tag{8.34}$$

It means that only main contribution to thermo-mechanical coupling is taken into consideration.

The evolution equation for temperature again yields

$$\dot{\vartheta} = \mathbf{M} : L_{\boldsymbol{v}}\boldsymbol{\tau} + \mathbf{N} : \mathbf{d}, \tag{8.35}$$

where now

$$\mathbf{M} = \frac{\chi(\tilde{\boldsymbol{\tau}}:\mathbf{P})\mathbf{P}}{Hc_p - \chi(\tilde{\boldsymbol{\tau}}:\mathbf{P})\pi}, \qquad \mathbf{N} = \frac{\chi(\tilde{\boldsymbol{\tau}}:\mathbf{P})}{Hc_p - \chi(\tilde{\boldsymbol{\tau}}:\mathbf{P})\pi}(\mathbf{P}\cdot\tilde{\boldsymbol{\tau}} + \tilde{\boldsymbol{\tau}}\cdot\mathbf{P}) \tag{8.36}$$

and the second order terms which describe the cross coupling effects are neglecting.

(ii) By analogy with the infinitesimal theory of elasticity we postulate

$$(\mathcal{L}^e)^{ijkl} = G(\delta^{ik}\delta^{jl} + \delta^{il}\delta^{jk}) + (K - \frac{2}{3}G)\delta^{ij}\delta^{kl} + \tau^{jl}\delta^{ik}, \tag{8.37}$$

where G and K denote the shear and bulk modulus, respectively.

(iii) Assume that

$$\overset{-1}{\mathcal{L}^e}\cdot\mathcal{L}^{th} = \theta\mathbf{I}, \tag{8.38}$$

where θ is the thermal expansion in elastic range.

We shall consider two separate particular cases. First by assuming

(iv)

$$L_{\boldsymbol{v}}\boldsymbol{\tau} \cong \dot{\boldsymbol{\tau}}, \qquad L_{\boldsymbol{v}}\boldsymbol{\mu} \cong \dot{\boldsymbol{\mu}}, \tag{8.39}$$

i.e. postulating that the spatial covariance terms represented by the matrix $\mathcal{C}$ have no influence on criteria for localization along shear bands.

Second by replacing the simplification (i) by the stronger requirement
(v)

$$\vartheta = \text{const}, \tag{8.40}$$

i.e. the process considered is assumed to be isothermal.

Each particular case allows to discuss the influence of important effects on criteria for localization.

In the first case the discusion of thermo-mechanical coupling when there is no spatial covariance effects is given, while the second case is devoted to the investigation of the spatial covariance effects for assumed isothermal process.

Let first discuss superposition of (i) – (iv).

The evolution equation for the Kirchhoff stress tensor $\boldsymbol{\tau}$ takes now the form [7]

$$\dot{\boldsymbol{\tau}} = \bar{\mathbb{L}} : \mathbf{d}, \tag{8.41}$$

where

$$\bar{\mathbb{L}} = \bar{\mathcal{L}} + \bar{\mathcal{T}} \tag{8.42}$$

and

$$\begin{aligned}
(\bar{\mathcal{L}})^{ijkl} &= G(\delta^{ki}\delta^{lj} + \delta^{kj}\delta^{il}) + (K - \frac{2}{3}G)\delta^{ij}\delta^{kl} - \frac{1}{H + G + 9KA^2}\left(\frac{G}{\sqrt{\tilde{J}_2}}\tilde{\tau}'^{ij}\right. \\
&\left. +3KA\delta^{ij}\right)\left(\frac{G}{\sqrt{\tilde{J}_2}}\tilde{\tau}'^{kl} + 3KA\delta^{kl}\right), \\
(\bar{\mathcal{T}})^{ijkl} &= -\left\{G\left[(\Pi - 6A\Xi)(G\tilde{\tau}'^{ij} + 3KA\sqrt{\tilde{J}_2}\delta^{ij}) + 2\Xi(H + G + \right.\right. \\
&\left.\left. +9KA^2)\sqrt{\tilde{J}_2}\delta^{ij}\right](G\tilde{\tau}'^{kl} + 3KA\sqrt{\tilde{J}_2}\delta^{kl})\right\}\left\langle \tilde{J}_2(H \right. \\
&\left. +G + 9KA^2)\left(H + G + 9KA^2 - \Pi G + 6GA\Xi\right)\right\rangle^{-1},
\end{aligned} \tag{8.43}$$

with the denotations

$$\Pi = \frac{\chi(\sqrt{\tilde{J}_2} + A\tilde{J}_1)\pi}{c_p G}, \qquad \Xi = \frac{3\theta K\chi(\sqrt{\tilde{J}_2} + A\tilde{J}_1)}{2c_p G}. \tag{8.44}$$

[7] Since the simplification (iv) is assumed in which the terms of the order of magnitude $\sqrt{\tilde{J}_2}$ are neglected, then for this case in the matrix $\mathcal{L}^e$ the term $\tau^{jl}\delta^{ik}$ (cf. Eg.(8.37)) is disregarded.

Finally the fundamental matrix $\bar{\mathbb{L}}$ has a particular form as follows

$$(\bar{\mathbb{L}})^{ijkl} = G(\delta^{ki}\delta^{lj} + \delta^{kj}\delta^{il}) + (K - \tfrac{2}{3}G)\delta^{ij}\delta^{kl} - \frac{1}{H+G+9KA^2-\Pi G+6GA\Xi}\left[\frac{G}{\sqrt{\tilde{J}_2}}\tilde{\tau}'^{ij} + (3KA + 2G\Xi)\delta^{ij}\right]\left(\frac{G}{\sqrt{\tilde{J}_2}}\tilde{\tau}'^{kl} + 3KA\delta^{kl}\right). \tag{8.45}$$

The necessary condition for a localized shear band to be formed is

$$\det[\,\bar{\mathbb{L}}^{2jk2}] = 0. \tag{8.46}$$

Substituting the matrix $\bar{\mathbb{L}}$ into (8.46) gives

$$\begin{aligned} \frac{H_{cr}}{G} &= -\frac{1+\nu}{2}\left(T + 2A + \frac{1-2\nu}{1+\nu}\Xi\right)^2 + \frac{(1-2\nu)^2}{1-\nu^2}\Xi^2 + \Pi, \\ \tan\Theta &= \left(\frac{S - T_{\min}}{T_{\max} - S}\right)^{\frac{1}{2}}, \end{aligned} \tag{8.47}$$

where Θ denotes the angle between the vector $\mathbf{n}$ and the $\tilde{\tau}_{III}$-direction, and the following denotations

$$\begin{aligned} S &= -(1-\nu)T + 2(1+\nu)A + (1-2\nu)\Xi, \\ T_{\max} &= \frac{\tilde{\tau}'_I}{\sqrt{\tilde{J}_2}}, \qquad T = \frac{\tilde{\tau}'_{II}}{\sqrt{\tilde{J}_2}}, \qquad T_{\min} = \frac{\tilde{\tau}'_{III}}{\sqrt{\tilde{J}_2}}, \\ \nu &= \frac{3K - 2G}{2(3K + G)}, \end{aligned} \tag{8.48}$$

are introduced.

It is noteworthy that more general results have been obtained by DUSZEK-PERZYNA, PERZYNA and STEIN [1989]. They considered the case when

$$\mathbf{P} \neq \mathbf{Q} \tag{8.49}$$

i.e. $k_2 \neq 0$ and $A \neq B$, the normality rule does not apply.

For this general case they obtained

$$\frac{H_{cr}}{G} = -\frac{1+\nu}{2}\left(T + A + B + \frac{1-2\nu}{1+\nu}\Xi\right)^2 + \frac{1+\nu}{1-\nu}\left(A - B + \frac{1-2\nu}{1+\nu}\Xi\right)^2 + \Pi, \tag{8.50}$$

and the direction of the shear band is determined by $(8.47)_2$ with

$$S = -(1-\nu)T + (1+\nu)(A+B) + (1-2\nu)\Xi. \tag{8.51}$$

The result (8.50) for the critical hardening modulus rate $\frac{H_{cr}}{G}$ as function of T is shown in Fig.12.

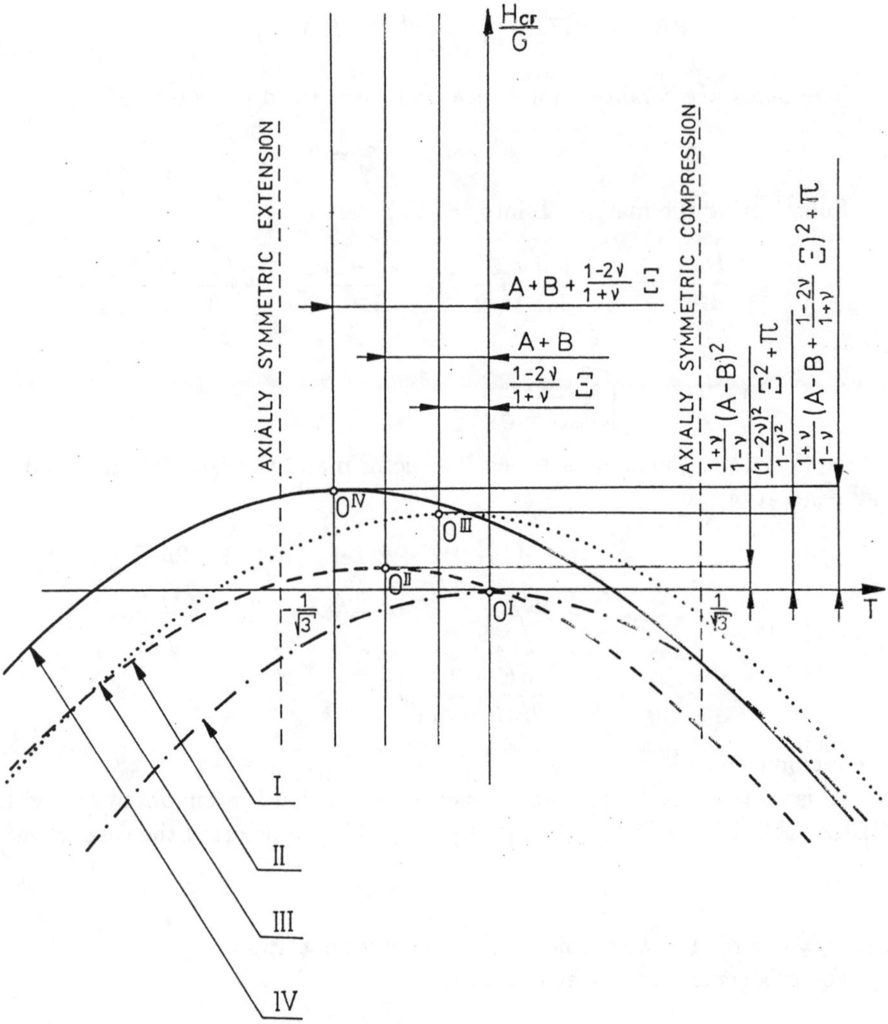

Figure 12: The critical hardening modulus rate H_{cr}/G as function of T (for $A \neq B$)

As second case let discuss superposition (ii), (iii) and (v) and additionally assume that there is no micro-damage process, i.e. $A = 0$. Then the evolution equation for the Kirchhoff stress tensor $\boldsymbol{\tau}$ is as follows

$$L_{\boldsymbol{v}}\boldsymbol{\tau} = \ \mathbb{L}^{\#} : \mathbf{d} \tag{8.52}$$

where

$$\mathbb{L}^{\#} = \mathcal{L}^{\#} + \mathcal{C}^{\#} \tag{8.53}$$

and

$$\begin{aligned} \mathcal{L}^{\#ijkl} &= G(\delta^{ik}\delta^{jl} + \delta^{il}\delta^{jk}) + \left(K - \frac{2}{3}G\right)\delta^{ij}\delta^{kl} \\ &\quad + \tau^{jl}\delta^{ik} - \frac{G^2}{(H+G)J_2}\tau'^{ij}\tau'^{kl}, \end{aligned} \tag{8.54}$$

$$\mathcal{C}^{\#ijkl} = -\frac{G}{(H+G)J_2}\tau'^{ij}\tau'^{kr}\tau'^{lr}\delta_{rs}. \tag{8.55}$$

The necessary condition for localization (8.32) takes now the form

$$\det[n_i(\mathcal{L}^{\#ijkl} + \mathcal{C}^{\#ijkl})n_l + n_i\tau^{il}n_l\delta^{ik}] = 0. \tag{8.56}$$

The last condition can be written as

$$\det[n_i(\mathcal{L}^{\#ijkl} + \mathcal{K}^{\#ijkl})n_l] = 0, \tag{8.57}$$

where the matrix

$$\mathcal{K}^{\#ijkl} = \mathcal{C}^{\#ijkl} + \tau^{il}\delta^{jk} \tag{8.58}$$

describes the influence of the spatial covariance terms on the criterion for localization.
The condition (8.57) gives the value of the rate hardening modules H in the form

$$H = H_0 + H_1, \tag{8.59}$$

where

$$H_0 = \frac{G}{J_2}\left[\frac{1-2\nu}{2(1-\nu)}\tau'^2_{22} + \tau^2_{12} - J_2\right] \tag{8.60}$$

is the value of rate hardening modulus which consists of the terms of the order of magnitude G, computed by neglecting the influence of the spatial covariance terms (assuming $\mathcal{K}^{\#} = 0$) and the terms of the order of magnitude $\sqrt{J_2}$ (concerned with the term $\tau^{ji}\delta^{ik}$ in Eq.(8.54)), and

$$H_1 = \frac{1}{J_2}\left[\frac{\nu(1-2\nu)}{2(1-\nu)^2}\tau'^2_{22}\tau_{22} + \tau^2_{12}\tau'_{11}\right] + 0\left(\frac{J_2}{G}\right) \tag{8.61}$$

which describes the higher order effects. These effects are implied by the spatial covariance terms and of the other terms of the order of magnitude $\sqrt{J_2}$. However we neglect the terms of the order of magnitude $\frac{J_2}{G}$ and smaller (it means that we neglect terms of the order of magnitude $\frac{\sqrt{J_2}}{G}$ in compparison to 1).

Taking into account the orientation of the plane within which the shear band localization first occurs we obtain the final result in the form as follows

$$\begin{aligned} \frac{H_{cr}}{G} &= -\frac{1+\nu}{2}T^2 + \left\{ \frac{(\nu-1)(2-\nu)}{2}T^3 + \left[\frac{\nu(1-2\nu)}{6}\frac{J_1}{\sqrt{J_2}} \right. \right. \\ & \left. \left. +(1-\nu+\nu^2)T_{\min} \right] T^2 + (1-\nu)T - T_{\min} \right\} \frac{\sqrt{J_2}}{G} + 0\left(\frac{J_2}{G^2}\right). \end{aligned} \tag{8.62}$$

The value in brackets { } represents again the main contribution to the value of H implied by the spatial covariance terms and the other higher order terms.

8.5 Discussion of different effects on localization criteria

The results $(8.47)_1$ for the critical hardening modulus rate $\frac{H_{cr}}{G}$ as function of the state of stress T may be represented by the parabola IV (cf. Fig.13). This parabola when compared with the parabola I (for $A = 0, \Xi = 0$ and $\Pi = 0$) is translated up by $\frac{(1-2\nu)^2}{1-\nu^2}\Xi^2 + \Pi$ and is shifted left by $2A + \frac{1-2\nu}{1+\nu}\Xi$. The translation up is caused by both thermal effects, i.e. by thermal expansion (represented by Ξ) and thermal plastic softening (represented by Π), while the shifting left is implied by the thermal expansion and the micro-damage process. The translation up means that the material is more inclined to instability by localization along the shear band and the shifting left shows that the inclination to instability for the axially symmetric compression is different from that for the axially symmetric tension. For tension material is more sensitive to localization than for compression.

Both these effects have been observed experimentally by ANAND and SPITZIG [1980]. They performed experimental investigations of the initation of localization along shear band in a maraging steel under condition of quasi-static plane strain. They observed that shear band localization occurs when the hardening modulus rate is decidedly positive but small and detected small diferences between the values of the hardening modulus rate for tension and compression, cf. Fig.14 .

By assuming the isothermal approximation of the process considered we can investigate the influence of micro-damage effects on criteria for localization. Then the necessary conditions for localization along the shear band are as follows (cf. DUSZEK and PERZYNA [1988])

$$\frac{H_{cr}}{G} = -\frac{1+\nu}{2}(T+2A)^2 \tag{8.63}$$

and the direction of localization is determined by Eq.$(8.47)_2$ with

$$S = -(1-\nu)T + 2(1+\nu)A. \tag{8.64}$$

(ii) In the second case it has been assumed that the strain rate $\dot{\varepsilon}^P_M$ varies in time.
Let consider the strain rate versus time (or strain) curve for typical dynamical test. For $\ddot{\varepsilon}^P_M > 0$ the yield stress increases and the rate sensitivity term $\mathbb{K}\ddot{\varepsilon}^P_M$ has stabilizing influence on the flow process considered , while for $\ddot{\varepsilon}^P_M < 0$ the yield stress decreases and the rate sensitivity term $\mathbb{K}\ddot{\varepsilon}^P_M$ implies some kind of softening effect, and as a result promotes localization.

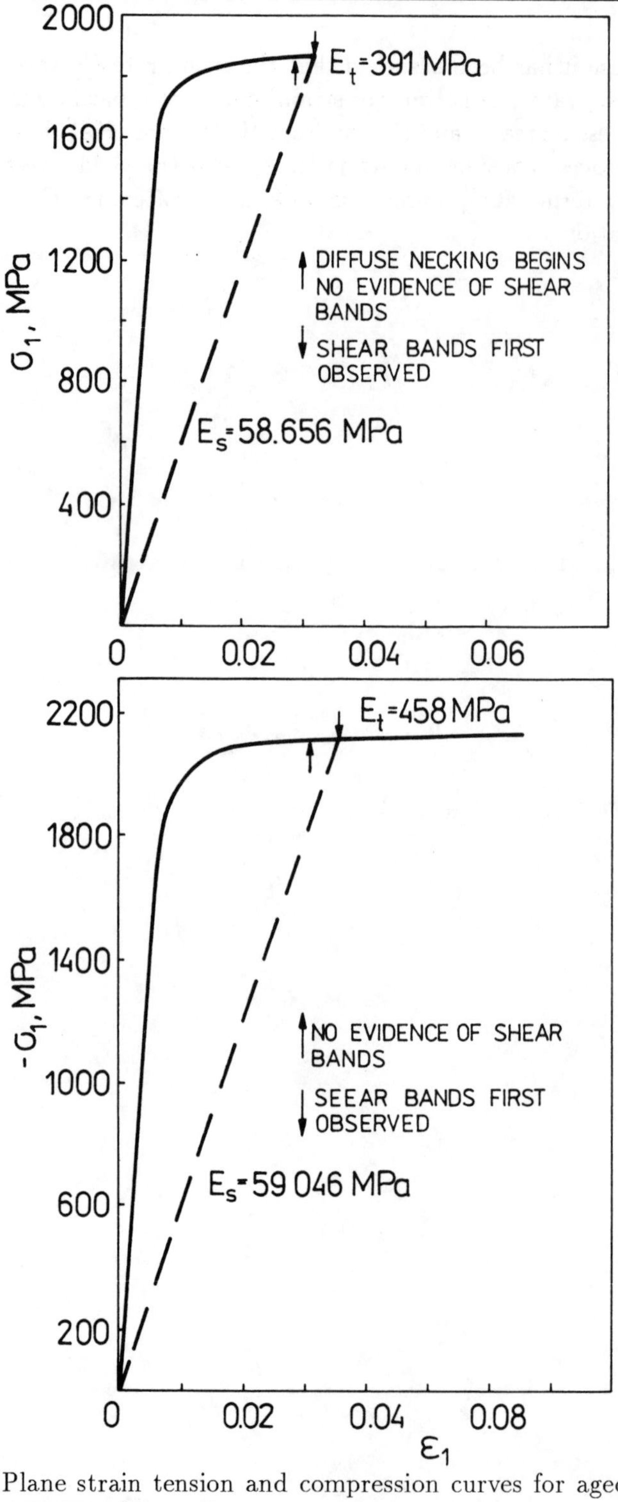

Figure 14: Plane strain tension and compression curves for aged margin steel (after ANAND and SPITZIG [1980])

The result (8.63) for the critical hardeming modulus rate $\frac{H_{cr}}{G}$ as function of the state of stress T may be represented by the parabola II, as it has been plotted in Fig.13 by broken line. The parabola II is shifted left by $2A$.

To investigate the influence of thermo-mechanical couplings only on criteria for localization let postulate that there is no micro-damage effects ($A = 0$) for adiabatic process. Then the necessary conditions for shear band localization are as follows (cf. DUSZEK and PERZYNA [1991])

$$\frac{H_{cr}}{G} = -\frac{1+\nu}{2}\left(T + \frac{1-2\nu}{1+\nu}\Xi\right)^2 + \frac{(1-2\nu)^2}{1-\nu^2}\Xi^2 + \Pi, \tag{8.65}$$

and the direction of localization is again given by Eq.(8.47) with

$$S = -(1-\nu)T + (1-2\nu)\Xi, \tag{8.66}$$

The result (8.65) is plotted in Fig.13 as the parabola III by dotted line. This parabola, when compared with the parabola I, is translated up by $\frac{(1-2\nu)^2}{1-\nu^2}\Xi^2 + \Pi$ and is shifted left by $\frac{1-2\nu}{1+\nu}\Xi$.

The translation up is caused by both thermal effects, i.e. by thermal expansion (represented by Ξ), and thermal plastic softening (represented by Π), while the shifting left is implied by thermal expansion only.

As it has been already pointed out by DUSZEK and PERZYNA [1991] the experimental results concerning the shear band localization phenomenon reported by ANAND and SPITZIG [1980] can be properly explained by taking account of the influence of thermo-machanical couplings only. Indeed, considering the influence of thermal expansion and thermal plastic softening we can have shear band localization when the hardening modulus rate is positive and we have also differences between the velues of the hardening modulus rate for axially symmetric compression and tension, cf. Fig.13.

DUSZEK-PERZYNA, PERZYNA and STEIN [1989] developed the estimation procedure for comparison of the influence of micro-damage and thermo-machanical coupling effects on initation of localization along shear bands.

Basing on the experimental results concerning the measurement of the temperature profile during shear band fromation in steels deforming at high strain rates reported by HARTLEY, DUFFY and HAWLEY [1987], MARCHAND and DUFFY [1988] and MARCHAND, CHO and DUFFY [1988] it has been estimated that increase of temperature at the instant of the initiation of localized shear bands is about 80°C.

A summary of the material parameters used in the calculations is presented in Table 1.

Table 1. Mild steel (0.18%C) material parameters

ρ_M	=	7850 kg/m^3	ξ^F	=	0.25
c_p	=	465 J/kg °C	ξ_0	=	0.0005
ν	=	0.3	b	=	const
E	=	200 000 MPa	n_1	=	0
κ_0	=	400 MPa	n_2	=	$n_2\bar{\vartheta}$
κ_0/κ_1	=	0.5	n_0	=	0.69
θ	=	$3\times10^{-5}/°C$	k_1	=	const
ϑ^0	=	20°	k_2	=	const
h_1	=	const			

χ has been postulated constant during the process considered and is assumed to be 0.85 .

The results obtained are as follows

(i) The micro-damage term (represented in Eq.(8.47)$_1$ by $2A$) is of the same order as the thermal expansion term (represented in Eq.(8.47)$_1$ by $\frac{1-2\nu}{1+\nu}\Xi$).

(ii) The dominated role plays the thermal plastic softeming term Π, which at initiation of localization is 2.5 times higher than the thermal expresion term $\frac{(1-2\nu)^2}{1-\nu^2}\Xi^2$.

To diuscuss the influence of the covarince terms in the condition for shear band localization let consider the matrix $\mathcal{K}^{\#}$.

The matrix $\mathcal{K}^{\#}$ consists of two different terms which have the magnitude of a stress component divided by an elastic modulus G. These terms are generally small compare to unity. This justifies the approximation procedure assumed in the determination of $\frac{H_{cr}}{G}$ (cf. Eq.(8.62)).

However, it is reasonable to point out that the rate of hardening modulus H decreases in value with ongoing plastic deformation and all terms in H_0 become small and can be comparable to terms in H_1. This supports the conjecture that the influence of terms which arise from the difference between the Lie derivative and the material rate of the Kirchhoff stress tensor are important and may play a dominated role when the inception of localization phenomenon is expected to take place for small values of H, that is near to the maximum stress attained during the process.

In fact, the experimental results obtained for AISI 1018 cold rolled steel by MARCHAND, CHO and DUFFY [1988], and for a low alloy structural steel (HY-100) by MARCHAND and DUFFY [1988], and performed by using a torsional Kolsky bar (split-Hopkinson bar) to impose a rapid deformation rate in a short thin-walled tubular specimen, showed that the inception of localization along shear band takes place in the stage of the plastic deformation process when a range of nominal strains is from about 15% to 45% for CRS and 25% to 50% for HY-100 and corresponds approximately to the maximum stress attained during the test. These results confirmed that with continued deformation the strain distribution is no longer homogeneous. During this stage of the process there is observed a continuous increase in the magnitude of localized strain along a narrow shear band. As the nominal strain withn this stage increases the localized strain increases over 150% for CRS and 170% for HY-steel. In this stage of deformation the value of the hardening modulus rate and flow stress level do not vary greatly, cf. Fig.15 .

9 THERMODYNAMIC RATE DEPENDENT PLASTIC FLOW PROCESS FOR POLYCRYSTALLINE SOLIDS

9.1 Formulation of the process

For a thermodynamic elastic – viscoplastic flow process the mapping $\mathbb{T}_{(\cdot)}$ is defined by the initial – boundary – value problem as follows:
Find $\varphi = (\phi, \boldsymbol{v}, \vartheta, \boldsymbol{\mu})$ as function of $\mathbf{x}$ and t such that
(i) the field equations

$$\begin{aligned}
\rho\dot{\boldsymbol{v}} &= \mathrm{div}\boldsymbol{\sigma}, \\
L_{\boldsymbol{v}}\boldsymbol{\tau} &= \mathcal{L}^e\cdot\mathbf{d} - \mathcal{L}^{th}\dot{\vartheta} - \frac{\lambda}{\beta}\left\langle\Phi(f-\kappa)\right\rangle\mathcal{L}^e\cdot\mathbf{P}, \\
L_{\boldsymbol{v}}\boldsymbol{\mu} &= \mathrm{m}(s)\frac{\lambda}{\beta}\left\langle\Phi(f-\kappa)\right\rangle, \\
\rho c_p\dot{\vartheta} &= \mathrm{div}(k\ \mathrm{grad}\ \vartheta) + \rho\chi\frac{\lambda}{\beta}\left\langle\Phi(f-\kappa)\right\rangle + \frac{\rho}{\rho_{Ref}}\vartheta\frac{\partial\boldsymbol{\tau}}{\partial\vartheta}:\mathbf{d};
\end{aligned} \tag{9.1}$$

(ii) the boundary conditions
traction $(\boldsymbol{\tau}\cdot\mathbf{n})^i$ are prescribed on $\partial\mathcal{B}$
temperature ϑ is prescribed on $\partial\mathcal{B}$;
(iii) the initial conditions
$\phi, \boldsymbol{v}, \vartheta$ and $\boldsymbol{\mu}$ are given at $X \in \mathcal{B}$ at $t = 0$; are satisfied.

9.2 Alternative formulation

To study the influence of the strain rate effects on plastic flow localization it is convenient to take advantage of the alternative constitutive equations for elastic-viscoplastic damaged solids. Then, to obtain an alternative formulation of the process considered we have to replace the last three field equations in (9.1) by

$$\begin{aligned}
L_{\boldsymbol{v}}\boldsymbol{\tau} &= \mathcal{L}:\mathbf{d} - \mathbf{z}\dot{\vartheta} + \mathcal{K}\dot{\sigma}_M, \\
L_{\boldsymbol{v}}\boldsymbol{\mu} &= \mathrm{m}(s)\left\langle\frac{1}{H}\{\mathbf{P}:[\dot{\boldsymbol{\tau}} - (\mathbf{d}\cdot\boldsymbol{\alpha} + \boldsymbol{\alpha}\cdot\mathbf{d})] + \pi\dot{\vartheta} - k\dot{\sigma}_M\}\right\rangle \\
\rho c_p\dot{\vartheta} &= \mathrm{div}(k\ \mathrm{grad}\ \vartheta) + \vartheta\frac{\rho}{\rho_{Ref}}\frac{\partial\boldsymbol{\tau}}{\partial\vartheta}:\mathbf{d} \\
&\quad + \rho\bar{\chi}\left\langle\frac{1}{H}\{\mathbf{P}:[\dot{\boldsymbol{\tau}} - (\mathbf{d}\cdot\boldsymbol{\alpha} + \boldsymbol{\alpha}\cdot\mathbf{d})] + \pi\dot{\vartheta} - k\dot{\sigma}_M\}\right\rangle.
\end{aligned} \tag{9.2}$$

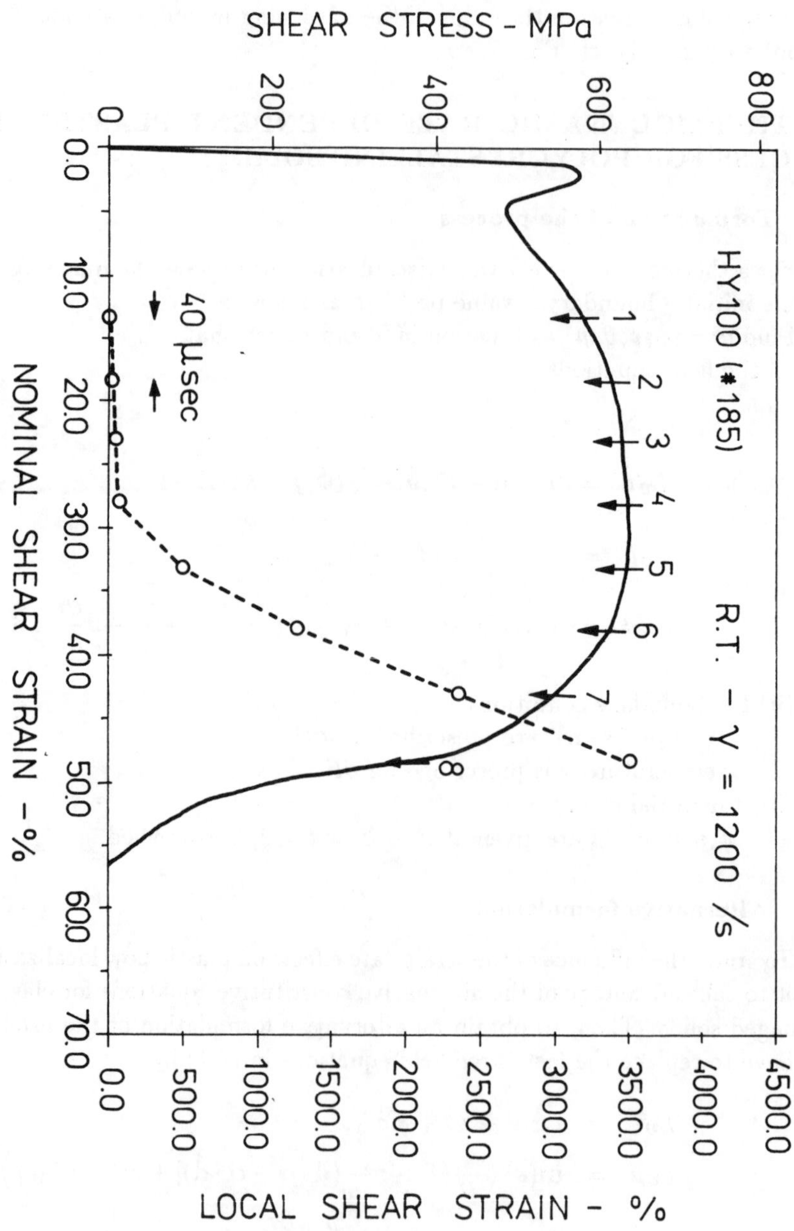

Figure 15: The shear stress – nominal shear strain curve and the local shear strain – nominal shear strain curve for HY–100 steel (after CHO, CHI and DUFFY [1988])

9.3 Quasi static and adiabatic approximation

Let consider the quasi-static and adiabatic approximation for the alternative formulation of the thermodynamic flow process.

The mapping $\mathbb{T}_{(\cdot)}$ is defined as follows:

Find $\varphi = (\phi, \boldsymbol{v}, \vartheta, \boldsymbol{\mu})$ as function of $\mathbf{x}$ and t such that

(i) the field equations

$$
\begin{aligned}
\operatorname{div}\boldsymbol{\sigma} &= 0, \\
L_{\boldsymbol{v}}\boldsymbol{\tau} &= \mathcal{L} : \mathbf{d} - \mathbf{z}\dot{\vartheta} + \mathcal{K}\dot{\sigma}_M, \\
L_{\boldsymbol{v}}\boldsymbol{\mu} &= \mathbf{m}(s)\left\langle \frac{1}{H}\{\mathbf{P} : [\dot{\boldsymbol{\tau}} - (\mathbf{d}\cdot\boldsymbol{\alpha} + \boldsymbol{\alpha}\cdot\mathbf{d})] + \pi\dot{\vartheta} - k\dot{\sigma}_M\}\right\rangle, \\
\rho c_p \dot{\vartheta} &= \frac{\rho}{\rho_{Ref}}\frac{\partial \boldsymbol{\tau}}{\partial \vartheta} : \mathbf{d} + \rho\bar{\chi}\left\langle \frac{1}{H}\{\mathbf{P} : [\dot{\boldsymbol{\tau}} - (\mathbf{d}\cdot\boldsymbol{\alpha} + \boldsymbol{\alpha}\cdot\mathbf{d})] + \pi\dot{\vartheta} - k\dot{\sigma}_M\}\right\rangle;
\end{aligned}
\tag{9.3}
$$

(ii) traction $(\boldsymbol{\tau}\cdot\mathbf{n})^i$ are prescribed on $\partial\mathcal{B}$;

(iii) $\phi, \boldsymbol{v}, \boldsymbol{\mu}$ and ϑ are given at $X \in \mathcal{B}$ at $t = 0$; are satisfied.

Taking advantage of the results obtained in section 7.5 the field equations can be reduced to

$$
\begin{aligned}
\operatorname{div}\boldsymbol{\sigma} &= 0 \\
L_{\boldsymbol{v}}\boldsymbol{\tau} &= \mathbb{L} : \mathbf{d} + \mathbb{K}\ddot{\varepsilon}^P_M
\end{aligned}
\tag{9.4}
$$

with prescribed tractions $(\boldsymbol{\tau}\cdot\mathbf{n})^i$ on $\partial\mathcal{B}$.

9.4 Discussion of the shear band localization criteria

To investigate the criteria for shear band localization let consider two particular dynamic loading processes.

(i) In the first case it is has been assumed that the strain rate is constant, i.e. $\dot{\varepsilon}^P_M$ =const and then $\ddot{\varepsilon}^P_M = 0$.

This kind of dynamical process is very frequently used in experimental investigation of properties of a material as well as in technological applications, particularly in metal forming technology.

Then the fundamental rate equation $(9.4)_2$ reduces to the form as follows

$$
L_{\boldsymbol{v}}\boldsymbol{\tau} = \mathbb{L} : \mathbf{d}. \tag{9.5}
$$

This result allows to use in the examination of the conditions for localization the standard bifurcation method.

In other words, for this kind of dynamical process the criteria for shear band localization are practically the same as for rate independent plastic flow process previously investigated in Chapter 9.

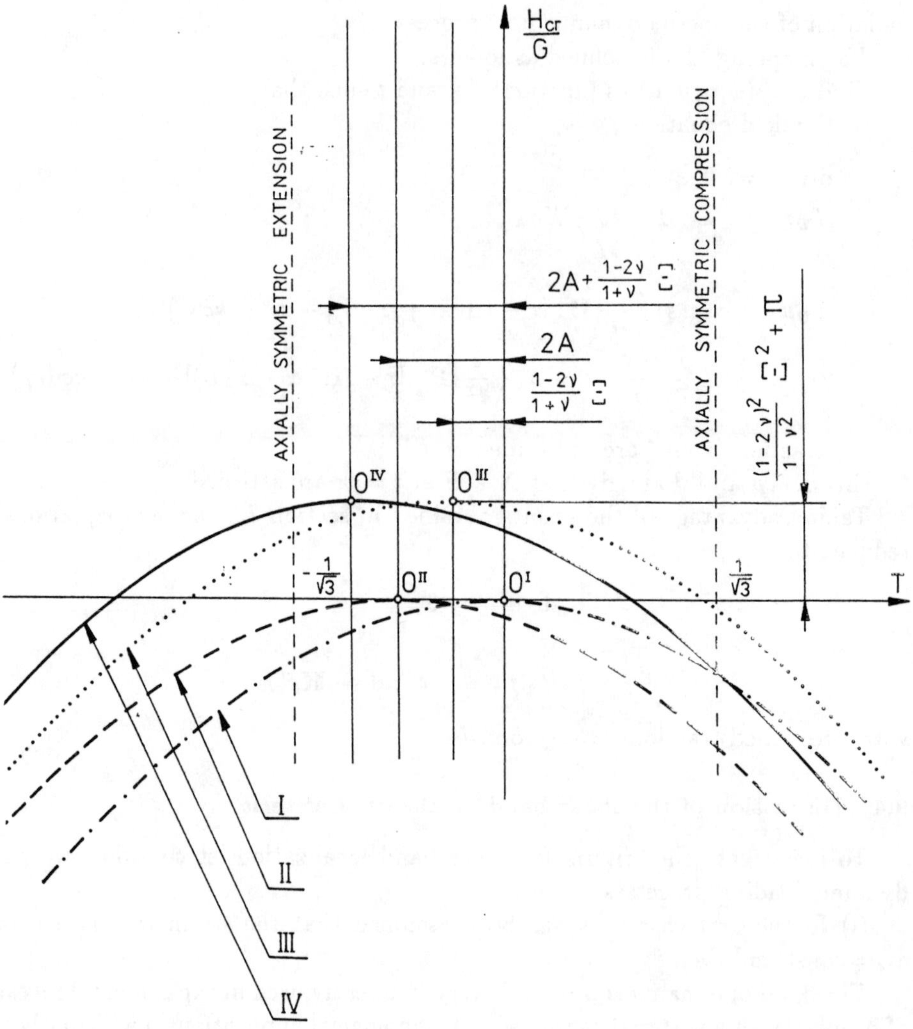

Figure 13: The critical hardening modulus rate H_{cr}/G as function of T (for $A = B$)

REFERENCES

1. Abraham, R., Marsden, J.E., Foundations of Mechanics, Second Edition, Addison-Wesley, Reading Mass., 1978

2. Abraham, R., Marsden, J.E., Ratiu, T., Manifolds, Tensor analysis and Applications, Addison-Wessley, Reading, MA, 1983

3. Agah-Tehrani, A., Lee, E.H., Mallett, R.L., Onat, E.H., The theory of elastic-plastic deformation at finite strain with induced anisotropy modelled as combined isotropic-kinematic hardening, J.Mech.Phys.Solids, 35, 519-539,1987

4. Anand, L., Spitzig, W.A., Initiation of localized shear band in plane strain, J.Mech.Phys. Solids, 28, 113-128, 1980

5. Asaro, R.J., Crystal plasticity, ASME, 50, 921-934, 1983

6. Asaro, R.J., Micromechanics of crystals and policrystals, Advances in Applied Mechanics, 23, 1-115, 1983

7. Asaro, R.J., Needleman, A., Texture development and strain hardening in rate dependent policrystals, Acta Metall., 33, 923-953, 1985

8. Asaro, R.J., Rice, R.J., Strain localization in ductile single crystals, J.Mech.Phys.Solids, 25, 309-338, 1977

9. Ashby, M.F., Frost, H.J., in Constitutive Equation in Plasticity (ed. A.S.Argon),MIT Press, Cambridge, Mass., 1975

10. Bingham, E.C., Fluidity and Plasticity, McGraw Hill, New York, 1922

11. Campbell, J.D., Dynamic plasticity macroscopic and microscopic aspects, Material Sci.Engng., 12, 3-12, 1973

12. Campbell, J.D., Ferguson, W.G., The temperature and strain-rate dependence of the shear strength of mild steel, Phil.Mag., 81, 63-82, 1970

13. Chang, Y.W., Asaro, R.J., An experimental study of shear localization in aluminum-copper single crystals, Acta Metall., 29, 241-257, 1981

14. Chang, Y.W., Asaro, R.J., Lattice rotations and localized shearing in single crystals, Arch.Mechanics, 32,369-401, 1980

15. Cho, K., Chi, Y.C., Duffy, J., Microscopic obserwations of adiabatic shear bands in three different steels, Brown University Report, Sept. 1988

16. Cox, T.B., Low, J.R. An investigation of the plastic fracture of AISI 4340 and 18 Nickel-200 grade maraging steels, Met. Trans., 5, 1457-1470, 1974

17. Dowling, A.R., Harding, J., Campbell, J.D., The dynamic punching of metals, J.Inst. of Metals, 98, 215-224, 1970

18. Duszek, M.K., Perzyna, P., Influence of the kinematic harden- ing on the plastic flow localization in damaged solids, Arch. Mechanics, 40, 595-609, 1988

19. Duszek, M.K., Perzyna, P., On combined isotropic and kinematic hardening effects in plastic flow processes, Int. J. Plasticity 1991 (in print)

20. Duszek, M.K., Perzyna, P., Plasticity of damaged solid and shear band localization, Ing.-Archiv, 58, 380-392, 1988

21. Duszek, M.K., Perzyna, P., The localization of plastic deformation in thermoplastic solids, Int.J.Solids and Structures, 27, 1419-1443,1991

22. Duszek-Perzyna, M. Perzyna, P., Stein, E., Adiabatic shear band localization in elastic-plastic damaged solids,The Second Int. Symposium on Plasticity and its current Applications, July 31- August 4, 1989, Mie University; Int.J. Plasticity (in print)

23. Duszek, M.K., Perzyna, P., Adiabatic shear band localization in elastic-plastic crystals, 1991 (in preparation to publication)

24. Evans, A.G., Rawlings, R.D., The thermally activated deformation of crystalline materials, Phys.Stat.Sol., 34, 9-31, 1969

25. Gilbert, J.E., Knops, R.J., Stability of general systems, Arch.Rat.Mech.Anal., 25, 271-284, 1967

26. Green, A.E., Naghdi, P.M., Some remarks on elastic-plastic deformation at finite strain, Int.J.Engng.Sci., 9, 1219-1229, 1971

27. Green, A.S., Naghdi, P.M., A general theory of an elastic-plastic continuum, Arch. Rat. Mech. Anal., 18, 1964

28. Gurson, A.L., Continuum theory of ductile rupture by void nucleation and growth , Part I. Yield criteria and flow rules for porous ductile media , J.Eng.Mater.Technol., 99, 2-15, 1977

29. Gurtin, M.E., Thermodynamics and stability, Arch. Rat. Mech. Anal., 59, 63-96, 1975

30. Gurtin, M.E., Thermodynamics and the energy criterion for stability, Arch.Rat. Mech.Anal., 52, 93-103, 1973

31. Hadamard, J., Lecons sur la propagation des ondes et les equations de l'hydrodynamique, chap.6, Paris, 1903

32. Hartley, K.A., Duffy, J., Hawley, R.H., Measurement of the temperature profile during shear band formation in steels deforming at high strain rates, J.Mech.Phys.Solids, 35, 283-301, 1987

33. Hauser, F.E., Simmons, J.A., Dorn, J.E., Strain rate effects in plastic wave propagation, in Response of Metals to High Velocity Deformation, Wiley (Interscience) New York, 93-114, 1961

34. Hill, R., Acceleration wave in Solids, J.Mech.Phys.Solids, 10, 1-16, 1962

35. Hill, R., Aspects of invariance in solids mechanics, Adv.Appl.Mech., 18, 1-75, 1987

36. Hill, R., The essential structure of constitutive laws for metal composites and policrystals, J.Mech.Phys.Solids, 15, 255-262, 1967

37. Hill, R., Rice, J.R., Constitutive analysis of elastic- plastic crystals at arbitrary strain, J.Mech.Phys.Solids, 20, 401-413, 1972

38. Hill, R., Rice, J.R., Elastic potentials and the structure of enelastic constitutive laws, SIAM J.Appl.Math., 25, 448-461, 1973

39. Hohenemser, K., Prager, W., Uber die Ansatze der Mechanik isotroper Kontinua, ZAMM, 12, 216-226, 1932

40. Hutchinson, J.W., Bounds and self-consistent estimates for creep of polycrystalline materials, Proc.Royal Soc. London, Sec.A 348, 101-127, 1976

41. Iwakuma, T., Nemat-Nasser, S., Finite elastic-plastic deformation of polycrystalline metals, Proc.Royal Soc. London, A 394, 87-119, 1984

42. Knops, R.J., Wilkes, E.W., Theory of elastic stability, Handbuch der Physik, VI a/3, Berlin, Heidelberg, New York, Springer, 1973

43. Kocks, U.F., Argon, A.S., Ashby, M.F., Thermodynamics and Kinetics of Slip, Pergamon Press, 1975

44. Kumar, A., Hauser, F.E., Dorn, J.E., Viscous drag on dislocations in aluminum at high strain rates, Acta Metall., 16, 1189-1197, 1968

45. Kumar, A., Kumble, R.G., Viscous drag on dislocations at high strain rates in copper, J.Appl.Physics, 40, 3475-3480, 1969

46. Le Roy, G., Embury, J.D., Edward, G., Ashby, M.F., A model of ductile fracture based on the nucleation and growth of voids, Acta Metall., 29, 1509-1522, 1981

47. Lee, E.H., Elastic-plastic deformations at finite strains, J. Appl. Mech., 36, 1-6, 1969

48. Lindholm, U.S., Some experiments with the split Hopkinson pressure bar, J.Mech.Phys.Solids, 12, 317-335, 1964

49. Lindholm, U.S., in: Mechanical Behaviour of Materials under Dynamic Loads, (ed. U.S.Lindholm) , Springer Verlag, 77-95, 1968

50. Lippmann, H., Velocity field equations and strain localization, Int.J.Solids Structure, 22, 1399-1409, 1986

51. Malvern, L.E., The propagation of longitudinal waves of plastic deformation in a bar of material exhibiting, a strain-rate effects, J.Appl.Mech., 18, 203-208, 1951

52. Mandel, J., Conditions de stabilite et postulat de Drucker, in: Rheology and Soil Mechanics, eds. J.Kravtchenko and P.M.Sirieys, Springer-Verlag, 58-68, 1966

53. Marchand, A., Cho, K., Duffy, J., The formation of adiabatic shear bands in an AISI 1018 cold-rolled steel, Brown University Report, September 1988

54. Marchand, A., Duffy, J., An experimental study of the formation process of adiabatic shear bands in a structural steel, J.Mech.Phys.Solids, 36, 251-283, 1988

55. Marsden, J.E., Hughes, T.J.R., Mathematical Foundations of Elasticity, Prentice-Hall, Englewood Cliffs, NJ, 1983

56. Mear, M.E., Hutchinson, J.W. Influence of yield surface curvature on flow localization in dillatant plasticity, Mechanics and Materials, 4, 395-407, 1985

57. Mehrabadi, M.M., Nemat-Nasser, S., Some basic kinematical relations for finite deformations of continua, Mechanics of Materials, 6, 127-138, 1987

58. Meyers, M.A., Aimone, C.T., Dynamic fracture (spalling) of metals, Prog. Mater. Sci., 28, 1-96, 1983

59. Nabarro, F.R.N., Theory of crystal dislocations, Oxford , 1967

60. Naghdi, P.M., A critical review of the state of finite plasticity, ZAMP, 41, 315-394, 1990

61. Needleman, A., Rice, J.R., Limits to ductility set by plastic flow localization, in: Mechanics of Sheet Metal Forming (ed. Koistinen, D.P. and Wang, N.-M.), Plenum, New York, 237-267, 1978

62. Needleman, A., Tvergaard, V., Limits to formability in, rate sensitive metal sheets, Mechanical Behaviour of Materials-IV (ed.J.Carlsson and N.G.Ohlson), 51-65, 1984

63. Nemat-Nasser, S., Decomposition of strain measures and their rates in finite deformation elastoplasticity, Int.J.Solids Structures, 15, 155-166, 1979

64. Nemat-Nasser, S., Chung, D.-T., Taylor, L.M., Phenomenological modelling of rate-dependent plasticity for high strain rate problems, Mechanics of Materials, 7, 319-344, 1989

65. Nemat-Nasser, S., Obata, M., Rate dependent, finite elasto-plastic deformation of polycrystals, Proc.Royal Soc. London, A 407, 343-375, 1986

66. Nemes, J.A., Eftis, J., Randles, P.W., Viscoplastic constitutive modeling of high strain-rate deformation, material damage and spall fracture, ASME J.Appl.Mech., 57, 282-291, 1990

67. Oldroyd, J., On the formulation of rheological equations of state, Proc.Roy.Soc. (London), Ser. A 200, 523-541, 1950

68. Pan, J., Rice, J.R., Rate sensitivity of plastic flow and implications for yield surface vertices, Int. J. Solids Structures, 19, 973-987, 1983

69. Pan, J., Saje, M., Needleman, A., Localization of deformation in rate sensitive porous plastic solids, Int.J.Fracture, 21, 261-278, 1983

70. Peirce, D., Asaro, R.J., Needleman, A., An analysis of non-uniform and localized deformation in ductile single crystals, Acta Metall., 30, 1087-1119, 1982

71. Peirce, D., Asaro, R.J., Needleman, A., Material rate dependence and localized deformation in crystalline solids, Acta Metall., 31, 1951-1976, 1983

72. Perzyna, P., The constitutive equations for rate sensitive plastic materials, Quart. Appl. Math., 20, 321-332, 1963

73. Perzyna, P., Fundamental problems in viscoplasticity, Advances in Applied Mechanics, vol.9, 243-377, 1966

74. Perzyna, P., Thermodynamic theory of viscoplasticity, Advances in Applied Mechanics, vol.11 313-354, 1971

75. Perzyna, P., Coupling of dissipative mechanisms of viscoplastic flow, Arch. Mechanics, 29, 607-624, 1977

76. Perzyna, P., Modified theory of viscoplastcity. Application to advanced flow and instability phenomena, Arch. Mechanics, 32, 403-420, 1980

77. Perzyna, P., Thermodynamics of dissipative materials, in Recent Developments in Thermodynamics of Solids,Eds. Lebon, G. and Perzyna, P., Springer, Wien 1980, 95-220, 1980

78. Perzyna, P., Stability phenomena of dissipative solids with internal defects and imperfections, XV-th IUTAM Congress, Toronto, August 1980, Theoretical and Applied Mechanics, Proc. ed. F.P.J. Rimrott and B. Tabarrok, North-Holland, pp. 369-376, Amsterdam 1981

79. Perzyna, P., Stability problems for inelastic solids with defects and imperfections, Arch.Mech., 33, 587-602, 1981

80. Perzyna, P., Application of dynamical system methods to flow processes of dissipative solids, Arch.Mech., 34, 523-539, 1982

81. Perzyna, P., Stability of flow processes for dissipative solids with internal imperfections, ZAMP, 35, 848-867, 1984

82. Perzyna, P., Constitutive modelling for brittle dynamic fracture in dissipative solids, Arch. Mechanics, 38, 725-738, 1986

83. Perzyna, P., Constitutive equations of dynamic plasticity, Post Symposium Short Course, August 4-5,1989, Nagoya,Japan

84. Perzyna, P., Temperature and rate dependent theory of plasticity of polycrystalline solids, The Second Int. Symposium on Plasticity and its current Applications, July 31- August 4, 1989, Mie University, Tsu, Japan

85. Perzyna, P., Duszek-Perzyna, M.K., Stein, E., Analysis of the influence of different effects on criteria for shear band localization,28th Polish Solids Mech.Conf., Kozubnik, 1990

86. Perzyna, P., Adiabatic shear band localization in rate dependent plastic solids,1991 (in preparation to publication)

87. Prager, W., Introduction to Mechanics of Continua, Gin and Co., New York, 1961

88. Rice, J., Plasticity of soil mechanics, in Proc. of the Symposium of the Role of Plasticity in Soil Mechanics, Cambridge, England 1973, ed. by Palmer, A.C., 263-275, 1973

89. Rice, J.R., Continuum mechanics and thermodynamics of plasticity in relation to microscale deformation mechanisms, in Constitutive Equations in Plasticity,(ed.A.S.Argon), The MIT Press, Cambridge, 23-75, 1975

90. Rice, J.R., Inelastic constitutive relations for solids: an internal variable theory and its application to metal plasticity, J.Mech.Phys.Solids, 19, 433-455, 1971

91. Rice, J.R., On the structure of stress-strain relation for time-dependent plastic deformation in metals, J.Appl.Mech., 37, 728-737, 1970

92. Rice, J.R., The localization of plastic deformation, Theoretical and Applied Mechanics, ed. Koiter, W.T., North-Holand, 207-220, 1976

93. Rice, J.R., Rudnicki, J.W., A note on some features of the theory of localization of deformation, Int.J.Solids Structure, 16, 597-605, 1980

94. Rudnicki, J.W., Rice, J.R., Conditions for the localization of deformation in pressure-sensitive dilatant materials, J. Mech. Phys. Solids, 23, 371-394, 1975

95. Seeger, A., The generation of lattice defects by moving dislocations and its application to the temperature dependence of the flow-stress of f.c.c. crystals, Phil.Mag., 46, 1194-1217, 1955

96. Shima, S., Oyane, M., Plasticity theory for porous solids, Int.J.Mech.Sci., 18, 285-291, 1976

97. Shioiri, J., Satoh, K., Nishimura, K., in High Velocity Deformation in Solids, IUTAM Symp.Proc. (ed.K.Kawata and J.Shioiri), Springer, 50-66, 1979

98. Shockey, D.A., Seaman, L., Curran, D.R., The microstatistical fracture mechanics approach to dynamic fracture problems, Int. J. Fracture, 27, 145-157, 1985

99. Simo, J.C., A framework for finite strain elasto-plasticity based on maximum plastic dissipation and the multiplicative decomposition: Part II, Computational aspects, Comput.Meths.Appl.Mech.Engng., 68, 1-31, 1988

100. Simo, J.C., A framework for finite strain elastoplasticity based on maximum plastic dissipation and the multiplicative decomposition: Part I. Continuum formulation, Comput. Meth. Appl. Mech. Engng. , 66, 199-219, 1988

101. Spitzig, W.A., Deformation behaviour of nitrogenated Fe-Ti-Mn and Fe-Ti single crystals, Acta Metall., 29, 1359-1377, 1981

102. Stein, E., Duszek-Perzyna, M.K., Perzyna, P., Influence of thermal effects on shear band localization in elastic-plastic damaged solids, Proc. Euromech Colloquium 255,Paderborn,1989

103. Teodosiu, C., Sidoroff, F., A theory of finite elastoplasticity of single crystals, Int.J.Engng.Sci., 14, 165-176, 1976

104. Thomas, T.Y., Plastic Flow and Fracture in Solids, New York, Academic Press, 1961

105. Truesdell, C., Noll, W., The nonlinear field theories, in: Handbuch der Physik, Band III/3, Springer, Berlin, 1965

106. Tvergaard, V., Effects of yield surface curvature and void nucleation on plastic flow localization, J.Mech.Phys.Solids, 35, 43-60, 1987

107. Tvergaard, V., Needleman, A., Effect of material rate sensitivity on failure modes in the Charpy V-notch test, J. Mech. Phys. Solids, 34, 213-241, 1986

108. Willems, J.C., Dissipative dynamical systems, Part I: General Theory, Part II: Linear systems with quadratic supply rates, Arch.Rat.Mech.Anal., 45, 321-393, 1972

109. Ziegler, H., A modification of Prager's hardening rule, Quart. Appl. Math., 17, 55-65, 1959

Table of Contents

NUMERICAL SIMULATION OF PLASTIC LOCALIZATION USING FE-MESH REALIGNMENT

R. Larsson, K. Runesson and A. Samuelsson
Chalmers University of Technology, Göteborg, Sweden

ABSTRACT

Proper design of the computational algorithm is absolutely essential in order to account for the failure mechanisms that are responsible for the development of a strongly localized mode of deformation. In this paper we discuss how to simulate numerically localized behavior of the deformation due to incorporation of non–associated plastic flow and/or softening behavior in the elasto–plastic material model. The development of a localization zone of a slope stability problem is captured by the use of a FE–mesh adaptation strategy, which aims at realigning the inter–element boundaries so that the most critical kinematical failure mode is obtained. Based on the spectral properties of the characteristic material operator we define a criterion for discontinuous bifurcation. As a by–product from this criterion, we obtain critical bifurcation directions which are used to realign the element mesh in order to enhance the ability of the model to describe properly the failure kinematics. Moreover, a successful algorithm also includes consideration of stability properties of the elasto–plastic solution.

INTRODUCTION

It is a well established fact that a smooth and continuously varying deformation field of an elastic–plastic solid may give way to an inhomogeneous highly localized deformation mode that is sometimes known (with somewhat loose terminology) as a *shear band*. At such a "critical" state of stress and strain the incremental solution may become *non–unique*. A classical example of localization of deformation is the existence of so–called Lüders' lines in thin steel plates subjected to plane stress states, c.f. Nadai[1] and Thomas[2]. Another example of localized deformation, which is of importance in geotechnical engineering, is the progressive development of a slip–line along which a soil slope fails. Although this problem has received considerable attention in the past classical methods have a limited ability to model the progressive development of the localization zone, which has a significant influence on the ultimate bearing capacity. Therefore, considerable attention has recently been paid to the design of efficient finite element procedures for resolving the problem of how to simulate the deformation localization in elastic–plastic solids.

It seems to be quite generally accepted that localization of the deformation is associated with discontinuous bifurcations of the incremental elasto–plastic solution, c.f. Bazant[3], Leroy and Ortiz[4,5], Ortiz and Quigley[6]. Upon introducing the concept of a characteristic surface, across which the strain or, possibly, the displacement increments may become discontinuous, a necessary condition for the existence of a discontinuous solution is that the characteristic, or acoustic, material tensor becomes singular. However, since the characteristic material tensor signals the possibility for discontinuous bifurcation only at the constitutive level it is of considerable interest to characterize stable equilibrium paths of the structural behavior, as discussed by Bazant[7] and Runesson *et* al.[8,9]. This seems particularly important in view of the fact that the incremental boundary value problem has a non–unique solution, and it is of interest to identify solutions representing localized deformation. Such a characterization shows that localization is strongly associated with stability of the incremental solution. In this context we note that stable equilibrium states can be defined by a minimizing property for a range of so–called "simple" plasticity models upon integrating the rate equations by an implicit method, such as the backward Euler method or, equivalently, the closest–point–projection–method, c.f. Runesson *et* al.[8,9]. However, when no minimizing property (no variational principle) exists other arguments are required in order to define stability. To this end, it seems reasonable to adopt Hill's stability criterion[10], which states that the equilibrium state is stable whenever the second order work is positive. We note that this stability criterion applied to a "simple" plasticity model yields identical stability results as compared to those corresponding to the minimizing property.

The feasibility of capturing localized plastic deformation using the FE–method has been demonstrated by a number of investigators for both finite and small deformations in various fields of application. For instance, in metal plasticity, analyses have been carried out by Tvergaard *et* al.[11] and Triantafyllidis *et* al.[12] concerning the incrementally non–linear J_2 –corner theory. A comprehensive survey of these analyses is given by Needleman and Tvergaard[13]. Soil plasticity has been considered by Prevost[14], de Borst[15,16], Bardet[17], Bardet and Mortazavi[18]. Development of cracks in concrete has also been studied by Nilsson and Oldenburg[19], Rots *et* al.[20], de Borst[21], Dahlbom and Ottosen[22], Glemberg[23]. However, such analyses tend to be quite sensitive to details of mesh design.

A computational algorithm that captures the solution which possesses the physically realistic deformation pattern is conveniently based on a FE–mesh adaptation strategy, that utilizes critical bifurcation directions corresponding to the singularity points of the characteristic material tensor. Two concepts have been adopted in the literature: The first one stems, typically, from conventional FE–analysis with a fixed mesh that is being enriched by local basis functions, as described by Ortiz *et* al.[4,24]. The other concept is to adopt a remeshing procedure which aims at realigning the element sides successively as the loading proceeds, as discussed by Larsson[25]. In practice the realignment (or enrichment) of the elements has to be carried out based on a diagnostic analysis since the updated state is not known in advance. In this context, we note that some care is needed when bifurcation results are converted from the rate to the incremental behavior. Moreover, it is known that failure of a structure that obeys an elasto–plastic material law may not generally be associated with the development of a shear band. For example, if a non–associated flow rule is adopted "spurious" hardening/softening as compared to elastic behavior may occur, whereby structural softening effects are possible even though the material is formally hardening, as discussed by Runesson and Mroz[26]. It is, therefore, interesting to link this failure phenomenon to the development of shear bands.

Furthermore, for a given finite element mesh it is important to design the incremental/iterative algorithm for obtaining equilibrium solutions in such a way that the stable equilibrium path is traced, as described by Larsson *et* al.[27] and de Borst[15]. For example, it does not seem feasible to use true Newton iterations to ensure convergence towards a stable solution. We note that different techniques can be conceived based on the adopted integration rule. When implicit integration techniques are used, for instance, it is natural to assess the stability properties at the updated time, whereas for explicit time integration it seems natural consider the stability properties at the current time.

A possible drawback of using classical plasticity theory is that it contains no inherent information about the width of the shear bands. Moreover, the question which material model that most properly describes localization from a physical viewpoint has been subjected to keen debate. Traditionally, the band width has been explained as a mesh scaling effect in finite element calculations, like in the smeared crack approach for concrete, whereby mesh objective material parameters should be utilized. To properly model the shear band width and to regularize the discontinuous problem, other material descriptions have been proposed in the literature, such as visco–plasticity, non–local theories and Cosserat theory. To this end, it seems that a feasible approach for enhancing the FE–discretization would be to include regularized displacement discontinuities, as discussed by Johnson and Scott[28]. Furthermore, it is believed that the localization phenomenon is inherently dynamic in nature, i.e. the structure will undergo vibrating motions when the deformation field changes abruptly into a localized deformation mode. Thus, a proper treatment of the progression of the localization zone would require a dynamic analysis. Such an analysis has quite recently been carried out by Leroy and Ortiz[5].

PLASTICITY PROBLEM

Constitutive Equations

The constitutive relation in plasticity may be written

$$\dot{\epsilon} = \mathbf{C}^e : \dot{\sigma} + \dot{\epsilon}^p \tag{1}$$

$$F(\sigma, \kappa) \leq 0 \quad , \quad \dot{\lambda} \geq 0 \quad , \quad F(\sigma, \kappa)\dot{\lambda} = 0 \tag{2}$$

where $\dot{\epsilon}$ and $\dot{\sigma}$ are strain and stress rates, κ is the hardening softening variable, $\mathbf{C}^e = (\mathbf{D}^e)^{-1}$ is the (constant) tensor of elastic flexibility, $\dot{\lambda}$ is a plastic multiplier and F is the (convex) yield function. It is also assumed that the hardening/softening of the yield surface is represented only with a scalar κ, which is defined by the rate law

$$\dot{\kappa} = h(\dot{\epsilon}^p) = \dot{\lambda} h(\mathbf{g}) \tag{3}$$

where $\dot{\epsilon}^p$ is the plastic portion of the strain rate and h is a homogeneous (but generally non–linear) function of $\dot{\epsilon}^p$. Furthermore, in (3) we have assumed the non–associated flow rule

$$\dot{\epsilon}^p = \dot{\lambda}\mathbf{g} \tag{4}$$

where $\mathbf{g}$ is the flow direction and $\dot{\lambda} \geq 0$ is a plastic multiplier. In the case of an associated flow rule we have $\mathbf{g} = \mathbf{f}$, where $\mathbf{f} = \partial F / \partial \sigma$. In order to describe the non–associated flow rule it is convenient to relate the flow direction to the gradient of the yield surface by the transformation $\mathbf{A}(\sigma, \kappa)$

$$\mathbf{g} = \mathbf{A}^{-1} : \mathbf{f} \tag{5}$$

An important special case of $\mathbf{A}$ is when it is state independent and when it expresses non–associative flow only in the volumetric part of the plastic flow direction. In such a case $\mathbf{A}$ can be expressed as

$$\mathbf{A} = \mathbf{I} + \frac{1}{3}\beta\delta\delta \quad , \quad \mathbf{A}^{-1} = \mathbf{I} - \frac{1}{3}\frac{\beta}{1+\beta}\delta\delta \tag{6}$$

where β is a dilatancy parameter, $\mathbf{I}$ is the fourth order identity tensor and δ is Kronecker's delta. An associated flow rule is clearly obtained when $\beta = 0$.

Subsequently, the bilinear Continuum Tangent Stiffness (CTS) tensor $\mathbf{D}_T$, that pertinent to a strain driven format, will be used. This tensor is defined by

$$\dot{\sigma} = \mathbf{D}_T : \dot{\epsilon} \quad , \quad \mathbf{D}_T = \mathbf{D}^e - \frac{1}{K_T}\mathbf{D}^e : \mathbf{gf} : \mathbf{D}^e \tag{7}$$

In (7) K_T is the generalized plastic modulus

$$K_T = H + \mathbf{f} : \mathbf{D}^e : \mathbf{g} > 0 \quad , \quad H = -\frac{\partial F}{\partial \kappa} h(\mathbf{g}) \tag{8}$$

where it must be required that $K_T > 0$ in order that uniqueness of the response can be assured, as discussed by Klisinski *et* al.[29].

Incremental Relations

Closest Point Projection Method

In the computational application, the constitutive relations of plasticity must be integrated in discrete (load) time steps to simulate the response of the elastic–plastic material for a given load. Classical integration schemes employ a tangent stiffness formulation and are traditionally explicit in character. On the other hand, algorithms that are based on implicit integration schemes have gained in popularity because a consistent return to the yield surface is obtained in the sense that incremental objectivity can be assured, c.f. Runesson *et* al.[30]. The most widely used implicit integration scheme is undoubtedly the Backward Euler (BE) method, which is also known as the Closest–Point–Projection–Method (CPPM). The notion "Closest–Point" stems from that the stress projection in the stress deviator subspace is obtained in Euclidean norm for isotropic elastic and plastic material properties. The CPPM can be obtained by applying BE differences directly to the constitutive (rate) equations. As an alternative, the CPPM can be retrieved by application of the BE–method to the proper constitutive inequality that defines the flow rule, as described by Johnson[31,32,33].

Applied to (1) and (2) the CPPM gives, for isotropic elastic and plastic material properties with the shear modulus G and the bulk modulus K, at the time $t_{n+1} = t_n + \Delta t$

$$p_{n+1} = p^e - \lambda \frac{K}{1+\beta} \frac{\partial F(\sigma_{n+1}, \kappa^*)}{\partial p} \tag{9a}$$

$$q_{n+1} = q^e \cos(\theta^e - \theta) - \lambda 3G \frac{\partial F(\sigma_{n+1}, \kappa^*)}{\partial q} \tag{9b}$$

$$q_{n+1} q^e \sin(\theta^e - \theta) - \lambda 3G \frac{\partial F(\sigma_{n+1}, \kappa^*)}{\partial \theta} = 0 \tag{9c}$$

which holds for a yield surface fixed in the stress space with $\kappa = \kappa^*$. In (9) the stress invariants p, q and θ are defined as

$$p = -\frac{1}{3}\mathrm{tr}(\sigma) \quad , \quad q = \frac{\sqrt{3}}{\sqrt{2}}|\mathrm{s}| \quad , \quad \cos 3\theta = \frac{27}{7}\frac{\det(\mathrm{s})}{(q)^3}$$

where $\Delta\epsilon$ is the strain increment, $\sigma^e = \sigma_n + \mathbf{D}^e : \Delta\epsilon$ is the "elastic" stress, $\mathbf{s} = \sigma + p\delta$ is the stress deviator, and $|\cdot|$ denotes the Euclidean length. Important special cases of (9) are the yield criteria of von Mises and Drucker–Prager, which offers explicit stress solutions from (9) for every fixed yield surface. However, in the general case the non-linear problem (9) has to be solved iteratively.

Remark: Equation (9) is an extremal of the following constrained minimization problem

$$\min_{\tau \in B(\kappa^*)} \left\{ \frac{1}{2} |\sigma^e - \tau|^2_{ca} \right\} = \min_{\tau \in B(\kappa^*)} \left\{ \frac{1}{2} (\sigma^e - \tau) : \hat{\mathbf{C}}^e : (\sigma^e - \tau) \right\}$$

where $\hat{\mathbf{C}}^e = \mathbf{A} : \mathbf{C}^e$ is a modified elastic flexibility tensor. Moreover, we have introduced the convex set $B(\kappa^*)$ of plastically admissible stresses

$$B(\kappa^*) = \{\sigma \mid F(\sigma, \kappa^*) < 0\}$$

In the case that σ^e is plastically admissible, i.e. $\sigma^e \in B(\kappa_n)$, we obtain an unconstrained minimization problem and the stress solution is simply $\sigma_{n+1} = \sigma^e$. If, on the other hand, σ^e violates the yield criterion, i.e. $\sigma^e \notin B(\kappa_n)$, a constrained minimization problem is obtained. The stress σ_{n+1} is then obtained as the closest projection of σ^e onto the set $B(\kappa^*)$ measured in the adjusted complementary energy norm $|\cdot|_{ca}$ ■

In order to define the solution of the complete incremental problem we supplement (9) with the incremental form of (4). The true stress solution is then obtained when the yield criterion is satisfied for the updated value $\kappa(\Delta\epsilon) = \kappa_n + \Delta\kappa(\Delta\epsilon)$, which implies that

$$F(\sigma(\Delta\epsilon), \kappa(\Delta\epsilon)) = 0 \tag{10}$$

Algorithmic Tangent Stiffness

The algorithmic tangential behavior of the updated stress $\sigma_{n+1}(\Delta\epsilon)$, pertinent to the CPPM, is governed by the Algorithmic Tangent Stiffness (ATS) tensor $\mathbf{D}_A$ defined as

$$\mathbf{D}_A(\Delta\epsilon) = \frac{\partial \sigma_{n+1}}{\partial(\Delta\epsilon)} \tag{11}$$

It has been shown elsewhere, e.g. Runesson *et* al.[9], that the ATS–tensor has the structure

$$\mathbf{D}_A(\Delta\epsilon) = \mathbf{D}^e{}_A - \frac{1}{K_A} \mathbf{D}^e{}_A : \hat{\mathbf{g}}\hat{\mathbf{f}} : \mathbf{D}^e{}_A \tag{12}$$

where the algorithmic elastic stiffness moduli $\mathbf{D}^e{}_A$ is defined by

$$\mathbf{D}^e{}_A(\Delta\epsilon) = \left[\mathbf{C}^e + \lambda \frac{\partial \mathbf{g}}{\partial \sigma} + \lambda^2 \left(1 - \lambda \frac{\partial h}{\partial \kappa}\right)^{-1} \frac{\partial \mathbf{g}}{\partial \kappa} \frac{\partial h}{\partial \sigma} \right]^{-1} \tag{13}$$

In (12) we have introduced the modified gradient $\hat{\mathbf{f}}$ and the modified flow direction $\hat{\mathbf{g}}$, which are given by

$$\hat{\mathbf{f}} = \mathbf{f} + \lambda\left(1 - \lambda\frac{\partial h}{\partial \kappa}\right)^{-1}\frac{\partial F}{\partial \kappa}\frac{\partial h}{\partial \sigma} \tag{14a}$$

$$\hat{\mathbf{g}} = \mathbf{g} + \Delta\kappa\left(1 - \lambda\frac{\partial h}{\partial \kappa}\right)^{-1}\frac{\partial \mathbf{g}}{\partial \kappa} \tag{14b}$$

Furthermore, the generalized algorithmic hardening modulus K_A becomes

$$K_A = \hat{H} + \hat{\mathbf{f}} : \mathbf{D}^e{}_A : \hat{\mathbf{g}} \ , \quad \hat{H} = \left(1 - \lambda\frac{\partial h}{\partial \kappa}\right)^{-1} H \tag{15}$$

General and "Simple" Plasticity Models

It was demonstrated in Runesson *et* al.[8,9] that it is quite simple to produce an incremental energy density $\Delta w(\Delta\epsilon)$ that serves as a potential for a uniaxial stress state with isotropic hardening/softening characteristics. For multiaxial stress states it must be required that the Hessian tensor $\mathbf{D}_H$ be symmetric for $\Delta w(\Delta\epsilon)$ to exist, i.e.

$$\sigma_{n+1}(\Delta\epsilon) = \frac{\partial(\Delta w(\Delta\epsilon))}{\partial(\Delta\epsilon)} \quad \text{and} \quad \mathbf{D}_H(\Delta\epsilon) = \frac{\partial^2(\Delta w(\Delta\epsilon))}{\partial(\Delta\epsilon)\partial(\Delta\epsilon)} \tag{16}$$

However, the tangential behaviour of the updated stress is generally non–symmetric as displayed by the structure of $\mathbf{D}_A$. Hence, we conclude that $\Delta w(\Delta\epsilon)$ exists only in special cases since the structure of $\mathbf{D}_A$ reveals, in fact, three sources of non–symmetry. The first one is the choice of an non–associated flow rule, where $\mathbf{g} \neq \mathbf{f}$. The second and third sources of non–symmetry are the choices of general hardening/softening ($\hat{\mathbf{f}} \neq \mathbf{f}$) and state dependent plastic flow ($\hat{\mathbf{g}} \neq \mathbf{g}$). Note that in the case of an associated flow rule the non–symmetry of $\mathbf{D}_A(\Delta\epsilon)$ still prevails due to general hardening/softening and state dependent plastic flow. In order to preserve symmetry, which leads to $\mathbf{D}_A = \mathbf{D}_H$, it must be required that $\partial h/\partial\sigma = 0$, $\partial \mathbf{g}/\partial\kappa = 0$ and that $\mathbf{g} = \mathbf{f}$; such plasticity models are henceforth termed "simple". This issue will be further elaborated on in the subsequent chapter.

Secant Relations

It is possible to establish an Algorithmic Secant Stiffness (ASS) tensor $\mathbf{D}_S$, which defines the incremental (algorithmic) constitutive behavior. Upon fully implicit integration of the relations (1) and (2) we obtain

$$\Delta\sigma = \mathbf{D}^e : \Delta\epsilon - \lambda\mathbf{D}^e : \mathbf{g}_{n+1} \tag{17}$$

The secant hardening modulus H_S can be calculated from

$$\lambda = \frac{1}{K_S}\mathbf{f}_{n+1} : \mathbf{D}^e : \Delta\epsilon \tag{18}$$

where K_S is the generalized secant plastic modulus

$$K_S = H_S + \mathbf{f}_{n+1} : \mathbf{D}^e : \mathbf{g}_{n+1} \tag{19}$$

By combining (18) and (19) we may calculate H_S as

$$H_S = \frac{1}{\lambda} \mathbf{f}_{n+1} : \Delta\sigma \tag{20}$$

The ASS–tensor $\mathbf{D}_S$ is then obtained from (17) and (18)

$$\Delta\sigma = \mathbf{D}_S(\Delta\epsilon) : \Delta\epsilon \tag{21}$$

where

$$\mathbf{D}_S(\Delta\epsilon) = \mathbf{D}^e - \frac{1}{K_S} \mathbf{D}^e : \mathbf{g}_{n+1}\mathbf{f}_{n+1} : \mathbf{D}^e \tag{22}$$

Thus, the ASS–tensor has the same (polyadic) structure as the CTS–tensor. However, the ASS–tensor is incrementally non–linear, and an explicit expression can be obtained only for special cases, whereas the CTS–tensor is incrementally bilinear with respect to the linearized loading and unloading criteria. We note in this context that conditions for uniqueness, stability and localized deformation are traditionally evaluated for the rate response as defined by the CTS–tensor. Such analyses results in the establishment of criteria, in terms of bounds on the "physical" hardening modulus H, that defines well posed boundary value problems. However, in the computational application it is $\mathbf{D}_S$ (rather than $\mathbf{D}_T$) that defines the incremental behavior. Therefore, it is of considerable value to establish the general relationships between the "physical" modulus H and the "computational" secant modulus H_S. For example, in the case of a perfectly plastic material ($H = 0$) it is simple to find such a relation. From convexity of the yield surface it follows in this case that $\mathbf{f}_{n+1} : \Delta\sigma \geq 0$, which together with the condition $\lambda \geq 0$ gives $H_S \geq 0$ from (20).

Furthermore, quite independent of the physical hardening modulus or the computational secant modulus the constitutive behavior may exhibit "spurious" hardening/softening characteristics in certain loading directions when non–associated plastic flow is considered, as discussed by (for instance) Runesson and Mroz[26]. To this end we consider the directional stiffness ratio defined as

$$R_\epsilon = \frac{S_\epsilon}{S^e_\epsilon} \quad , \quad S_\epsilon = \frac{\Delta\sigma^* : \Delta\epsilon}{|\Delta\epsilon|^2} \quad , \quad S^e_\epsilon = \frac{\Delta\sigma^e : \Delta\epsilon}{|\Delta\epsilon|^2} \tag{23}$$

which for a strain controlled loading process defines the linearized incremental stiffness at different loading directions. It is noted that this stiffness may not generally be positive in all loading directions for a hardening material. It is, therefore, of considerable interest to evaluate the bounds of R_ϵ as compared to the hardening modulus H. For this purpose, the incremental constitutive behavior is linearized at time $t = t_n$ such that

$$\Delta\sigma^* = \mathbf{D}_{T,n} : \Delta\epsilon \quad , \quad \mathbf{D}_{T,n} = \mathbf{D}^e - \frac{1}{K_{T,n}} \mathbf{D}^e : \mathbf{g}_n\mathbf{f}_n : \mathbf{D}^e \tag{24}$$

where $\Delta\sigma^*$ is the tangential stress increment that is obviously different from the true stress increment $\Delta\sigma$ which is governed by the ASS–tensor. However, for very small strain increments one may approximate the incremental behavior by (24). The reason for this approximation is that it is not feasible to determine bounds on R_ϵ for the true incremental behavior (as defined by the ASS–tensor) since this behavior is incrementally highly non–linear. Thus, in order to evaluate the bounds of R_ϵ it is appropriate to evaluate the Rayleigh quotient

$$R_\epsilon = \frac{\Delta\epsilon : \mathbf{D}_{T,n} : \Delta\epsilon}{\Delta\epsilon : \mathbf{D}^e : \Delta\epsilon} \quad \forall \Delta\epsilon \tag{25}$$

where it is noted that only the symmetric part of the CTS–tensor $\mathbf{D}_{T,n}$ (denoted $\mathbf{D}^s{}_{T,n}$) will contribute to the directional stiffness ratio R_ϵ. The bounds of R_ϵ are, thus, controlled by the lowest and the highest eigenvalues pertinent to the eigenvalue problem

$$\mathbf{D}^s{}_{T,n} : \mathbf{x}_k = \mu_k \mathbf{D}^e : \mathbf{x}_k \quad , \quad k = 1.2, \ldots p \quad , p = \mathrm{Dim}(\mathbf{D}^e) \tag{26}$$

These eigenvalues can be obtained explicitly and according to Runesson and Mroz[26] we summarize as follows: The magnitude of the directional stiffness ratio is bounded by

$$\mu_p \le R_\epsilon \le \mu_1 \tag{27}$$

where μ_1 and μ_p becomes

$$\mu_1 = \frac{1}{K_{T,n}}\left(H_n + \mathbf{f}_n : \mathbf{D}^e : \mathbf{g}_n \frac{1}{2}\left(1 + \frac{1}{\cos a}\right)\right) \tag{28a}$$

$$\mu_p = \frac{1}{K_{T,n}}\left(H_n + \mathbf{f}_n : \mathbf{D}^e : \mathbf{g}_n \frac{1}{2}\left(1 - \frac{1}{\cos a}\right)\right) \tag{28b}$$

In (28ab) the angle a denotes the deviation between $\mathbf{f}_n$ and $\mathbf{g}_n$ measured in the elastic energy norm, i.e.

$$\cos a = \frac{\mathbf{f}_n : \mathbf{D}^e : \mathbf{g}_n}{|\mathbf{f}_n|_e |\mathbf{g}_n|_e} \quad , \quad |\mathbf{x}|_e = (\mathbf{x} : \mathbf{D}^e : \mathbf{x})^{\frac{1}{2}} \tag{29}$$

For a non–associated flow rule ($\cos a \neq 1$), we note in view of (28b) that a certain amount of hardening ($H_n \ge 0$) is always requires in order that the minimum value is positive $\mu_p > 0$; otherwise "spurious" softening may occur. The aforementioned characteristics of the incremental plasticity problem along with a non–associated flow rule have a significant impact on the structural behavior, which may exhibit "structural" softening even though the material is formally hardening. However, for an associated flow rule we have $\cos a \equiv 1$ and the smallest eigenvalue μ_p is always positive for $H_n \ge 0$. Thus, structural softening requires material softening for an associated flow rule.

Incremental Boundary Value Problem

Preliminaries

For uniaxial stress states it is quite simple to establish an incremental energy that serves as a potential for the incremental boundary value problem. This potential becomes non–convex for softening material behavior, and multiple equilibrium solutions (extremum points) exist in general. By inspection of the deformation patterns for simple truss problems, it was noted in Runesson *et* al. [8] that the solution with the most pronounced localization of plastic deformation represents the global minimum of the incremental potential. It is also possible to put forward arguments, Runesson *et* al. [9], that the global minimum of the incremental potential, which has multiple stationary points in general, corresponds to the stable (=physical) solution.

For a plasticity model that displays non-symmetry of the ATS-tensor no incremental potential exists. It is possible, however, to support the hypothesis that the symmetric part of the Algorithmic Tangent Stiffness operator is positive definite for a stable incremental solution, although this situation cannot be associated with any extremum property. This condition is the discrete (incremental) equivalent to the classical requirement for stability by Hill[10] that the second order rate of work involving the CTS-tensor be positive definite, which will be discussed subsequently.

Characteristics – Stability

By constructing a finite-dimensional space V_h, which within the FE-solution of the incremental boundary value problem is sought, of constant strain triangles

$$V_h \subset C^0(\Omega), \quad C^0(\Omega) = \{ \ \mathbf{v} : v_i \text{ is cont. on } \Omega \text{ and } \nabla v_i \text{ is piecewise continuous on } \Omega \ \} \tag{30}$$

the finite element solution $\mathbf{u}_h \in V_h$ can be represented as a linear combination of basis functions, which are defined locally on each triangle via the conventional finite element expansion

$$\mathbf{u}_h = \mathbf{\Phi}(\mathbf{x})\mathbf{p} \tag{31}$$

where $\mathbf{\Phi}$ is a matrix containing the basis functions and $\mathbf{p}$ is a vector of nodal displacement variables. At this point it is appropriate to make a comment on the finite element formulation used for the spatial discretization: Since $V_h \subset C^0(\Omega)$, discontinuous velocities (or slip surfaces) can not appear, but the strain rate can indeed be discontinuous across the inter-element boundaries. Consequently, the FE-discretization is able to represent a discontinuous deformation state only in a restricted sense.

Remark: It should be born in mind that the actual plasticity model determines the appropriate function space to which the solution belongs. For example, in the case of a perfectly plastic von Mises material, Johnson and Scott[34] introduced an extended function space, instead of the usual Hilbert spaces, in order to prove convergence of a finite element method. The discrete finite element analogue of the extended function space, termed "functions of bounded deformation", consists of piecewise discontinuous polynomials. However, such extensions are beyond the scope of this Paper ■

Consider a finite element discretized elastic-plastic solid subjected to applied loads $\mathbf{P}_n$ and prescribed displacements $\mathbf{r}_n$ corresponding to free displacements $\mathbf{p}_n$ at the time t_n. During the next time increment the load is changed to $\mathbf{P}_{n+1} = \mathbf{P}_n + \Delta\mathbf{P}$ and the prescribed displacements to $\mathbf{r}_{n+1} = \mathbf{r}_n + \Delta\mathbf{r}$, which results in a displacement increment $\Delta\mathbf{p}$ and reaction increment $\Delta\mathbf{R}$.

A virtual work formulation of the equilibrium equations at the updated state gives

$$\mathbf{N}_{n+1}(\Delta\mathbf{p}, \Delta\mathbf{r}) = \mathbf{P}_{n+1} \tag{32}$$

where $\mathbf{N}_{n+1}$ represents the internal nodal forces. With the strain matrix $\mathbf{B}$ defining the part of $\Delta\epsilon$ produced by $\Delta\mathbf{p}$, we obtain the usual expression

$$\mathbf{N}_{n+1} = \int_V \mathbf{B}^T\sigma_{n+1}(\Delta\epsilon)dV \tag{33}$$

It seems reasonable to adopt Hill's criterion for stability[10], i.e. we postulate that the updated state is stable whenever the second order work caused by kinematically admissible variations of $\Delta\mathbf{p}$ is non-negative. Hence, stability requires

$$d(\mathbf{N}_{n+1})^T d(\Delta\mathbf{p}) \geq 0 \tag{34}$$

Since, by definition,

$$d(\mathbf{N}_{n+1}) = \mathbf{K}_A d(\Delta\mathbf{p}) \tag{35}$$

where $\mathbf{K}_A$ is the (generally non-symmetric) algorithmic stiffness matrix defined as

$$\mathbf{K}_A = \frac{\partial(\mathbf{N}_{n+1})}{\partial(\Delta\mathbf{p})} = \int_V \mathbf{B}^T\mathbf{D}_A(\Delta\epsilon)\mathbf{B}dV \tag{36}$$

it follows that the stability criterion (34) can be reformulated as

$$d(\Delta\mathbf{p})^T\mathbf{K}_A d(\Delta\mathbf{p}) \geq 0 \tag{37}$$

This criterion is equivalent to the requirement that the symmetric part of $\mathbf{K}_A$ must be positive definite.

For "simple" plasticity one may define a (global) potential in terms of the incremental energy density $\Delta w(\Delta\epsilon)$ and the finite element approximation as

$$W(\Delta\mathbf{u}_h) = \int_\Omega \Delta w(\epsilon(\Delta\mathbf{p}))dV - \Delta\mathbf{p}^T\mathbf{P}_{n+1} \quad , \quad \epsilon = \mathbf{B}\Delta\mathbf{p} \tag{38}$$

where $\mathbf{B}$ is the strain–displacement matrix and $\Delta\mathbf{p} = \mathbf{p}_{n+1} - \mathbf{p}_n$. The first variation of this incremental potential may be expressed as

$$W'(\Delta\mathbf{p};\Delta\mathbf{p}') = \Delta\mathbf{p}'^T(\mathbf{N}_{n+1}(\Delta\mathbf{p},\Delta\mathbf{r}) - \mathbf{P}_{n+1}) = 0 \quad \forall\ \mathbf{u}_h(\Delta\mathbf{p}') \in V_h \tag{39}$$

which corresponds to the equilibrium equation (33). The character of a specific finite element solution of (39) can now be checked (at lest superficially) by considering the second variation of $W(\Delta\mathbf{u}_h)$, i.e.

$$W''(\Delta\mathbf{p};\Delta\mathbf{p}') = \Delta\mathbf{p}'^T\mathbf{K}_H(\Delta\mathbf{p})\Delta\mathbf{p}' \quad \forall\ \mathbf{u}_h(\Delta\mathbf{p}') \in V_h \tag{40}$$

where it is clear that the sign of $W''(\Delta\mathbf{p};\Delta\mathbf{p}')$, in a certain direction $\Delta\mathbf{p}'$, is reflected by the spectral properties of the Hessian stiffness matrix

$$\mathbf{K}_H(\Delta\mathbf{p}) = \int_\Omega \mathbf{B}^T\mathbf{D}_H(\epsilon(\Delta\mathbf{p})\mathbf{B}dV \tag{41}$$

It is possible to guarantee that the incremental solution $\Delta\mathbf{p}$ of the finite element problem is unique and corresponds to a minimum point of $W(\Delta\mathbf{p})$ only in the case that $W(\Delta\mathbf{p})$ is strictly convex, i.e. $\mathbf{K}_H(\Delta\mathbf{p})$ is positive definite for all $\mathbf{u}_h(\Delta\mathbf{p}) \in V_h$. On the other hand, if $\mathbf{K}_H(\Delta\mathbf{p})$ is semi-positive definite we may have the possibility that no distinct solution exists, i.e. the limit load has been attained. In the case that $\mathbf{K}_H(\Delta\mathbf{p})$ has one or more negative eigenvalues, there exists the additional possibility of multiple distinct solutions of (39). Whether a specific solution corresponds to a minimum, maximum or saddle point can, in practice, be checked from the signs of the diagonal elements of the factorized matrix $\mathbf{K}_H(\Delta\mathbf{p})$.

The discussed concepts were illustrated in Runesson *et* al.[8,9] with the aid of explicit expressions of the incremental energy density function for a uniaxial stress problem and for a multiaxial stress problem with von Mises yield criterion along with an isotropic hardening/softening law. It is shown that localization is strongly linked to stability and global minimization, i.e. the "most localized" solution corresponds to a global minimum of $W(\Delta\mathbf{p})$ and the Hessian stiffness matrix is of course positive definite at this point.

DISCONTINUOUS BIFURCATION OF THE INCREMENTAL SOLUTION

Preliminaries

The capability of capturing localized plastic deformation using the FE–method has been demonstrated by a number of investigators for both finite and small deformations in various fields of application. There is, however, a tendency for such analyses to be highly sensitive to details of mesh design. In order to remedy this situation, it appears to be important to employ a mesh adaptation strategy in order to enhance the possibility for a localized solution. For example, in order to accommodate a shear band by a FE–mesh of constant strain triangles, it is certainly desirable that the element sides are aligned along the shear band, and it is useful to determine the orientation of the shear band via a diagnostic bifurcation analysis of the pertinent material operators. Recent developments of analytical results and numerical search algorithms for quite general plasticity models facilitate a computer implementation of such diagnostic analysis, c.f. Ottosen and Runesson[35], Runesson *et* al.[36], Ortiz *et* al.[24], Molenkamp[37], Sobh[38], Willam and Sobh[39] and Sabban[40].

Concept of characteristic surface

An important concept in the analysis of localized deformation is the characteristic surface C, as discussed by Thomas[2], across which the secondary (bifurcated) solution $\dot{\mathbf{u}}_2$ differs abruptly from the primary (continuous) field $\dot{\mathbf{u}}_1$, as shown in Figure 1. The difference (discontinuity) is denoted $[\dot{\mathbf{u}}] = \dot{\mathbf{u}}_2 - \dot{\mathbf{u}}_1$. If the primary field $\dot{\mathbf{u}}_1$ is continuous everywhere, then C defines locations within the elastic–plastic body where the bifurcated solution $\dot{\mathbf{u}}_2$ is discontinuous. In the case that no characteristic surface exists, we have $[\dot{\mathbf{u}}] = 0$.

The concept of a characteristic surface does also imply that the discontinuity $[\dot{\mathbf{u}}]$ is preserved constant along C. This condition may be expressed as

$$d[\dot{\mathbf{u}}] = \frac{\partial[\dot{\mathbf{u}}]}{\partial \mathbf{x}} \cdot d\mathbf{x} = 0 \quad or \quad [\dot{u}_{i,j}]dx_j = 0 \quad \forall \quad d\mathbf{x} \;//\; C \tag{42}$$

where $d\mathbf{x}$ is an arbitrary differential tangent vector of C. With $\mathbf{n}$ being the (only) vector normal to all $d\mathbf{x}$ of C we also have

$$\mathbf{n} \cdot d\mathbf{x} = 0$$

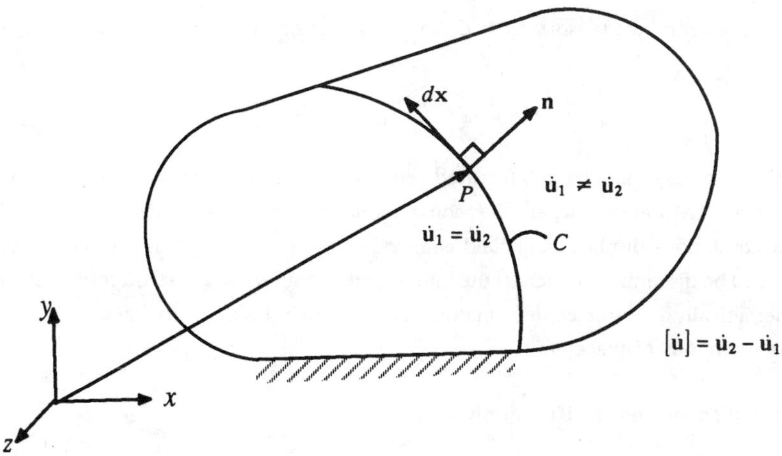

Figure 1 Singular surface C of a deforming solid

It appears that the problem of finding the general solution of (42) becomes quite complicated since the determinant of a 3×3 matrix containing the components of the jump of the rate of deformation tensor $\partial[\dot{\mathbf{u}}]/\partial\mathbf{x}$ must be evaluated. This means that the solution represents the condition that the tensor $\partial[\dot{\mathbf{u}}]/\partial\mathbf{x}$ becomes singular with respect to $d\mathbf{x}$. This problem can, however, be simplified pointwise along C by chosing a local coordinate system of basis vectors such that $\mathbf{e}^*_1 = \mathbf{n}$ and $\mathbf{e}^*_2$, $\mathbf{e}^*_3$ are in the tangent plane of C, as shown in Figure 1.

Since $\mathbf{e}^*_2$ and $\mathbf{e}^*_3$ are now in the tangent plane of C, we have $d\mathbf{x} = dx_k\mathbf{e}^*_k$, $dx_1 = 0$. Equation (42) may then be rephrased as

$$d[\dot{\mathbf{u}}] = [\dot{u}_{i,k}](\mathbf{e}^*_i\mathbf{e}^*_k) \cdot (dx_k\mathbf{e}^*_k) = [\dot{u}_{i,k}]dx_k\mathbf{e}^*_i = 0 \tag{43}$$

In view of (43), it follows that it must be required that $[\dot{u}_{i,2}] = [\dot{u}_{i,3}] = 0$ since dx_2 and dx_3 are arbitrary. Moreover, since we have trivially $dx_1 = 0$, a consequent non–trivial solution of (43) implies $[\dot{u}_{i,1}] \neq 0$. As a result the general solution of (42) takes the form

$$\frac{\partial[\dot{\mathbf{u}}]}{\partial\mathbf{x}} = \boldsymbol{\xi}\mathbf{n} \quad or \quad [\dot{u}_{i,j}] = \xi_i n_j \tag{44}$$

In the specific case that the components $[\dot{u}_{i,j}]$ of (44) are represented in the coordinate system $\mathbf{e}^*_1 = \mathbf{n}$, $\mathbf{e}^*_2$, $\mathbf{e}^*_3$, it follows that the components $\xi_i = [\dot{u}_{i,1}]$. Furthermore, in this coordinate system the rate of deformation field always bifurcates in the normal direction $\mathbf{n}$ of C, i.e. it is only the components $[\dot{u}_{i,1}]$ that are discontinuous.

The jump of the strain rate tensor across C, in the case of small deformations, can now be obtained via (44) as

$$[\dot{\epsilon}] = \frac{1}{2}(\xi\mathbf{n} + \mathbf{n}\xi) \quad or \quad [\dot{\epsilon}_{ij}] = \frac{1}{2}(\xi_i n_j + \xi_j n_i) \tag{45}$$

In conclusion, we may state that a bifurcated velocity field does also impose a discontinuous rate of deformation field. Moreover, when the conventional finite element approximation (30–31) is employed, a continuous displacement field is introduced a priori, whereas the rate of deformation field may indeed be discontinuous across the inter element boundaries. It is, therefore, still possible to describe kinematically a bifurcated deformation state although this deformation state is not associated with any discontinuous displacements.

Condition for discontinuous bifurcation

The fact that discontinuous bifurcations may occur across the characteristic surface, as defined in (44), is used as a constitutive criterion for localization. The properties of the boundary value problem that are of special interest in this context are non–associated flow rules and strain–softening behaviour. We note that a discontinuous solution is possible even in the hardening regime provided that a non–associated flow rule is adopted. Non–associated plastic flow rules are of relevance especially for the modeling of granular materials, like sand, silt and clay, with regard to dilatancy–contractancy properties, as discussed by Lade[41] and Sture *et* al.[42]. Experimental observations for granular materials indicate that non–associated plastic flow is associated with the occurrence of shear bands, as described by Lade[43] and Vardoulakis[44].

Let us now assume that the characteristic surface C exists, whereby it must be required that a "critical" state of stress has been attained. Since static equilibrium across the characteristic surface must be satisfied, the traction rate $\dot{\mathbf{t}}$ across the characteristic surface must be continuous as shown in Figure 2, i.e.

$$\dot{\mathbf{t}}^{(2)} - \dot{\mathbf{t}}^{(1)} = 0 \quad or \quad [\dot{\sigma}_{ij}]n_j = 0 \tag{46}$$

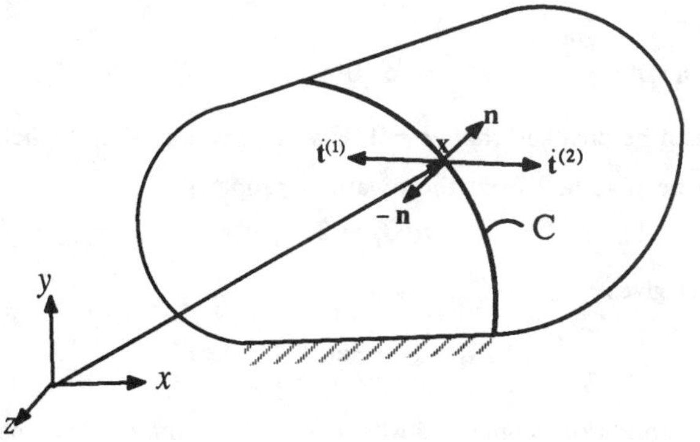

Figure 2 Body in static equilibrium

The relations

$$[\dot{\sigma}]\cdot\mathbf{n} = 0 \;\;,\;\; \dot{\sigma} = \mathbf{D}_T : \dot{\epsilon} \;\;,\;\; [\dot{\epsilon}] = \frac{1}{2}(\mathbf{n}\xi + \xi\mathbf{n})$$

may now be combined to give

$$\mathbf{Q}_T \cdot \xi = 0 \;\;,\;\; \mathbf{Q}_T = \mathbf{n}\cdot\mathbf{D}_T\cdot\mathbf{n} \tag{47}$$

where $\mathbf{Q}_T$ is the Characteristic Tangent Stiffness (ChTS) tensor. In (47) we assume that the material responds plastically on both sides of the characteristic surface. That this assumption is in order has been shown by Ottosen and Runesson[35].

In view of (47), we have a continuous strain state when $\xi \equiv 0$, in which case $\mathbf{Q}_T$ may be non-singular. On the other hand, a discontinuous strain state, i.e. $\xi \neq 0$, is possible whenever

$$\det(\mathbf{Q}_T) = 0 \tag{48}$$

At this point we emphasize that it is traditional to consider bifurcations in terms of the continuous (or rate) response, whereas the computational algorithm deals with truly finite increments. Since finite increments should be considered, this means that $\mathbf{D}_T$ in (47) is replaced by the ASS-tensor $\mathbf{D}_S$ to yield

$$\det(\mathbf{Q}_S) = 0 \;\;,\;\; \mathbf{Q}_S = \mathbf{n}\cdot\mathbf{D}_S\cdot\mathbf{n} \tag{49}$$

where $\mathbf{Q}_S$ is the Characteristic Secant Stiffness (ChSS) tensor, which defines the possibility for a discontinuous bifurcation of the incremental response.

In order to express the singularity condition (49) explicitly it is appropriate to study the following eigenvalue problem

$$\mathbf{Q}_S\cdot\mathbf{y} = \mu\mathbf{Q}^e\cdot\mathbf{y} \;\;,\;\; \mathbf{Q}^e = \mathbf{n}\cdot\mathbf{D}^e\cdot\mathbf{n} \tag{50}$$

whereby it is possible to rewrite (50) as

$$\mathbf{G}\cdot\mathbf{y} = \mu\mathbf{y} \;\;,\;\; \mathbf{G} = \delta - \frac{1}{K_S}\mathbf{P}^e\cdot\mathbf{b}(\mathbf{n})\mathbf{a}(\mathbf{n}) \tag{51}$$

where

$$\mathbf{b}(\mathbf{n}) = \mathbf{n}\cdot\mathbf{D}^e : \mathbf{g}_{n+1}\ , \quad \mathbf{a}(\mathbf{n}) = \mathbf{n}\cdot\mathbf{D}^e : \mathbf{f}_{n+1}\ , \quad K_S = H_S + \mathbf{f}_{n+1} : \mathbf{D}^e : \mathbf{g}_{n+1} \tag{52}$$

and $\mathbf{P}^e = (\mathbf{Q}^e)^{-1}$. It can be checked that $\mu = 1$ is an eigenvalue of multiplicity of two, whereby the third eigenvalue can be obtained from the invariant property

$$tr(\mathbf{G}) = 2 + \mu_3 \tag{53}$$

which in view of (51) gives

$$\mu_3 = 1 - \frac{1}{K_S}\mathbf{a}(\mathbf{n})\cdot\mathbf{P}^e\cdot\mathbf{b}(\mathbf{n}) \tag{54}$$

Since the singularity condition is obtained whenever $\mu_3 = 0$ the critical secant hardening modulus can be expressed

$$H_b = -\mathbf{f}_{n+1} : \mathbf{D}^e : \mathbf{g}_{n+1} + \mathbf{a}(\mathbf{n})\cdot\mathbf{P}^e\cdot\mathbf{b}(\mathbf{n}) \tag{55}$$

which is now a function of the orientation $\mathbf{n}$ for a given state of stress and strain. We conclude that discontinuous bifurcation becomes possible for the first time when

$$H^{max}{}_b = -\mathbf{f}_{n+1} : \mathbf{D}^e : \mathbf{g}_{n+1} + \max_{|\mathbf{n}| \equiv 1} \mathbf{a}(\mathbf{n})\cdot\mathbf{P}^e\cdot\mathbf{b}(\mathbf{n}) \tag{56}$$

A similar analysis may also be carried out for the eigenvalue μ_3 , which in fact takes its minimum corresponding to the same critical directions as (56) since

$$\mu_3{}^{min} = 1 - \frac{1}{K_S}\max_{|\mathbf{n}| \equiv 1} \mathbf{a}(\mathbf{n})\cdot\mathbf{P}^e\cdot\mathbf{b}(\mathbf{n}) \tag{57}$$

Since the extremal properties of (56) and (57) are identical we may, thus, evaluate the value of the minimum of μ_3 by combining (56) and (57) to obtain

$$\mu_3{}^{min} = \frac{1}{K_S}(H_S - H^{max}{}_b) \tag{58}$$

From (58) we conclude that the singularity condition is obtained for the first time when the secant hardening modulus becomes equal to the critical hardening modulus. In the case that $\mu_3{}^{min}$ is further reduced below zero discontinuous bifurcation can occur also in other directions than the critical ones that correspond to the maximum problem (56).

Analytical evaluation of critical bifurcation directions for plane strain states

An efficient computer implementation of the solution to the maximum problem (56) is conveniently based on analytical results, as described by Ottosen and Runesson[35], Runesson *et* al.[36]. We note that this information can be (and it has indeed been) obtained via numerical search algorithms, e.g. Ortiz *et* al.[24], Molenkamp[37], Sobh[38], Willam and Sobh[39] and Sabban[40]. However, since the bifurcation analysis has to be carried out in a large number of material points (typically the Gaussian integration points), it seems attractive to utilize analytical bifurcation results in order to save computer effort.

The critical hardening modulus $H^{max}{}_b$ for a frictional material was first determined by Mandel[45], who employed Mohr–Coulomb's yield criterion with a non-associated flow rule to describe dilatancy.

In the pioneering papers by Rudnicki and Rice[46] and Rice[47] explicit analytical expressions of the critical hardening modulus and the critical bifurcation directions were derived for Drucker–Prager's yield criterion in the unrestricted three–dimensional stress state. Only the volumetric part of the plastic flow was assumed to be non–associated. Recently, Ottosen and Runesson[35] retained the assumption of non–associativity only in the volumetric part of the plastic flow and generalized the results of Rudnicki and Rice[46] to arbitrary three invariant yield criteria. More recently Runesson *et* al.[36] considered the cases of plane stress and plane strain for quite general yield criteria and plastic potentials and derived *analytical* solutions of the critical hardening modulus and the corresponding critical bifurcation directions. The only restriction is that the yield function and the plastic potential function must have the same principal directions and that two of these are located in the plane of interest.

Of particular relevance for the numerical examples presented in this Paper are the bifurcation results obtained by Runesson *et* al.[36] for plane strain bifurcations, although a slightly different approach for obtaining the pertinent bifurcation results is presented subsequently, where non–associated plastic flow is allowed only in the volumetric part of the flow direction. In the case of plane strain states the strain increment must satisfy the condition

$$\Delta\epsilon_{i3} = 0 \ , \quad i = 1,2,3 \quad \textit{plane strain} \tag{59}$$

which implies that the elastic stress strain relation can be expressed by the in plane components as

$$\Delta\sigma_{\alpha\beta} = D^e{}_{\alpha\beta\gamma\kappa}\Delta\epsilon^e{}_{\gamma\kappa} \ , \qquad \alpha,\beta = 1,2 \tag{60}$$

In (60) Greek indices refer to the in–plane components, whereas Latin subscript letters refer to unrestricted stress or strain states. Concerning the structure of the ASS–tenor $\mathbf{D}_S$ for plane states we note that

$$D_{S,\alpha\beta\gamma\kappa} = D^e{}_{\alpha\beta\gamma\kappa} - \frac{1}{K_S} D^e{}_{\alpha\beta ij} g_{ij} D^e{}_{\gamma\kappa kl} f_{kl} \ , \quad K_S = H_S + f_{ij} D^e{}_{ijkl} g_{kl} \tag{61}$$

where it is tacitly assumed that the components f_{ij} and g_{ij} refers to the updated state at time $t = t_{n+1}$. Moreover, for plane strain bifurcations we note the following:

$$n_i \rightarrow n_\alpha \ , \qquad y_l \rightarrow y_\alpha$$

$$Q^e{}_{il} \rightarrow Q^e{}_{\alpha\beta} \ , \qquad P^e{}_{il} \rightarrow P^e{}_{\alpha\beta} \ , \qquad Q_{il} \rightarrow Q_{\alpha\beta}$$

$$b_i \rightarrow b_\alpha = n_\beta D^e{}_{\alpha\beta pq} g_{pq} \ , \qquad a_l \rightarrow a_\alpha = n_\beta D^e{}_{\beta\alpha mn} f_{mn}$$

where it is noted in particular that (59) implies that no out–of–plane bifurcations can occur. It was concluded that the condition $\det(Q_{il}) = 0$ is obtained whenever $\mu_3 = 0$, whereby the hardening modulus H_b for plane strain states may be expressed as

$$H_b = -f_{ij} D^e{}_{ijkl} g_{kl} + n_\beta D^e{}_{mn\beta\alpha} f_{mn} P^e{}_{\alpha\beta} g_{pq} D^e{}_{\beta\kappa pq} n_\kappa \tag{62}$$

which is clearly a function of the normal vector $\mathbf{n}$ of C .

In order to obtain explicit results we restrict the analysis to isotropic elasticity as defined by the tensor

$$D^e_{ijkl} = 2G\left(\delta_{ijkl} - \frac{\delta_{ij}\delta_{kl}}{3}\right) + K\delta_{ij}\delta_{kl} \tag{63}$$

An explicit expression for the elastic characteristic stiffness tensor Q^e_{il} and its inverse P^e_{il} can now obtained as

$$P^e_{\alpha\beta} = \frac{1}{G}\left(-\left(\frac{G+3K}{4G+3K}\right)n_\alpha n_\beta + \delta_{\alpha\beta}\right) \tag{64}$$

Furthermore, the gradient of the yield function and the flow direction are split into their deviatoric and volumetric parts as

$$f_{ij} = f'_{ij} + \frac{f_{kk}}{3}\delta_{ij} \quad , \quad g_{ij} = g'_{ij} + \frac{g_{kk}}{3}\delta_{ij} \tag{65}$$

By using the transformation **A** of (6), which defines non–associativity only in the volumetric part of the flow rule, we obtain

$$g'_{ij} = f'_{ij} \quad , \quad g_{ii} = f_{ii}\frac{1}{1+\beta} \tag{66}$$

whereby it is possible (after some manipulations) to rewrite (62) as

$$\frac{H_b}{2G} = n_\alpha\left(2f'_{\alpha\gamma}f'_{\alpha\gamma} + \frac{2+\beta}{1+\beta}\frac{3K}{4G+3K}f_{kk}f'_{\alpha\beta}\right)n_\beta - 2\frac{G+3K}{4G+3K}\left(n_\alpha f'_{\alpha\beta}n_\beta\right)^2 - \frac{1}{1+\beta}\frac{2K}{4G+3K}f_{kk}^2 - f'_{ij}f'_{ij} \tag{67}$$

Moreover, in order to solve the maximum problem stated in (56) it is convenient to represent **f'** in terms if its principal components f'_1, f'_2 and f'_3. The in–plane components $f'_{1,2}$ are ordered so that $f'_1 \geq f'_2$, while the magnitude of the out–of–plane component f'_3 is not related to those which are in–plane.

We may now rewrite (67) as the simpler expression

$$\frac{H_b}{2G} = \left(2f'^2_1 + rf'_1\right)\bar{n}_1^2 + \left(2f'^2_2 + rf'_2\right)\bar{n}_2^2 - 2\psi\left(f'_1\bar{n}_1^2 + f'_2\bar{n}_2^2\right)^2 - k \tag{68}$$

where the "bifurcation parameters" r, ψ, k are shown in Table 1, and where $\bar{n}_\alpha$ are the components of **n** in the principal coordinate system.

Table 1 "Bifurcation parameters" for plane strain

r	ψ	k
$\frac{2+\beta}{1+\beta}\frac{3K}{4G+3K}f_{kk}$	$\frac{G+3K}{4G+3K}$	$\left(f'^2_1 + f'^2_2 + f'^2_3\right) + \frac{1}{1+\beta}\frac{2K}{4G+3K}f_{kk}^2$

Due to the restriction a priori to two dimensions the problem (56) can be reduced to finding the maximum of a scalar-valued function by introducing the constraint $\bar{n}^2_2 = 1 - \bar{n}^2_1$ directly into (68). A maximum value of H_b is clearly obtained when

$$\frac{1}{2G}\frac{\partial H_b}{\partial(\bar{n}_1)^2} = (f'_1 - f'_2)\left[2(f'_1 + f'_2) - 4\psi(f'_2 + \bar{n}^2_1(f'_1 - f'_2)) + r\right] = 0 \tag{69}$$

since

$$\frac{1}{2G}\frac{\partial^2 H_b}{\partial^2(\bar{n}_1)^2} = -\,4\psi(f'_1 - f'_2)^2 < 0 \tag{70}$$

Whenever the solution of (69) satisfies $0 \le \bar{n}^2_1 \le 1$, the maximum is unrestricted and corresponds to the critical bifurcation defined by

$$\bar{n}^2_1 = \frac{1}{2\psi}\frac{f'_1 + (1-2\psi)f'_2 + r/2}{f'_1 - f'_2}$$
$$\bar{n}^2_2 = -\frac{1}{2\psi}\frac{f'_2 + (1-2\psi)f'_1 + r/2}{f'_1 - f'_2} \tag{71}$$

By inserting (71) into (68) one may evaluate the critical hardening modulus $H^{max}{}_b$ as

$$\frac{H^{max}{}_b}{2G} = \frac{1}{2\psi}\left(f'_1 + f'_2 + \frac{1}{2}r\right)^2 - 2f'_1 f'_2 - k \tag{72}$$

In the case that $f'_1 + (1-2\psi)f'_2 + r/2 < 0$, a restricted maximum is obtained, that is defined by

$$\bar{n}^2_1 = 0 \;,\quad \bar{n}^2_2 = 1 \;,\quad \frac{H^{max}{}_b}{2G} = 2(1-\psi)f'^2_2 + rf'_2 - k \tag{73}$$

and in the case that $f'_2 + (1-2\psi)f'_1 + r/2 > 0$, we obtain

$$\bar{n}^2_1 = 1 \;,\quad \bar{n}^2_2 = 0 \;,\quad \frac{H^{max}{}_b}{2G} = 2(1-\psi)f'^2_1 + rf'_1 - k \tag{74}$$

DIAGNOSTIC BIFURCATION ANALYSIS - MESH ADAPTATION STRATEGY

General

The importance of using a "proper" mesh adaptation strategy is motivated by the fact that the FE–approximation must be designed in such a way that it represents kinematically the pertinent localized deformation mode. This problem has been pointed out by several investigators, e.g. Hsu *et* al.[48], Ortiz *et* al.[24], and Leroy and Ortiz[4]. By a "properly adapted" FE–mesh we mean here a FE-mesh that is able to represent a strain discontinuity across the characteristic curve.

Several alternative methods can be conceived in order to define such a "properly adapted" FE-mesh; however, we note that merely densifying of the FE–mesh along the characteristic curve would not be optimal, since the crucial problem of aligning the element sides is not considered in this approach. Ortiz *et* al.[24] employed a "local alignment strategy" within the elements of the original mesh. The element approximation was then enhanced by linear shape functions that were defined locally on each element. The orientation of these shape functions was defined by critical bifurcation directions, which were evaluated by using a numerical search algorithm.

In this Paper, however, we adopt the approach of successively realigning (remeshing) the mesh along the anticipated characteristic curve whenever discontinuous bifurcation is expected. The analytical bifurcation results presented in the previous chapter are thereby utilized.

Mesh adaptation - realignment

At first thought, it might be considered natural to use the condition $\mu_3^{min} \leq 0$ of (58) as a means for diagnosing the possibility for localization in each material point (each Gaussian integration point). That this is not the case will be motivated as follows: The idea of using a diagnostic bifurcation analysis is, clearly, that it can be used for *predicting* the possibility for a localized solution in the subsequent load increment *before* this incremental solution has been calculated. However, as pointed out previously, μ_3^{min} is strongly dependent, not only on the "physical" material hardening/softening, but also on the integration method and the incremental solution. This means that μ_3^{min} cannot be calculated before the desired (localized) solution has been obtained. Consequently, it is not feasible to use μ_3^{min} as a diagnostic measure.

The discussed characteristics are illustrated in Figures 3–4 for a rectangular sheet in plane strain obeying von Mises yield criterion with an isotropic softening rule. Two different types of mesh design are used; the mesh in Figure 3 is unbiased with respect to localization, whereas the mesh in Figure 4 is biased. We note that the condition $\mu_3^{min} \leq 0$ is a reflection of the characteristics of the deformation state since the diffuse solution in Figure 3 corresponds to $\mu_3^{min} \geq 0$, whereas the localized one in Figure 4 corresponds to $\mu_3^{min} \leq 0$. These findings are in agreement with the discussion above.

Remark: Although μ_3^{min} is not suitable for predicting bifurcations, it can still be used for assessing the characteristics of a specific incremental solution when this solution is available. For example, say that two solutions have been found: one corresponding to diffuse and the other corresponding to localized deformation modes. Our experience is then quite consistently that only the localized solution will be represented by a portion of the plastic zone for which the condition $\mu_3^{min} \leq 0$ is satisfied ■

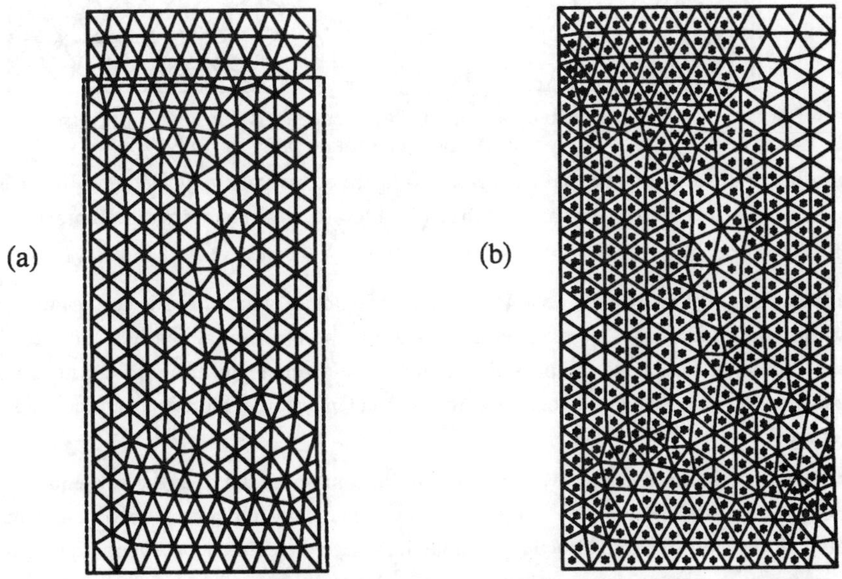

Figure 3 Rectangular sheet with unbiased mesh, (a) diffuse deformation pattern, (b) corresponding plastic zone (* means $\mu_3^{min} > 0$)

(a) (b)

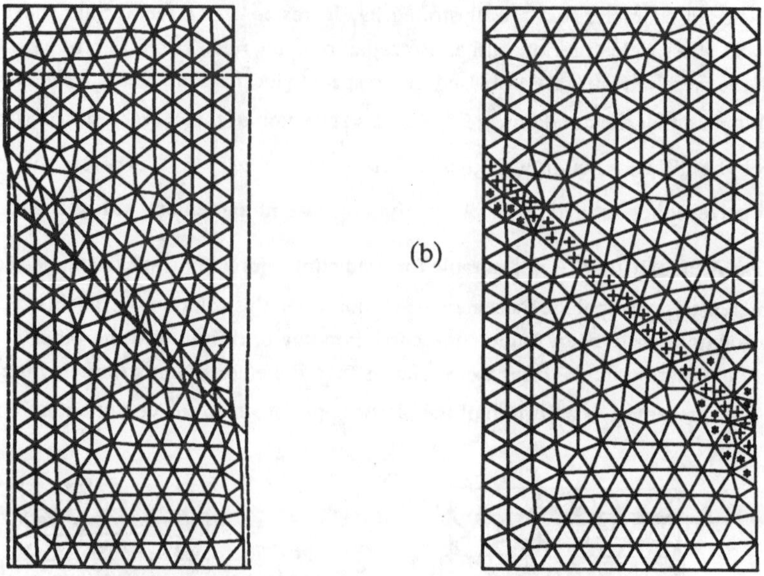

Figure 4 Rectangular sheet with biased mesh, (a) localized deformation pattern, (b) corresponding plastic zone (+ means $\mu_3^{min} < 0$)

It is possible to use a technique that aims at assessing the stress state in *all* elements that are located in the plastic zone. It is thereby emphasized that only a few of these elements will, in practice, eventually turn into a critical state (in the sense that $\mu_3^{min} \leq 0$) when a shear band develops.

The adopted strategy is to define an anticipated (approximate) characteristic curve among material points within the plastic zone. No explicit information is thereby used about how close to "critical" the stress level is, or how large the possibility is for these elements to turn into a critical state after mesh realignment. The objective is thereafter to realign element sides along the "anticipated" characteristic curve.

The anticipated characteristic curve serves as an "internal boundary" at the subsequent mesh regeneration, as discussed in Larsson[25]. This is shown schematically in Figure 5, for an element with the anticipated critical bifurcation direction **n** that has been calculated based on the actual stress state."Layers" of triangular elements, that are aligned with this internal boundary, seem to form an efficient mesh in terms of its localization capturing capability. This concept is utilized in the mesh generation package ADMESH, which was developed by Jin and Wiberg[49]. The mesh generation is performed from the external and the internal boundaries with a moving front technique. ADMESH also provides the option of specifying the rate of change of the width of element layers away from the internal boundary. A typical mesh realignment (remeshing), obtained from ADMESH, is depicted in Figure 6.

The described realignment procedure is performed in an iterative fashion within each load increment. Consequently, the internal boundary is redefined during the successive mesh realignments. Convergence is obtained when the anticipated characteristic curve changes insignificantly and the plastic zone is confined to one single element layer.

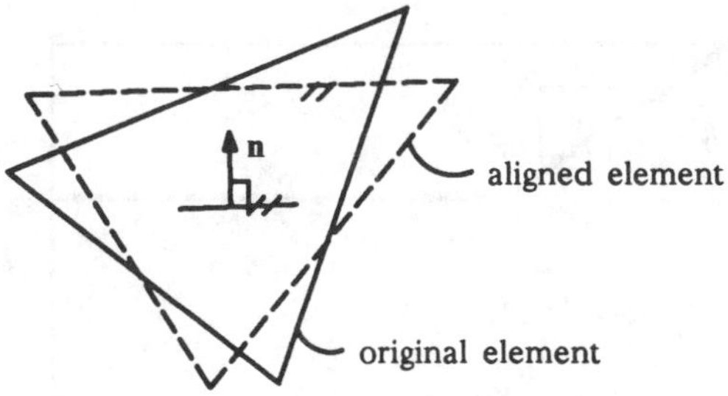

Figure 5 Alignment of an element

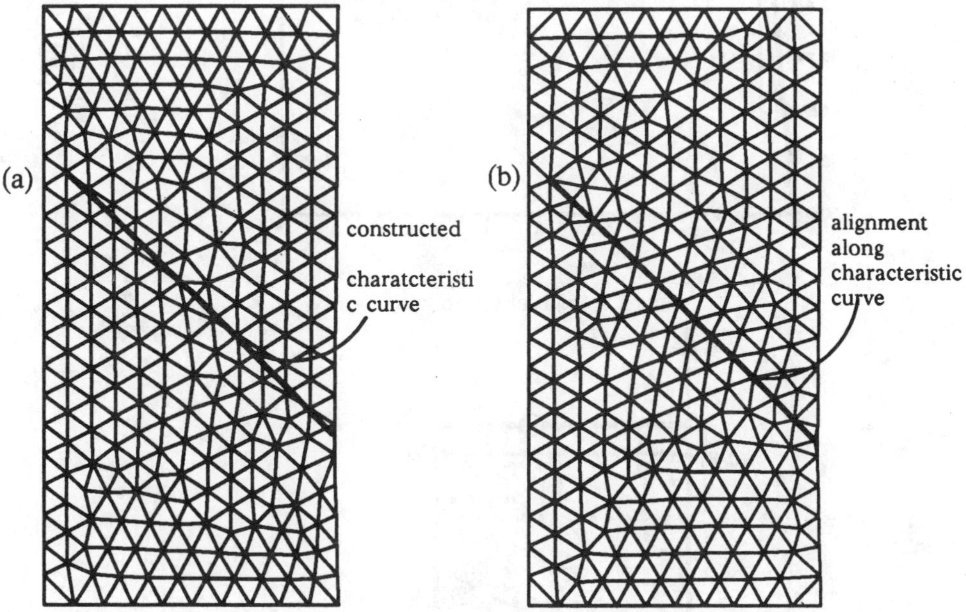

Figure 6 Remeshing of FE-mesh due to anticipated localization
(a) initial mesh, (b) realigned mesh

Numerical example: Extension of rectangular sheet

Development of plane strain bifurcations will be considered for von Mises yield criterion with an isotropic softening rule, as defined by the uniaxial test curve in Figure 7. Poisson's ratio is $\nu = 0.3$.

A sheet in a state of plane strain is subjected to a prescribed uniform extension r, as shown in Figure 8, which leads to a uniform stress state within the sheet in the absence of imperfections. However, in order to trigger the initiation of a shear band a very small geometric imperfection is introduced at point A, as indicated Figure 8. The loading is applied in such a way that the sheet is in the elastic state, but just before the onset of yielding, in the first load step. When the second load step is applied, almost the entire sheet may enters the plastic range if the deformation is uniform due to the fact that the stress state is only slightly inhomogeneous.

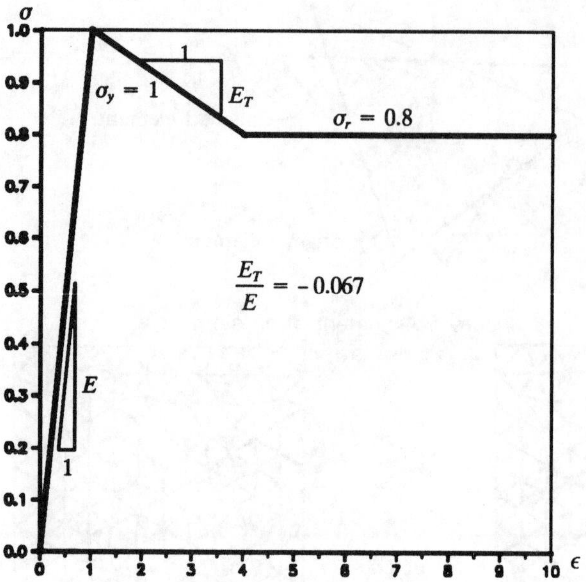

Figure 7 Uniaxial test curve

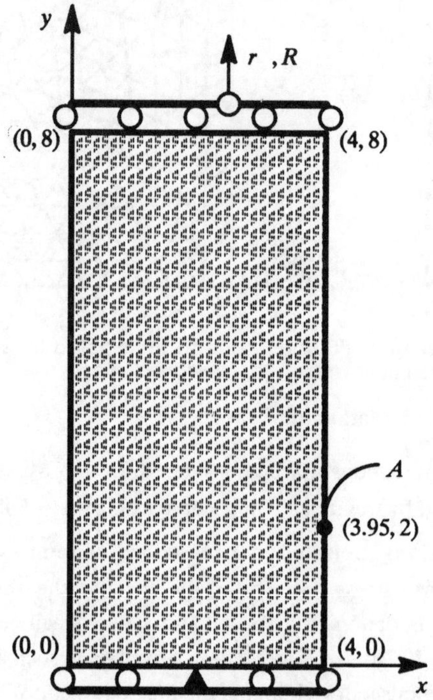

Figure 8 Analyzed sheet

The significance of the mesh adaptation strategy is demonstrated by choosing the initial mesh unbiased, as shown in Figure 9a. Since the first load step is elastic a symmetric deformation mode is obtained, as shown in Figure 9b. When the second (plastic) load was applied, convergence was obtained towards a non–localized and symmetric mode of deformation, as shown in Figure 10a. The solution involves 340 plastic elements and has the critical bifurcation directions that are shown in Figure 10b. When the distribution of the eigenvalue μ_3^{min} is considered in Figure 10c, we note that only one element that has reached a critical state defined by $\mu_3^{min} < 0$.

Since the initial mesh is not properly adapted to capture the pertinent localized deformation mode, the mesh is realigned by constructing a characteristic internal boundary on the basis of the critical bifurcation directions in Figure 10c; the resulting realigned mesh is shown in Figure 11a along with the symmetric elastic deformation mode corresponding to the first load step (Figure 11b).

The plastic solution due to the first realignment for the second load step is shown, in Figure 12, in terms of the deformation pattern, the critical bifurcation directions of the 57 plastic elements and the bifurcated or non–bifurcated elements. It is interesting to note that the diffuse mode of deformation obtained by the unbiased mesh has now turned into a localized and unsymmetrical deformation mode (as is displayed by Figure 12a). This demonstrates the significance of proper mesh design for capturing localized deformation.

Since the plastic zone is still somewhat diffuse a second realignment resulting in the mesh in Figure 13a was performed. In this case an almost perfect alignment with the bifurcation directions is obtained. The localized deformation mode is now even more localized as compared to the solution corresponding to the first realignment, as shown in Figure 14a.

It is appropriate to make a further comment on the advantage of using a realignment procedure as opposed to densifying the elements that are penetrated by the characteristic line in the previous mesh. Firstly, the triangles become more capable of representing a strain discontinuity in the critical bifurcation directions of the elements when the elements are realigned. Moreover, all converged solutions are stable in the sense that the Hessian stiffness matrix is positive definite.

(a)

(b)

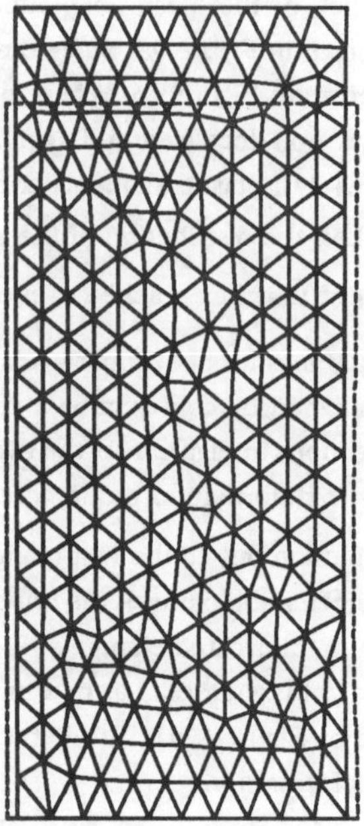

Figure 9 Plane strain: (a) unbiased mesh and (b) elastic solution

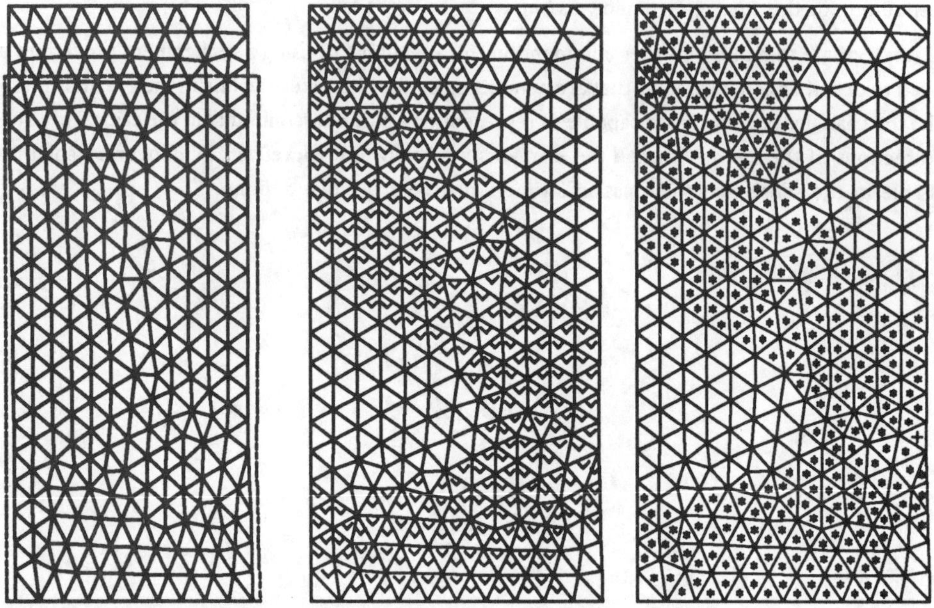

Figure 10 Plane strain: plastic solution characteristics for unbiased mesh, (a) displacement map, (b) bifurcation directions, (c) bifurcated elements (+ $\mu_3^{\min} \leq 0$, * $\mu_3^{\min} > 0$)

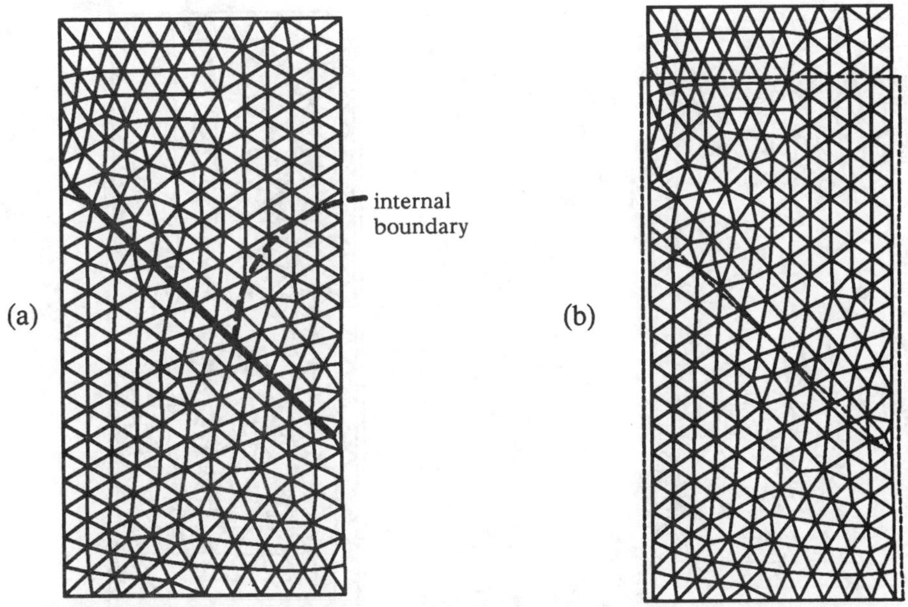

Figure 11 Plane strain: (a) first aligned mesh and (b) elastic solution

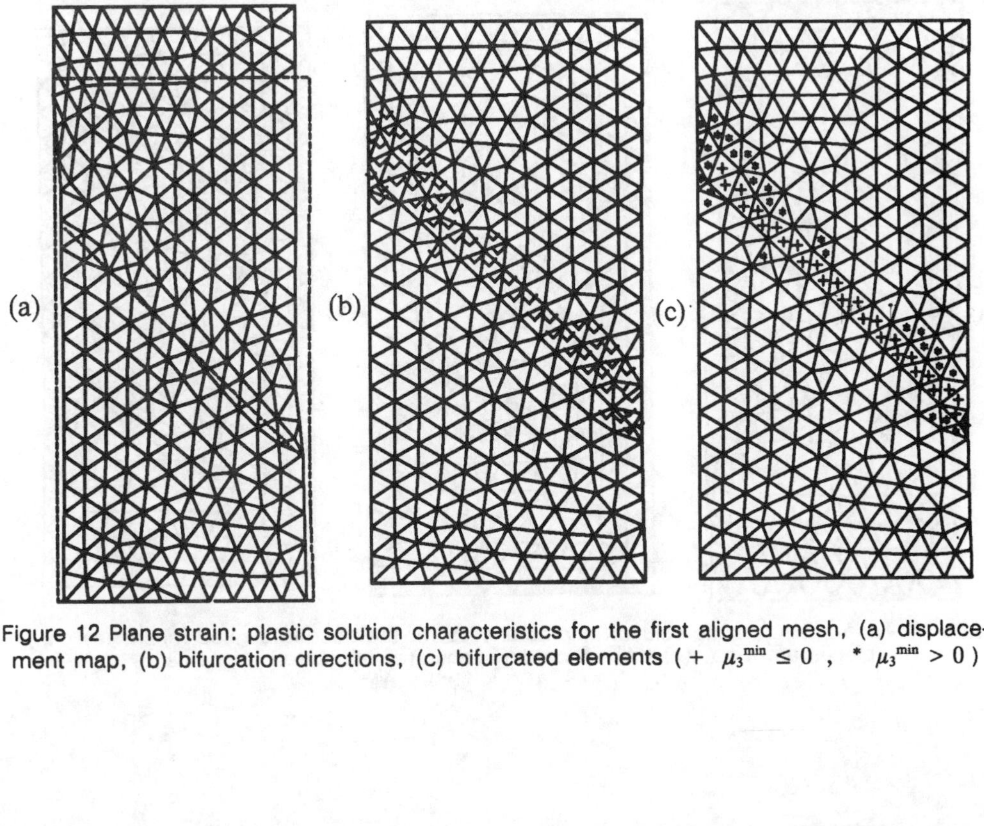

Figure 12 Plane strain: plastic solution characteristics for the first aligned mesh, (a) displacement map, (b) bifurcation directions, (c) bifurcated elements (+ $\mu_3^{min} \leq 0$, * $\mu_3^{min} > 0$)

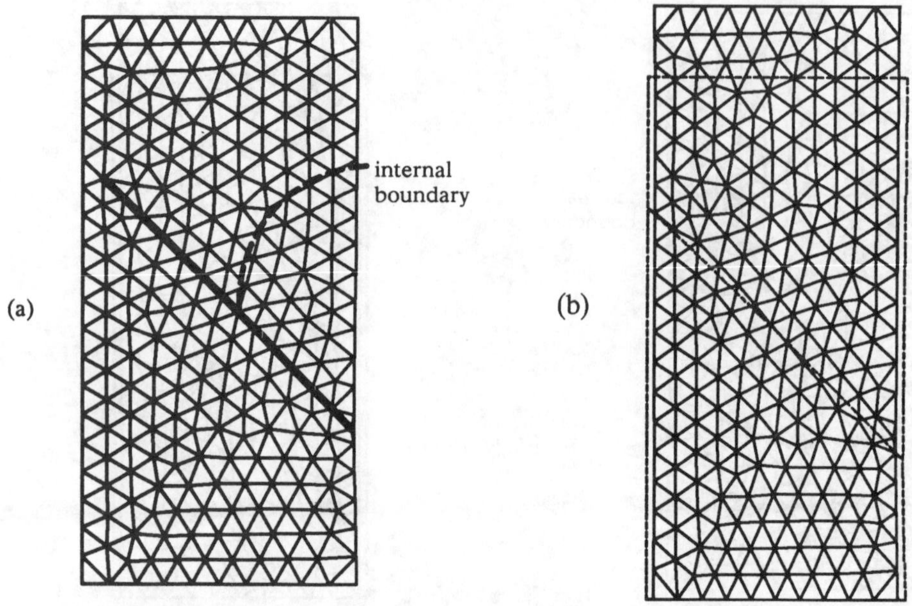

Figure 13 Plane strain: (a) second aligned mesh and (b) elastic solution

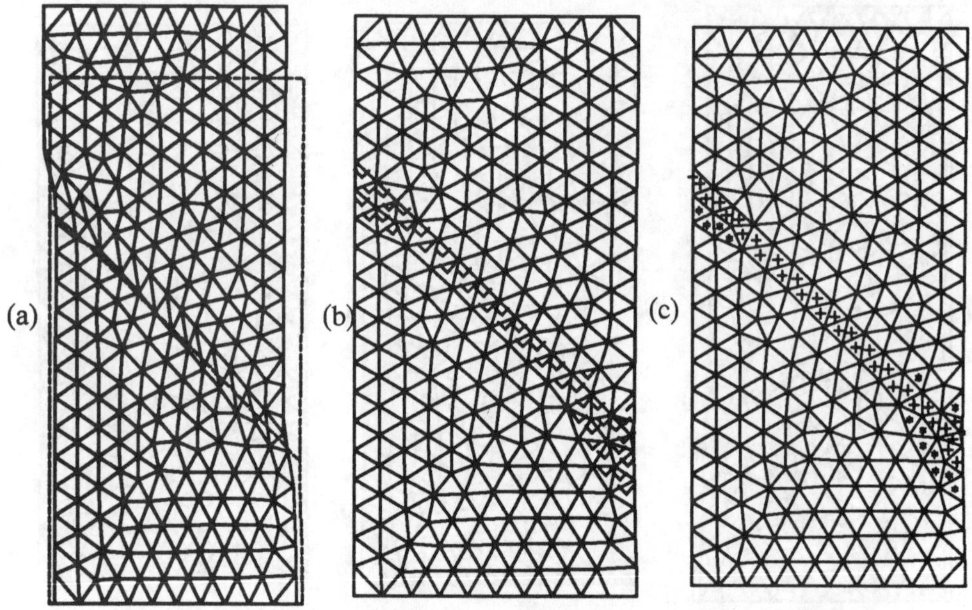

Figure 14 Plane strain: plastic solution characteristics for the second aligned mesh, (a) displacement map, (b) bifurcation directions, (c) bifurcated elements
(+ $\mu_3^{\min} \leq 0$, * $\mu_3^{\min} > 0$)

Numerical example: Stability of a Slope with a Footing

A slope with a footing resting on its crest is analyzed with respect to different types of mesh design; one biased and one unbiased mesh will be considered. As for the analysis of the sheet, it is assumed that the slope is obeys von Mises yield criterion with an isotropic softening rule, which may be thought of as representing undrained behavior of a heavily overconsolidated soil. Moreover, the footing, which is quite stiff as compared to the soil, is subjected to a prescribed displacement r at its center, whereby the footing is allowed to rotate. For simplicity, the contact between the footing and the soil is assumed to be completely adhesive. Plane strain is assumed and Poisson's ratio is $\nu = 0.3$. The prescribed displacement is applied in six load steps.

By considering the deformations modes at the final time step, in Figures 15a–16a, the same solution characteristics as as for the sheet are obtained in terms of the mesh design. In fact, the biased mesh displays a strongly localized plastic zone at this time step, whereas the plastic zone for the unbiased mesh is rather diffuse. Like in the analysis of the sheet, we note that all converged solutions are stable in the sense that the Hessian stiffness operator is positive definite. It is also noteworthy that the bifurcation directions indicated in Figures 15b–16b show a quite good alignment with the element sides for the biased mesh.

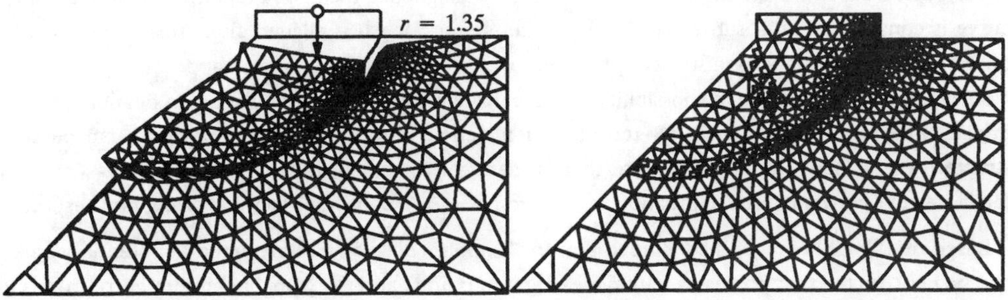

Figure 15 Plastic solution for biased mesh at final time step: (a) localized deformation mode, (b) Bifurcation directions in plastic zone

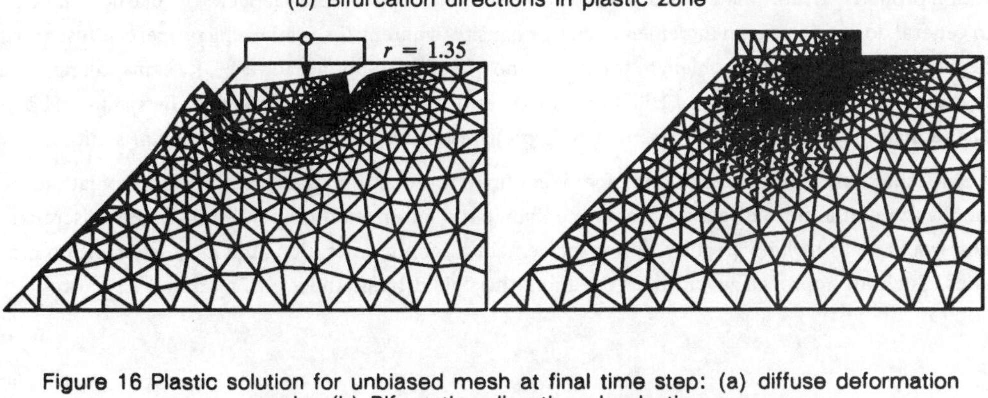

Figure 16 Plastic solution for unbiased mesh at final time step: (a) diffuse deformation mode, (b) Bifurcation directions in plastic zone

CONCLUDING REMARKS

In this paper we have demonstrated the importance of using finite element mesh realignment procedures for capturing strongly localized elastic–plastic solutions due to softening behavior and/or nonassociated plastic flow. However, the success of the numerical procedure is also dependent on the incremental/iterative algorithm for obtaining a stable equilibrium path.

The suggested mesh adaptation strategy is based on mesh realignments in those time steps where discontinuous bifurcation is anticipated. For the purpose of realigning the mesh, a bifurcation analysis is performed based on the incremental solution as a diagnostic means. A tentative "characteristic" curve is constructed and, subsequently, the finite element mesh is aligned along this curve. As opposed to the mesh adaptation strategy proposed by Ortiz et al.[24], the present approach represents a global means for enhancing the possibility for a localized solution. We note that a significant feature of implicit integration is that the ASS–tensor, which defines the incremental behaviour, can be constructed with a mathematical structure that is identical to that of the CTS–tensor. This similarity facilitates the conversion of bifurcation results from the rate behaviour to the incremental response.

Based on the CPPM, it was concluded that for "simple" plasticity an incremental potential energy can be constructed that possesses multiple extremum points, and the global minimum corresponds to the (most) stable and localized deformation pattern. The advantage of the existence of this potential is that the incremental solution can be characterized in terms of stability corresponding to the minimization property. Unfortunately, because of non–symmetry of the material operator, plasticity models in general do not possess an incremental energy density, whereby the minimizing property is lost and it becomes quite a delicate problem to trace the "most" stable solution. However, it seems still possible to define stability in the sense of Hill's stability criterion[10], whereby it follows that the symmetric part of the algorithmic tangent stiffness matrix is positive definite at a stable incremental solution.

Since the analysis presented in this Paper is confined to a continuous displacement interpolation, we emphasize that a more direct approach to enhance the localization properties of the FE–discretization would be to include (regularized) displacement discontinuities along the inter–element boundaries. Such an approach would then resemble the "yield hinge theory" for framed structures.

REFERENCES

1. A. Nadai, *Plasticity*, McGraw-Hill Book Company (1931)
2. T.Y. Thomas, *Plastic Flow and Fracture in Solids*, Academic Press (1961)
3. Z.P. Bazant, "Mechanics of distributed cracking", *Appl. Mech. Rev.*, **39**, 675–705 (1986)
4. Y. Leroy and M. Ortiz, "Finite element analysis of strain localization in frictional materials", *Int. J. Num. Anal. Meth. Geomech.*, **13**, 53–74 (1989)
5. Y. Leroy and M. Ortiz, "Finite element analysis of transient strain localization phenomena in frictional materials", *Int. J. Num. Anal. Meth. Geomech.*, **14**, 93–124 (1990)
6. M. Ortiz and J.J. Quigley, "Element design and adaptive remeshing in strain localization problems", in: D.R.J Owen, E. Hinton and E. Onate (eds.), *Computational Plasticity*, COMPLAS II pp. 213–236 Swansea: Pineridge Press (1989)
7. Z.P. Bazant, "Stable states and paths of structures with plasticity of damage", *J. Engng. Mech.*, ASCE, **114**, 2013–2034 (1988)
8. K. Runesson, R. Larsson and S. Sture, "Characteristics and computational procedure in softening plasticity", *J. Engng. Mech.*, ASCE, **115** 1628–1646, (1989)
9. K. Runesson, R. Larsson and N.S. Ottosen, "Extremal Properties of Incremental Solutions in Plasticity at Implicit Integration - Localization", submitted to ASCE, J. Engng Mech. (1991)
10. R. Hill, "A general theory of uniqueness and stability in elastic-plastic solids." *J. Mech. Phys. Solids*, **6**, 236–249 (1958)
11. V. Tvergaard, A. Needleman and K. K. Lo, "Flow localization in the plane strain tensile test", *J. Mech. Phys. Solids*, **29**, 115–142 (1981)
12. N. Triantafyllidis, A. Needleman and V. Tvergaard, "On the development of shear bands in pure bending", *Int. J. Solids Struct.*, **18**, 121–138 (1982)
13. A. Needleman and V. Tvergaard, "Finite element analysis of localization in plasticity", in: J.T. Oden and C.F. Carey (eds.), *Finite elements: Special problems in solid mechanics*, **5**, pp. 94–157 Prentice-Hall (1984)
14. J.H. Prevost, "Localization of deformations in elastic-plastic solids", *Int. J. Num. Anal. Meth. Geomech.*, **8**, 187–196 (1984)
15. R. de Borst, "Numerical methods for bifurcation analysis in geomechanics", *Ingenieur-Archiv*, **59**, 160–174 (1989)
16. R. de Borst, "Bifurcations in finite element methods with a non-associated flow law", *Int. J. Num. Anal. Meth. Geomech.*, **12**, 99–116 (1988)
17. J.P. Bardet, "A note on the finite element simulation of strain localization", in: O.Z. Zienkiewicz, G.N. Pande and J. Middleton (eds.), *Numerical Methods in Engineering: Theory and Applications*, NUMETA 87, Martinus Nijhoff (1987)
18. J.P. Bardet and S. M. Mortazavi, "Simulation of shear band formation in over consolidated", in: C.S. Desai, E. Krempl, P.D. Kiousis and T. Kundu (eds.), *Constitutive Laws for Engineering Materials and Applications*, pp. 805–812 Elsevier (1987)
19. L. Nilsson and M. Oldenburg, "Non-linear wave propagation in plastic fracturing materials", in: U. Nigel and J. Engelbrecht (eds.), *Nonlinear Deformation Waves*, IUTAM Symp., pp. 209–217 Berlin: Springer (1983)

20. J.G. Rots, P. Nauta, G.M.A. Kusters and J. Blaauwendraad, "Smeared crack approach and fracture localization in concrete", *Heron*, **30**, 1–48 (1985)

21. R. de Borst, "Computation of post–bifurcation and post–failure behavior of strain–softening solids", *Computers & Structures*, **25**, 211–224 (1987)

22. O. Dahlbom and N.S. Ottosen, "Smeared crack analysis using generalized fictitious crack model", *J. Engng. Mech.*, ASCE, **116**, 55–76 (1990)

23. R. Glemberg, *Dynamic analysis of concrete structures*, Publ. 84:1 (Ph. D. Dissertation), Dept. of Structural Mechanics, Chalmers University of Technology, (1984)

24. M. Ortiz, Y. Leroy and A. Needleman, "A finite element method for localized failure analysis", *Comp. Meth. Appl. Mech. Engng.*, **61**, 189–214 (1987)

25. R. Larsson, *Numerical Simulation of Plastic Localization*, Publ. 90:5 (Ph. D. Dissertation), Dept. of Structural Mechanics, Chalmers University of Technology, (1990)

26. K. Runesson and Z. Mroz, "A Note on Non–associated Plastic Flow Rules", *Int. J. Plasticity*, **5**, 639–658 (1989).

27. R. Larsson, K. Runesson and S. Sture, "Numerical simulation of localized plastic deformation", *Ingenieur–Archiv*, in print (1990)

28. C. Johnson and R. Scott, "A finite element method for problems in perfect plasticity using discontinuous trial functions", in: W. Wunderlich, E. Stein and K.–J. Bathe (eds.), *Non–linear Finite Element Analysis in Structural Mechanics*, pp. 307–324. Berlin: Springer (1981)

29. M. Klisinski, Z. Mroz and K. Runesson, "Structure of constitutive equations in plasticity for different choices of state and control variables", in print

30. K. Runesson, S. Sture and K. Willam, "Integration in computational plasticity", *Computers & Structures*, **30**, 119–130 (1987)

31. C. Johnson, "On finite element methods for plasticity problems", *Numer. Math.*, **26**, 79–84 (1976)

32. C. Johnson, "On plasticity with hardening", *J. Math. Anal. and Appl.*, **62**, 325–336 (1976)

33. C. Johnson, "A mixed finite element method for plasticity problems with hardening", *J. Numer Anal.*, SIAM, **14**, 575–583 (1977)

34. C. Johnson and R. Scott, "A finite element method for problems in perfect plasticity using discontinuous trial functions", in: W. Wunderlich, E. Stein and K.–J. Bathe (eds.), *NonLinear Finite Element Analysis in Structural Mechanics*, pp. 307–324 Springer–Verlag (1981)

35. N.S. Ottosen and K. Runesson, "Properties of bifurcation solutions in elasto–plasticity", *Int. J. Solids Struct.*, **27**, 401–421 (1991)

36. K. Runesson, N.S. Ottosen and D. Peric, "Discontinuous bifurcations of elastic–plastic solutions at plane stress and plane strain", *Int. J. Plasticity* (1990), in print

37. F. Molenkamp, "Comparison of frictional material models with respect to shear band initiation", *Geotechnique*, **35**, 127–143 (1985)

38. N.A. Sobh, *Bifurcation analysis of tangential material operators*, (Ph. D. Dissertation) Univ. of Colorado, Boulder (1987)

39. K. Willam and N. Sobh, "Bifurcation analysis of tangential material operators", in: O.Z. Zienkiewicz, G.N. Pande and J. Middleton (eds.), *Numerical Methods in Engineering: Theory and Applications*, NUMETA 87, Martinus Nijhoff (1987)

40. S.A. Sabban, *Property analysis and incremental formulation of J_2 elasto–plastic solids in plane stress*, (Ph. D. Dissertation) Univ. of Colorado, Boulder (1989)

41. P.V. Lade, "Effects of voids and volume changes on the behavior of frictional materials", *Int. J. Anal. Num. Meth. Geomech.*, **12**, 351–370 (1988)

42. S. Sture, K. Runesson and E. J. Macari–Pasqualino, "Analysis and calibration of a three–invariant plasticity model for granular materials", *Ingenieur–Archiv*, **59**, 253–266 (1989)

43. P.V. Lade, "Experimental observations of stability, instability and shear planes in granular materials", *Ingenieur–Archiv*, **59**, 114–123 (1989)

44. I. Vardoulakis, "Bifurcation analysis of the plane rectilinear deformation test on dry sand samples", *Int. J. Solids Struct.*, **17**, 1085–1101 (1981)

45. J. Mandel, "Conditions de stabilité et postulat de Drucker", in: J. Kravtchenko and P.M. Siryies (eds.), *Proc. IUTAM Symposium on Reology and Soil-Mechanics*, pp. 58–68. Berlin: Springer (1964)

46. J.W. Rudnicki and J.R. Rice, "Conditions for the localization of deformation in pressure-sensitive dilatant materials", *J. Mech. Phys. Solids*, **23**, 371–394 (1975)

47. J.R. Rice, "The localization of plastic deformation", in: W.T. Koiter (ed.), *Theoretical and Applied Mechanics*, Proc. 14th IUTAM, Congress pp. 207–220. Amsterdam: North–Holland (1977)

48. T.S. Hsu, J.F. Peters and S.K. Saxena, "Importance of mesh design for capturing strain localization", in: C.S. Desai, E. Krempl, P.D. Kiousis and T. Kundu (eds.), *Constitutive Laws for Engineering Materials and Applications*, pp. 857-864 Elsevier (1987)

49. H. Jin and N.E. Wiberg, "Two dimensional mesh generation, adaptive remeshing and refinement", *Int. J. Num. Meth. Engng*, **29**, 1501–1526 (1990)

RECENT DEVELOPMENTS IN THE NUMERICAL ANALYSIS OF PLASTICITY

J. C. Simo
Stanford University, Stanford, CA, USA

INTRODUCTION

The goal of this lectures is to survey some recent developments in the numerical analysis of classical plasticity and viscoplasticity. For the infinitesimal theory, the continuum mechanics aspects of the subject are currently well understood and firmly established. Classical expositions of the basic theory can be found in the work of HILL [1950], KOITER [1960] and others. On the mathematical side, classical plasticity experienced a significant development in the 70's and early 80's, starting with the pioneering work of DUVAUT & LIONS [1972]. The subsequent improvement of JOHNSON [1978], MATTHIES [1979], SUQUET [1979], TEMAM & STRANG [1980] and others produced at the beginning of the 80's a fairly complete mathematical picture of the theory.

From a computational perspective, the key contribution to the subject appears in the early work of WILKINS [1964] and MAENCHEN & SACKS [1964] with the formulation of the now classical radial return method for J_2-flow theory. By the early 70's this approach was widely used in the 'hydrocodes' employed in National Laboratories in the Unite States. Subsequent extensions to the basic method to account for various forms of hardening were made by a number of authors. In particular, the generalizations of KRIEG & KEY [1976] and BALMER ET AL. [1974] to linear kinematic/isotropic hardening, and more recent formulation for the plane stress problem in SIMO & TAYLOR [1987]. An equivalent approach from an entirely different and more general perspective is contained in the work MOREAU [1977], where classical plasticity is regarded as a convex minimization problem. Conceptually, the key result exploited in these algorithmic treatments of infinitesimal plasticity is the interpretation of the *local* evolution equations defining plastic flow as the optimality conditions of a local convex optimization problem. The implementation of this result, known as the principle of maximum dissipation (see HILL [1950, page 51]) leads to the *return mapping* algorithms.

In these lectures we shall review some recent developments in the are of computational plasticity. Part I addresses the numerical analysis of classical and recently proposed algorithms for infinitesimal plasticity, while Part II is concerned with the formal extension of these results to the finite strain regime.

1. Overview of Results in Part I

Currently, return mapping algorithms are widely use in many commercial finite element problems. The technique is remarkably simple. Within a typical time step one first computes a trial elastic state for prescribed strain increments by ignoring plastic flow. Then, the actual stress is defined as the closest-point projection of the trial state onto the elastic domain in the natural norm defined by the complementary (Helmholtz) energy. This projection is computed locally at each quadrature point of a typical finite element and depends exclusively on the functional form adopted by the yield criterion in stress space. Globally, the problem discretized via finite difference or finite element methods is typically solved by Newton or quasi-Newton iterative solution procedures. Up to the mid 80's, implementations of Newton's method used the standard elastoplastic moduli obtained by enforcing the consistency condition on the rate model; see e.g., STRANG, MATTHIES & TEMAM [1980]. SIMO & TAYLOR [1985] showed that the disappointing rates of convergence exhibited by the iteration arise from lack of consistency between the classical elastoplastic moduli and the return mapping algorithm. The exact closed-form linearization of the return map produces modified elastoplastic moduli, referred to as the consistent *algorithmic* moduli, which restore the quadratic rate of convergence of Newton's method and result in improved rates of convergence for quasi-Newton methods employing periodic refactorizations.

The first part of this lectures is devoted to the numerical analysis aspects related to return-mapping algorithms formulated within the framework of the infinitesimal theory. In the now classical approach, the convex minimization problem arises via a backward Euler integration of the plastic flow evolution equations. We shall consider an extension of this technique recently proposed in SIMO & GOVINDJEE [1990] which can achieve second order accuracy while preserving the symmetric of the algorithmic tangent moduli. The numerical analysis of this class of methods relies on the contractivity property of the semiflow in stress space generated by classical plasticity, relative to the natural norm induced by the complementary Helmholtz free energy. As recently noted in SIMO [1991], unconditional nonlinear B-stability of the return mapping algorithms (in the sense of BURRAGE & BUTCHER [1980]) follows from the observation that the closest-point-projection preserves the contractivity property relative to the natural norm. In Part I a summary of these results is given after the basic mathematical structure structure of the initial boundary value problem for dynamic plasticity is reviewed.

1. Overview of Results in Part II

It is apparent from the preceding discussion that the actual implementation of a numerical method for computational plasticity involves two key ingredients: (**i**) A finite element method suitable for quasi-incompressible infinitesimal elasticity, and (**ii**) A finite dimensional optimization algorithm to compute the closest projection of a point (the trial stress) onto a convex set (the elastic domain). The main objective in Part II of these lectures is to construct an exact counterpart in the finite strain regime of the algorithmic scheme outlined above in such a way that the closest-point-projection algorithms of the infinitesimal theory remain unchanged. It will be shown that this goal can be achieved by considering a formulation of plasticity at finite strains based on the multiplicative decomposition of the deformation gradient. The resulting class of algorithms constitute a significant departure from the approachs based on the so-called hypoelastic formulations, widely used up to the early 80's; see e.g. the reviews of NEEDLEMAN & TVERGAARD [1984] and HUGHES [1984].

Computational approaches based on the multiplicative decomposition appear to have been first proposed by ARGYRIS & DOLTSINIS [1979] within the context of the so-called *natural formulation.* Subsequently, however, these authors appear to favor hypoelastic rate models on the basis that multiplicative formulations "... lead in principle to non-symmetric relations between stress rates and strain rates" (see ARGYRIS, DOLTSINIS, PIMENTA & WUSTENBERG [1982, page 22]. SIMO & ORTIZ [1985] and SIMO [1985] proposed a computational approach entirely based on the multiplicative decomposition and pointed out the role of the intermediate configuration in a definition of the trial state via mere function evaluation of hyperelastic stress-strain relations. Extensions of classical volume/displacement mixed methods within the framework of the multiplicative decomposition, originally introduced for plasticity problems in NAGTEGAAL, PARK & RICE [1974], are presented in SIMO, TAYLOR & PISTER [1985].

In recent years, computational approaches based on the multiplicative decomposition have received considerable attention in the literature. SIMO [1988a,b] exploited a strain-space version of the principle of maximum dissipation to obtain the associative flow rule consistent with the multiplicative decomposition, and used a (covariant) backward method to derive a finite strain version of the return mapping algorithms. Subsequently, WEBER & ANAND [1990] and ETEROVICH & BATHE [1991] used the multiplicative decomposition in conjunction with a logarithmic stored energy function and an exponential approximation to the flow rule cast in terms of the full plastic deformation gradient. The use of logarithmic models is advocated in ROLPH & BATHE [1984]. The multiplicative decomposition along with a logarithmic stored energy function is also used in PERIC, OWEN & HONNOR [1989]. More recently, MORAN, ORTIZ & SHI [1990] addressed a number of computational aspects of multiplicative plasticity and presented explicit/implicit integration algorithms. Methods of convex analysis, again in the context of the multiplication decomposition, are discussed in EVE, REDDY & ROCKEFELLAR [1991]. Related approaches based on the theory of materials of grade N are presented in KIM & ODEN [1990].

The preceding survey, although by no means comprehensive, conveys the popularity gained in recent years by computational elastoplasticity based on the multiplicative decomposition. Despite their success these approaches involve modifications, and often a complete reformulation, of the standard closest-point projection algorithms of the infinitesimal theory. From a practical stand-point the implication is that the implementation of classical models needs to be considered on a case-by-case basis in the finite strain regime. The formulation described in Part II of these lectures, recently proposed in SIMO [1991], preserves the structure of the closest-point projection algorithms of the infinitesimal theory and retains full symmetry of the algorithmic moduli, which are computed exactly in closed-form. These properties are achieved with one further simplification: The closest-point projection algorithm is now formulated in principal (Kirchhoff) stresses.

I. NUMERICAL ANALYSIS ASPECTS OF CLASSICAL PLASTICITY

The goal in the first part of this lectures is to describe recent results on the numerical analysis of algorithms for static and dynamic plasticity, within the framework of the infinitesimal theory. In particular, a rigorous nonlinear stability analysis of a class of time discretizations of the weak form of the initial boundary value problem is presented, both for rate–independent and rate–dependent infinitesimal elastoplasticity. The material in this part follows closely the analysis recently presented in SIMO [1991]. These results are easily extended to the dynamic; see CORIGLIANO & PEREGO [1991] and the analysis below. Salient features of this approach are:

i. The stability analysis is performed directly on the system of variational equations discretized in time, and not on the algebraic system arising from both a temporal and a spatial discretization. The results carry over immediately to the finite dimensional problem obtained via a Galerkin (spatial) discretization.

ii. Previous stability analyses employ either the notion of A–stability, introduced by DAHLQUIST [1963] in the context of linear multistep methods for systems of ODE's, or the concept of *linearized stability*; see e.g., HUGHES [1982] and references therein. The results given below prove *nonlinear stability* in the sense that arbitrary perturbations in the initial data are attenuated by the algorithm relative to certain *algorithm–independent* norm associated with the continuum problem called the *natural norm.*

For nonlinear systems of ODE's this notion of nonlinear stability reduces to the concept of A–contractivity or B–stability introduced by BUTCHER [1975] in the context of implicit Runge–Kutta methods. A–contractivity is now widely accepted as the proper definition of nonlinear stability; see e.g., BURRAGE & BUTCHER [1979,1980], and DAHLQUIST & JELTSCH [1979]. For *linear* semigroups the definition of stability employed in this paper coincides with the notion of Lax–stability; see RICHTMYER & MORTON [1967]. A key step in the stability analysis given below is the identification of the natural norm for the continuum problem relative to which the crucial contractivity property holds. For infinitesimal elastoplasticity the natural norm is shown to be the norm induced by the complementary Helmholtz free energy function. A given algorithm is then said to be A–contractive if it inherits the contractivity property present in the continuum problem relative to the natural norm.

For infinitesimal elastoplasticity and viscoplasticity, it is shown that the system of variational inequalities associated with a class of return mapping algorithms based on the generalized mid–point rule, recently proposed in SIMO & GOVINDJEE [1990], is nonlinearly stable (A–contractive) for $\alpha \geq \frac{1}{2}$. This class of methods encompass the well–known return mapping algorithms based on the implicit backward Euler method ("catching–up" algorithms in the terminology of MOREAU [1977]). Algorithms of this type are now widely used in the numerical solution of the elastoplastic initial boundary value problem in conjunction with Newton (or quasi–Newton) methods. Convergence of the discrete incremental problem has been recently examined in MARTIN [1988], COMMI & MAIER [1989], and MARTIN & CADDEMI [1990]. We emphasize that these results apply to the generalized mid-point rule. It is well–known that A–contractivity cannot hold for the generalized trapezoidal rule (see WANNER [1976] for a counter–example).

The nonlinear stability proof given below also applies to the class of generalized mid–point rule algorithms in HUGHES & TAYLOR [1978] (perfect viscoplasticity) and SIMO & TAYLOR [1987] (plane stress elastoplasticity), but does not cover the class of methods in ORTIZ & POPOV [1988], whose stability properties remain an open question. The results in Section **3** for the full variational problem in elastoplasticity extend and complete the stability analysis in SIMO & GOVINDJEE [1990], which considers the system of ODE's obtained by assuming a strain driven problem and is restricted to smooth elastic domains defined by a single (smooth) yield condition.

1. Infinitesimal Elastoplasticity

The steps in the stability analysis of the time-discretization of the IBVP for infinitesimal plasticity and viscoplasticity by means of a generalized mid–point rule depend crucially on a formulation of plasticity as a variational inequality. The analysis proceeds according to the following steps:

i. Infinitesimal elastoplasticity is formulated in weak form as a variational problem of evolution. Assuming convexity of yield criterion and associativity of the flow rule, the solutions of this problem are shown to be contractive relative to a *natural norm* defined as the sum of the the kinetic energy and the complementary Helmholtz free energy function.

ii. The resulting time dependent variational inequality is discretized by means a one–parameter family of generalized mid–point rule algorithms depending on the parameter $\alpha \in [0, 1]$. This defines a class of generalized return mapping algorithms recently proposed in SIMO & GOVINDJEE [1990]. It is then shown that this time discrete problem of evolution inherits the contractivity property of the continuum problem provided that $\alpha \geq \frac{1}{2}$. Hence, this class of algorithms is nonlinearly B–stable (A–contractive).

The preceding analysis is performed without introducing any spatial discretization. An identical result holds for the semi–discrete problem of evolution obtained by a Galerkin finite element projection. The stability proof relies critically on the enforcement of the equilibrium condition at the mid-point value $t_{n+\alpha}$. The analysis presented below extends reproduces the results described in SIMO [1991]. The extension of this analysis to the time-dependent case was carried out by Maier and his co-workers; see CORIGLIANO & PEREGO [1991].

1.1. The Local Form of the Balance Laws

Let $\Omega \subset \mathsf{R}^{n_{\mathrm{dim}}}$, with $1 \leq n_{\mathrm{dim}} \leq 3$, be the reference placement of an elasto-plastic body, with smooth boundary $\partial\Omega$; let $\mathsf{I} \subset \mathsf{R}_{+}$ the time interval of interest, and denote by

$$\boldsymbol{u} : \overline{\Omega} \times \mathsf{I} \rightarrow \mathsf{R}^{n_{\mathrm{dim}}}, \qquad \text{and} \qquad \boldsymbol{\sigma} : \overline{\Omega} \times \mathsf{I} \rightarrow \mathsf{S} \tag{1.1}$$

the displacement field and the stress tensor, respectively. Here $\mathsf{S} \cong \mathsf{R}^{(n_{\mathrm{dim}}+1)\cdot n_{\mathrm{dim}}/2}$ is the vector space of symmetric rank-two tensors. Let $\boldsymbol{n}\colon \partial\Omega \rightarrow \mathrm{S}^2$ be the unit outward normal to the boundary. We shall denote by $\bar{\boldsymbol{u}}\colon \Gamma_u \times \mathsf{I} \rightarrow \mathsf{R}^{n_{\mathrm{dim}}}$ the prescribed boundary boundary displacement and designate by $\bar{\boldsymbol{t}}\colon \Gamma_\sigma \times \mathsf{I} \rightarrow \mathsf{R}^{n_{\mathrm{dim}}}$ is the prescribed boundary traction vector. As usual, we assume

$$\overline{\Gamma_u \cup \Gamma_\sigma} = \overline{\partial\Omega}, \quad \text{and} \quad \Gamma_u \cap \Gamma_\sigma = \emptyset, \tag{1.2}$$

(with the conventional interpretation). Therefore, denoting by $\boldsymbol{f} : \Omega \times \mathsf{I} \rightarrow \mathsf{R}^{n_{\mathrm{dim}}}$ the body force (per unit of volume) the local form of the momentum equations becomes

$$\left.\begin{aligned} \dot{\boldsymbol{u}} &= \boldsymbol{v} \\ \rho_0 \dot{\boldsymbol{v}} &= \operatorname{div}[\boldsymbol{\sigma}] + \boldsymbol{f} \end{aligned}\right\} \quad \text{in} \quad \Omega \times \mathsf{I}, \tag{1.3a}$$

where $\boldsymbol{v}(\cdot, t)$ is the velocity field. This balance law is supplemented by the boundary conditions:

$$\left.\begin{aligned} \boldsymbol{u} &= \bar{\boldsymbol{u}} \quad \text{on} \quad \Gamma_{\boldsymbol{u}} \times \mathsf{I} \\ \boldsymbol{\sigma} \cdot \boldsymbol{n} &= \bar{\boldsymbol{t}} \quad \text{on} \quad \Gamma_q \times \mathsf{I} \end{aligned}\right\} \tag{1.3b}$$

along with the initial conditions:

$$\boldsymbol{u}(\cdot, t)|_{t=0} = \boldsymbol{u}_0(\cdot) \quad \text{and} \quad \boldsymbol{v}(\cdot, t)|_{t=0} = \boldsymbol{v}_0(\cdot) \quad \text{in} \quad \Omega. \tag{1.3c}$$

Observe that the preceding equations are linear. The source of nonlinearity in this problem arises from the type of constitutive equation that relates the stress field and the displacement field, as discussed below.

1.2. Classical rate-independent plasticity

In addition to the stress tensor $\boldsymbol{\sigma}(\boldsymbol{x},t)$, one introduces a n_{int}–dimensional vector field $\boldsymbol{q}: \Omega \times \mathsf{I} \to \mathsf{R}^{n_{\text{int}}}$ $(n_{\text{int}} \geq 1)$ of phenomenological internal variables which, from a physical standpoint, characterize strain hardening in the material. For convenience, the following notation is adopted

$$\boldsymbol{\Sigma}(\boldsymbol{x},t) := (\boldsymbol{\sigma}(\boldsymbol{x},t),\, \boldsymbol{q}(\boldsymbol{x},t)), \quad \text{for } (\boldsymbol{x},t) \in \Omega \times \mathsf{I}. \tag{1.4}$$

One refers to $\boldsymbol{\Sigma}$ as the *generalized stress* which is constrained to lie within a *convex domain*, called the elastic domain, and denoted by E. Typically, E is defined in terms of smooth convex functions $\phi_\mu : \mathbb{S} \times \mathsf{R}^{n_{\text{int}}} \to \mathsf{R}$, with $\mu \in \{1,\cdots,m\}$, as the constrained *convex* set

$$\mathsf{E} := \{\boldsymbol{\Sigma} := (\boldsymbol{\sigma},\boldsymbol{q}) \in \mathbb{S} \times \mathsf{R}^{n_{\text{int}}} :\ \phi_\mu(\boldsymbol{\Sigma}) \leq 0, \text{ for } \mu = 1,\cdots,m\}. \tag{1.5}$$

The boundary $\partial\mathsf{E}$ of $\mathsf{E} \subset \mathbb{S} \times \mathsf{R}^{n_{\text{int}}}$ need not be smooth; in fact, in applications, $\partial\mathsf{E}$ typically exhibits 'corners'. A classical example is provided by the Tresca yield condition.

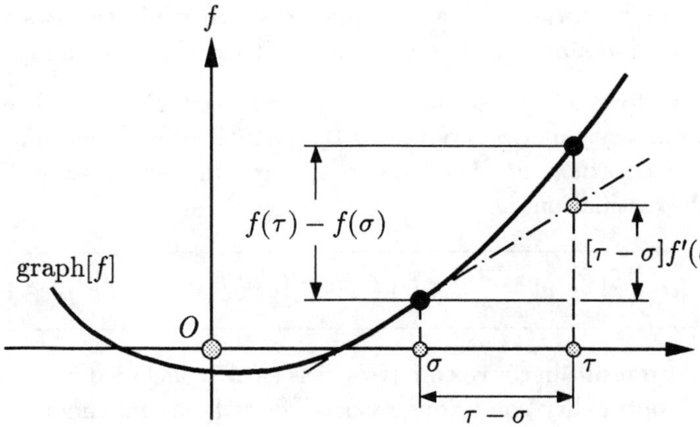

FIGURE 1.1. Illustration of the convexity property (1.7) for a smooth one dimensional function $f: \mathsf{R} \to \mathsf{R}$ $(n_{\text{dim}} = 1.)$.

Let $\mathbf{C}$ be the *elasticity tensor*, which is assumed to be constant in what follows. Further, let $\mathbf{H}$ denote a $n_{\text{int}} \times n_{\text{int}}$ given matrix, which is assumed to be positive definite, constant, and is referred to as the *generalized hardening moduli*. Under these assumptions, the complementary Helmholtz free energy function defined by

$$\Xi(\boldsymbol{\Sigma}) := \tfrac{1}{2}\boldsymbol{\sigma} \cdot \mathbf{C}^{-1}\boldsymbol{\sigma} + \tfrac{1}{2}\boldsymbol{q} \cdot \mathbf{H}^{-1}\boldsymbol{q} \tag{1.6}$$

is a strictly convex function on $\mathbb{S} \times \mathbb{R}^{n_{\rm int}}$. The convexity assumptions on the elastic domain and the Helmholtz free energy function are crucial for the stability analysis described below. Observe that the smoothness hypothesis on the functions $\phi_\mu(\boldsymbol{\Sigma})$ along with the convexity assumption yields the relation

$$\phi_\mu(\boldsymbol{T}) - \phi_\mu(\boldsymbol{\Sigma}) \geq [\boldsymbol{T} - \boldsymbol{\Sigma}] \cdot \nabla\phi_\mu(\boldsymbol{\Sigma}), \quad \forall \boldsymbol{T}, \boldsymbol{\Sigma} \in \mathbb{S} \times \mathbb{R}^{n_{\rm int}}. \tag{1.7}$$

An illustration of this property is given in Figure 1.1.

By definition, the *local dissipation function* $\mathcal{D}$ is the difference between the total stress power and the rate of change in free energy which, in the present setting, coincides with the complementary Helmholtz free energy. Accordingly, we have

$$\mathcal{D} := \boldsymbol{\sigma} \cdot \boldsymbol{\varepsilon}[\boldsymbol{v}] - \frac{d}{dt}\Xi(\boldsymbol{\Sigma}); \quad \text{in} \quad \Omega \times \mathbb{I}, \tag{1.8}$$

where $\boldsymbol{\varepsilon}[\cdot] = \text{sym}[\nabla(\cdot)]$ is the strain operator, and $\boldsymbol{v} = \dot{\boldsymbol{u}} := \frac{\partial}{\partial t}\boldsymbol{u}$ is the velocity field. Carrying out the time differentiation via the chain rule yields the expression

$$\mathcal{D} = \boldsymbol{\sigma} \cdot [\boldsymbol{\varepsilon}[\boldsymbol{v}] - \mathbf{C}^{-1}\dot{\boldsymbol{\sigma}}] + \boldsymbol{q} \cdot [-\mathbf{H}^{-1}\dot{\boldsymbol{q}}]. \tag{1.9}$$

One refers to $\boldsymbol{\epsilon}^p := \boldsymbol{\varepsilon}[\boldsymbol{u}] - \mathbf{C}^{-1}$ as the plastic strain while the strain-like variables $\boldsymbol{\xi} := -\mathbf{H}^{-1}\dot{\boldsymbol{q}}$ are the affinities conjugate to the stress-like variables $\boldsymbol{q}$.

The local form of the second law requires that $\mathcal{D} \geq 0$. The classical rate independent plasticity model is obtained by postulating the following *local principle of maximum dissipation*: For fixed rates $(\boldsymbol{\varepsilon}[\boldsymbol{v}], \dot{\boldsymbol{\sigma}})$, the actual state $\boldsymbol{\Sigma} = (\boldsymbol{\sigma}, \boldsymbol{q}) \in \mathbb{E}$ maximizes the dissipation; i.e:

$$\boxed{[\boldsymbol{\sigma} - \boldsymbol{\tau}] \cdot \boldsymbol{\varepsilon}[\boldsymbol{v}] - \frac{d}{dt}[\Xi(\boldsymbol{\Sigma}) - \Xi(\boldsymbol{T})] \geq 0, \quad \forall \boldsymbol{T} = (\boldsymbol{\tau}, \boldsymbol{p}) \in \mathbb{E}.} \tag{1.10}$$

Using a standard result in convex analysis, it is easily concluded from expression (1.9) that the local optimality conditions associated with the maximum principle (1.10) are given by the relations

$$\left.\begin{aligned} \boldsymbol{\varepsilon}[\dot{\boldsymbol{u}}] - \mathbf{C}^{-1}\dot{\boldsymbol{\sigma}} &= \sum_{\mu=1}^{m} \gamma^\mu \frac{\partial}{\partial \boldsymbol{\sigma}}\phi_\mu(\boldsymbol{\sigma}, \boldsymbol{q}), \\ -\mathbf{H}^{-1}\dot{\boldsymbol{q}} &= \sum_{\mu=1}^{m} \gamma^\mu \frac{\partial}{\partial \boldsymbol{q}}\phi_\mu(\boldsymbol{\sigma}, \boldsymbol{q}), \end{aligned}\right\} \tag{1.11}$$

$$\gamma^\mu \geq 0, \quad \phi_\mu(\boldsymbol{\sigma}, q) \leq 0 \text{ and } \textstyle\sum_{\mu=1}^{m} \gamma^\mu \phi_\mu(\boldsymbol{\sigma}, q) = 0, \quad \text{for } \mu = 1, \cdots, m.$$

One refers to $\gamma^\mu \geq 0$ as the *plastic consistency parameters*. The relations $(1.11)_3$ are called the Kuhn-Tucker complementarity conditions in the optimization literature.

The preceding evolution equations define the local form of the flow rule and the hardening law for multi-surface plasticity according to a prescription going back to KOITER [1960]. For the analysis given below, however, the global form of these relations embodied in the statement (1.10) of the principle of maximum dissipation are far more convenient.

2. The Weak Formulation of Dynamic Plasticity

The weak form of the momentum equations along with the consitutive equations formulated as a variational inequality comprise the weak formulation of the initial boundary value problem for classical plasticity and viscoplasticity. A key feature of this IBVP is the contractivity property of the solutions relative to certain natural norm for the problem, which motivates the notion of algorithmic stability employed in the subsequent analysis.

2.1. Variational Inequality Form of the Constitutive Equations

The weak form of the constitutive equations is simply the global formulation of the principle of maximum (plastic) dissipation (1.10) over the entire body. Accordingly, let (σ_{ij}, q_i) be components of $\boldsymbol{\Sigma} = (\boldsymbol{\sigma}, \boldsymbol{q})$ relative to a Cartesian orthonormal frame, and let

$$\mathcal{T} := \{\boldsymbol{\Sigma} = (\boldsymbol{\sigma}, \boldsymbol{q}) : \Omega \to \mathsf{S} \times \mathsf{R}^{n_{\text{int}}} : \ \sigma_{ij} \in L_2(\Omega) \text{ and } q_i \in L_2(\Omega)\}. \tag{2.1}$$

Further, define the bilinear forms $a(\cdot,\cdot)$ and $b(\cdot,\cdot)$ by the expressions

$$a(\boldsymbol{\sigma}, \boldsymbol{\tau}) := \int_\Omega \boldsymbol{\sigma} \cdot \mathbf{C}^{-1} \boldsymbol{\tau}\, d\Omega, \quad b(\boldsymbol{q}, \boldsymbol{p}) := \int_\Omega \boldsymbol{q} \cdot \mathbf{H}^{-1} \boldsymbol{p}\, d\Omega. \tag{2.2a}$$

Since the elastic moduli $\mathbf{C}$ and plastic hardening moduli $\mathbf{H}$ are positive definite on S and $\mathsf{R}^{n_{\text{int}}}$, respectively, it follows that $a(\cdot\,,\,\cdot)$, and $b(\cdot\,,\,\cdot)$ are coercive. Consequently, the bilinear form $A(\cdot,\cdot)$: $\mathcal{T} \times \mathcal{T} \to \mathsf{R}$ defined by

$$A(\boldsymbol{\Sigma}, \boldsymbol{T}) := a(\boldsymbol{\sigma}, \boldsymbol{\tau}) + b(\boldsymbol{q}, \boldsymbol{p}), \tag{2.2b}$$

induces an inner product on $\mathcal{T}$. Observe that relations (1.6) and (2.2) imply that the norm squared $A(\boldsymbol{\Sigma})$ is precisely twice the integral over the body of the complementary Helmholtz free energy function. With the this notation in hand, the dissipation over the entire body, denoted by $\mathcal{D}_\Omega$ and obtained by integration of (1.10) over the reference placement Ω, can be written as

$$\mathcal{D}_\Omega = A(\boldsymbol{\Sigma}, \dot{\boldsymbol{\Sigma}}^{\text{trial}}) - \tfrac{1}{2}\frac{d}{dt} A(\boldsymbol{\Sigma}, \boldsymbol{\Sigma}) = A(\dot{\boldsymbol{\Sigma}}^{\text{trial}} - \dot{\boldsymbol{\Sigma}}, \boldsymbol{\Sigma}) \geq 0, \tag{2.3}$$

where

$$\dot{\boldsymbol{\Sigma}}^{\text{trial}} := (\mathbf{C}\, \boldsymbol{\varepsilon}[\dot{\boldsymbol{u}}]\,,\, \mathbf{0})\,. \tag{2.4}$$

Note that $\dot{\boldsymbol{\Sigma}}^{\text{trial}}$ is the stress rate about $\boldsymbol{\Sigma} \in \mathsf{E}$ obtained by *freezing plastic flow* (i.e., by setting $\dot{\boldsymbol{\varepsilon}}^p = \mathbf{0}$ and $\dot{\boldsymbol{q}} = \mathbf{0}$) and, therefore, is often referred to as the rate of trial elastic stress. In view of (2.4), the global version of the principle of maximum dissipations leads to the variational inequality

$$A(\dot{\boldsymbol{\Sigma}}^{\text{trial}} - \dot{\boldsymbol{\Sigma}}, \boldsymbol{T} - \boldsymbol{\Sigma}) \leq 0, \quad \forall \boldsymbol{T} \in \mathsf{E} \cap \mathcal{T}, \tag{2.5}$$

which gives the weak formulation of the constitutive equation for rate–independent (hardening) elastoplasticity.

2.1.1. *The viscoplastic regularization.* Following DUVAUT & LIONS [1976], classical viscoplasticity is regarded as a (Yoshida) regularization of rate independent plasticity constructed as follows. Define the functional $J: \mathsf{S} \times \mathsf{R}^{n_{\text{int}}} \to \mathsf{R}$ by the constrained minimization problem

$$J(\boldsymbol{\Sigma}) := \min\{2\,\Xi(\boldsymbol{\Sigma} - \boldsymbol{T}), \text{ for all } \boldsymbol{T} \in \mathsf{E}\}. \tag{2.6}$$

Thus, $J(\boldsymbol{\Sigma})$ gives the (unique) distance measured in the complementary Helmholtz free energy $\Xi(\cdot)$ between a given point $\boldsymbol{\Sigma}$ and the convex set E. Clearly, $J(\boldsymbol{\Sigma}) \geq 0$ for any $\boldsymbol{\Sigma} \in \mathsf{S} \times \mathsf{R}^{n_{\text{int}}}$, and $J(\boldsymbol{\Sigma}) = 0$ iff $\boldsymbol{\Sigma} \in \mathsf{E}$. Now consider the following regularization of the dissipation function (2.3):

$$\mathcal{D}_{\Omega}^{\eta} := \mathcal{D}_{\Omega} + \frac{1}{\eta} \int_{\Omega} g(J(\boldsymbol{\Sigma}))\, d\Omega, \tag{2.7a}$$

where $\eta \in (0, \infty)$ is the regularization parameter, and $g(\cdot)$ is a *non–negative* convex function with the property

$$g(x) \geq 0, \quad \text{and} \quad g(x) = 0 \iff x = 0. \tag{2.7b}$$

By standard results in convex optimization; see e.g., LUENBERGER [1984], it follows that the problem of maximizing the regularized dissipation $\mathcal{D}_{\Omega}$ over all *unconstrained* stresses $\boldsymbol{\Sigma} \in \mathcal{T}$ is simply the penalty regularization of the classical *constrained* principle of maximum dissipation, with dissipation function $\mathcal{D}_{\Omega}$. The optimality condition associated with the regularized principle of maximum dissipation then yields the following inequality which characterizes the constitutive response of classical viscoplasticity

$$A(\dot{\boldsymbol{\Sigma}}^{\text{trial}} - \dot{\boldsymbol{\Sigma}}, \boldsymbol{T} - \boldsymbol{\Sigma}) \leq \frac{1}{\eta} \int_{\Omega} [g(J(\boldsymbol{T})) - g(J(\boldsymbol{\Sigma})]\, d\Omega, \quad \forall \boldsymbol{T} \in \mathcal{T}. \tag{2.8}$$

Observe that in contrast with rate–independent plasticity, neither the actual stress $\boldsymbol{\Sigma} \in \mathcal{T}$ nor the admissible stress variations $\boldsymbol{T} \in \mathcal{T}$ are constrained to lie in the elastic domain E. It is well–known, however, that as $\eta \to 0$, the stress $\boldsymbol{\Sigma}$ is constrained to lie in the elastic domain, and (2.8) reduces to inequality (2.5); see DUVAUT & LIONS [1976] and JOHNSON [1976,1978].

2.2. The Weak Formulation of the IBVP

We shall denote by $\mathcal{S}_t$ the solution space for the displacement field at (frozen) time $t \in \mathsf{I}$; i.e.,

$$\mathcal{S}_t = \{\boldsymbol{u}(\cdot,t) \in [H^1(\Omega)]^{n_{\text{dim}}} : \boldsymbol{u}(\cdot,t) = \bar{\boldsymbol{u}}(\cdot,t) \quad \text{on} \quad \Gamma_u\}. \tag{2.9}$$

Associated with the space $\mathcal{S}_t$ we have the space $\mathcal{V}$ of displacement test functions defined by standard expression

$$\mathcal{V} = \{\boldsymbol{\eta} \in [H^1(\Omega)]^{n_{\text{dim}}} : \boldsymbol{\eta} = \boldsymbol{0} \quad \text{on} \quad \Gamma_u\}. \tag{2.10}$$

In addition, with a slight abuse in notation, the same symbol $\langle\cdot,\cdot\rangle$ is used to denote the standard $L_2(\Omega)$ of either functions, vectors or tensors depending on the specific context. Since for fixed time $t \in \mathsf{I}$ the velocity field $\boldsymbol{v}(\cdot,t)$ lies in $\mathcal{V}$, the kinetic energy function $K(\boldsymbol{v})$ induces a natural inner product on $\mathcal{V}$. In summary, denoting by $\rho_0(\cdot) > 0$ is the reference density we have

$$K(\boldsymbol{v}) = \int_\Omega \rho_0 \boldsymbol{v}\cdot\boldsymbol{v}\, d\Omega \equiv \langle \rho_0 \boldsymbol{v}, \boldsymbol{v}\rangle.$$

With this notation in hand, the weak form of the momentum equations (1.3) along with the dissipation inequality (2.5) [or (2.8) for viscoplasticity] leads to the following variational problem:

Problem W_t: Find $t \in \mathsf{I} \mapsto \boldsymbol{u} \in \mathcal{S}_t$ and $t \in \mathsf{I} \mapsto \boldsymbol{\Sigma} = (\boldsymbol{\sigma}, \boldsymbol{q}) \in \mathsf{E} \cap \mathcal{T}$ such that

$$\left.\begin{aligned} \langle \rho_0(\boldsymbol{v} - \dot{\boldsymbol{u}}), \boldsymbol{\eta}\rangle &= 0 \quad \forall \boldsymbol{\eta} \in \mathcal{V}, \\ \langle \rho_0 \dot{\boldsymbol{v}}, \boldsymbol{\eta}\rangle + \langle \boldsymbol{\sigma}, \boldsymbol{\varepsilon}[\boldsymbol{\eta}]\rangle - \langle \boldsymbol{f}, \boldsymbol{\eta}\rangle - \langle \bar{\boldsymbol{t}}, \boldsymbol{\eta}\rangle_\Gamma &= 0 \quad \forall \boldsymbol{\eta} \in \mathcal{V}, \\ A(\dot{\boldsymbol{\Sigma}}^{\text{trial}} - \dot{\boldsymbol{\Sigma}}, \boldsymbol{T} - \boldsymbol{\Sigma}) &\le 0 \quad \forall \boldsymbol{T} \in \mathsf{E} \cap \mathcal{T}, \end{aligned}\right\} \tag{2.11a}$$

with $\boldsymbol{T} = (\boldsymbol{\tau}, \boldsymbol{p})$, subject to the initial conditions

$$\langle \boldsymbol{u}(\cdot,0), \boldsymbol{\eta}\rangle = \langle \boldsymbol{u}_0, \boldsymbol{\eta}\rangle \quad \text{and} \quad \langle \boldsymbol{v}(\cdot,0), \boldsymbol{\eta}\rangle = \langle \boldsymbol{v}_0, \boldsymbol{\eta}\rangle, \quad \forall \boldsymbol{\eta} \in \mathcal{V}. \tag{2.11b}$$

The weak form (2.11) of the dynamic elastoplastic problem involves the velocity and the generalized stress stress field. In the subsequent analysis it is convenient to use the notation

$$\boldsymbol{\chi} = (\boldsymbol{v}, \boldsymbol{\Sigma}) \in Z \quad \text{where} \quad Z = \mathcal{V} \times [\mathcal{T} \cap \mathsf{E}] \tag{2.12}$$

is the the space of admissible velocities and admissible generalized stress fields. The space Z is equipped with a natural inner product induced by the kinetic energy and the complementary Helmholtz free energy function, which is be denoted by $\langle\langle\cdot,\cdot\rangle\rangle$ and defined by

$$\langle\langle \boldsymbol{\chi}_1, \boldsymbol{\chi}_2\rangle\rangle := \langle \rho_0 \boldsymbol{\eta}_1, \boldsymbol{\eta}_2\rangle + A(\boldsymbol{T}_1, \boldsymbol{T}_2). \tag{2.13}$$

The associated natural is denoted by $|||\cdot|||:=\sqrt{\langle\langle\cdot,\cdot\rangle\rangle}$ and is interpreted as *twice* the sum of the kinetic and Helmolltz energies of the elastoplastic body. It will be shown below that the $|||\cdot|||:=\sqrt{\langle\langle\cdot,\cdot\rangle\rangle}$ is the *natural norm* for the elastoplastic problem.

Remarks 2.1.

1. The geometric interpretation of inequality $(2.11\text{a})_2$ is illustrated in Figure 2.1. The actual stress rate $\dot{\boldsymbol{\Sigma}}$ is the projection in the norm defined by the bilinear form $A(\cdot,\cdot)$ of the trial stress rate $\dot{\boldsymbol{\Sigma}}^{\text{trial}}$ onto the tangent plane at $\boldsymbol{\Sigma} \in \partial\mathsf{E}$. Convexity of E then implies that the angle measured in the inner product defined by $A(\cdot,\cdot)$ between $[\dot{\boldsymbol{\Sigma}}^{\text{trial}} - \dot{\boldsymbol{\Sigma}}]$ and $[\boldsymbol{\Sigma} - \boldsymbol{T}]$ is greater or equal that $\pi/2$; a condition equivalent to $(2.11\text{a})_2$.

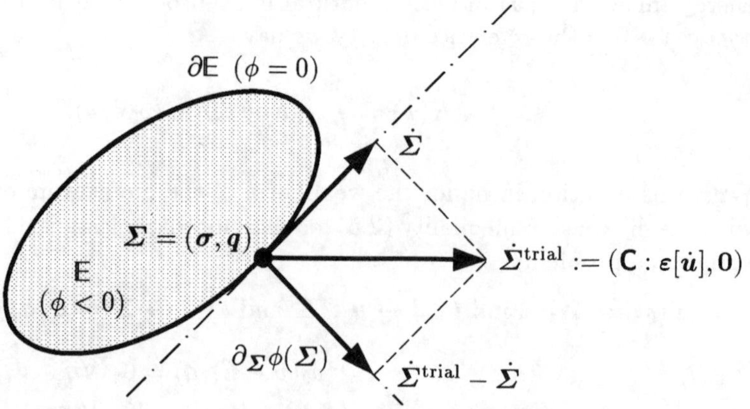

FIGURE 2.1. Geometric interpretation of the variational inequality $A(\dot{\boldsymbol{\Sigma}}^{\text{trial}} - \dot{\boldsymbol{\Sigma}}, \boldsymbol{\Sigma} - \boldsymbol{T}) \le 0$, for $\mathsf{E} := \{\boldsymbol{\Sigma} : \phi(\boldsymbol{\Sigma}) \le 0\}$.

2. The variational formulation of plasticity given by equations (2.11) goes back to the pioneering work DUVAUT & LIONS [1976] and has been considered by a number of authors, in particular, JOHNSON [1976,1978]. The assumption of hardening plasticity; i.e. the presence of the bilinear form $b(\cdot,\cdot)$ defined by $(2.2\text{a})_2$, is crucial if the functional analysis framework outlined above is to remain physically meaningful.

3. The situation afforded by perfect plasticity is significantly more complicated than hardening plasticity since the regularity implicit in the choice of $\mathcal{S}_t$ in (2.9) no longer holds. The underlying physical reason for this lack of regularity is the presence of strong discontinuities in the displacement field; the so-called *slip lines*, which rule out the use of standard Sobolev spaces. For perfect plasticity the appropriate choice appears to be $\mathcal{S}_t = BD(\Omega)$; i.e. the space of bounded deformations [displacements in $L_2(\Omega)$ with strain field a bounded measure] introduced by MATTHIES, STRANG & CHRISTIANSEN [1979], and further analyzed in TEMAM & STRANG [1980]. See MATTHIES [1978,1979], SUQUET [1979], STRANG, MATTHIES & TEMAM [1980], and the recent comprehensive summary in DEMENGEL [1989] for a detailed elaboration on these and related issues. ■

2.3. Contractivity of the Elastoplastic Flow

The following contractivity property inherent to problem W_t identifies the norm $|||\cdot|||$ defined by (2.2b) as the natural norm and plays, therefore, a crucial role in the subsequent algorithmic stability analysis. Let

$$\boldsymbol{\chi}_0 = (\boldsymbol{v}_0, \boldsymbol{\Sigma}_0) \in Z \qquad \text{and} \quad \tilde{\boldsymbol{\chi}}_0 = (\tilde{\boldsymbol{v}}_0, \tilde{\boldsymbol{\Sigma}}_0) \in Z \tag{2.14}$$

be two *arbitrary* initial conditions for the problem of evolution defined by (2.11) and denote by

$$t \in \mathbf{I} \mapsto \boldsymbol{\chi} = (\boldsymbol{v}, \boldsymbol{\Sigma}) \in Z \qquad \text{and} \qquad t \in \mathbf{I} \mapsto \tilde{\boldsymbol{\chi}} = (\tilde{\boldsymbol{v}}, \tilde{\boldsymbol{\Sigma}}) \in Z \tag{2.15}$$

the corresponding flows generated by (2.11) for the two initial conditions (2.14), respectively. For the quasi-static problem, following result is implicitly contained in MOREAU [1976,1977]; see also NGUYEN [1977].

Proposition 2.1. Relative to the norm $|||\cdot|||$ defined by (2.2) the following contractivity property holds:

$$\boxed{|||\boldsymbol{\chi}(\cdot,t) - \tilde{\boldsymbol{\chi}}(\cdot,t)||| \leq |||\boldsymbol{\chi}_0 - \tilde{\boldsymbol{\chi}}_0|||, \quad \forall t \in \mathbf{I}.} \tag{2.16}$$

Proof: By hypothesis, the flows in (2.14) satisfy the variational inequality $(2.11)_2$. In particular,

$$\left.\begin{aligned} A(\dot{\boldsymbol{\Sigma}}, \boldsymbol{\Sigma} - \tilde{\boldsymbol{\Sigma}}) &\leq \ A(\dot{\boldsymbol{\Sigma}}^{\text{trial}}, \boldsymbol{\Sigma} - \tilde{\boldsymbol{\Sigma}}), \\ -A(\dot{\tilde{\boldsymbol{\Sigma}}}, \boldsymbol{\Sigma} - \tilde{\boldsymbol{\Sigma}}) &\leq -A(\dot{\tilde{\boldsymbol{\Sigma}}}^{\text{trial}}, \boldsymbol{\Sigma} - \tilde{\boldsymbol{\Sigma}}). \end{aligned}\right\} \tag{2.17}$$

Adding these inequalities and using bilinearity, along with definitions $(2.2a)_1$ and (2.4) yields

$$A(\dot{\boldsymbol{\Sigma}} - \dot{\tilde{\boldsymbol{\Sigma}}}, \boldsymbol{\Sigma} - \tilde{\boldsymbol{\Sigma}}) \leq A(\dot{\boldsymbol{\Sigma}}^{\text{trial}} - \dot{\tilde{\boldsymbol{\Sigma}}}^{\text{trial}}, \boldsymbol{\Sigma} - \tilde{\boldsymbol{\Sigma}}) = \langle \boldsymbol{\varepsilon}[\boldsymbol{v} - \tilde{\boldsymbol{v}}], \boldsymbol{\sigma} - \tilde{\boldsymbol{\sigma}} \rangle. \tag{2.18}$$

Now observe that, for fixed (but arbitrary) $t \in \mathbb{I}$, $\boldsymbol{v}(\cdot,t) - \tilde{\boldsymbol{v}}(\cdot,t) \in \mathcal{V}$. Using the linearity of $\boldsymbol{\varepsilon}[\cdot]$ and the fact that the flows (2.15) satisfy the variational inequality $(2.11\text{a})_1$ gives

$$\langle \boldsymbol{\sigma}, \boldsymbol{\varepsilon}[\boldsymbol{v} - \dot{\tilde{\boldsymbol{v}}}]\rangle - \langle \tilde{\boldsymbol{\sigma}}, \boldsymbol{\varepsilon}[\boldsymbol{v} - \tilde{\boldsymbol{v}}]\rangle = -\langle \rho_0 \dot{\boldsymbol{v}} - \dot{\tilde{\boldsymbol{v}}}, \boldsymbol{v} - \tilde{\boldsymbol{v}}\rangle = -\frac{d}{dt} K(\boldsymbol{v} - \tilde{\boldsymbol{v}}), \tag{2.19}$$

so that the right hand side of (2.18) equals minus the rate of change in the kinetic energy of the difference $\boldsymbol{v} - \tilde{\boldsymbol{v}}$. Combining (2.18) and (2.19), along with the definition for the natural norm, yields

$$\frac{d}{dt} K(\boldsymbol{v} - \tilde{\boldsymbol{v}}) + \langle\langle \dot{\boldsymbol{\Sigma}} - \dot{\tilde{\boldsymbol{\Sigma}}}, \boldsymbol{\Sigma} - \tilde{\boldsymbol{\Sigma}} \rangle\rangle = \tfrac{1}{2} \frac{d}{dt} |||\boldsymbol{\chi} - \tilde{\boldsymbol{\chi}}|||^2 \leq 0, \tag{2.20}$$

which implies $\frac{d}{dt} |||\boldsymbol{\chi} - \tilde{\boldsymbol{\chi}}||| \leq 0$. Therefore, for any $t \in \mathbb{I}$,

$$|||\boldsymbol{\chi} - \tilde{\boldsymbol{\chi}}||| - |||\boldsymbol{\chi}_0 - \tilde{\boldsymbol{\chi}}_0||| = \int_0^t \frac{d}{d\tau} |||\boldsymbol{\chi}(\cdot,\tau) - \tilde{\boldsymbol{\chi}}(\cdot,\tau)||| \, d\tau \leq 0, \tag{2.21}$$

which proves the result. ■

Remark 2.2. An identical contractivity result holds for classical viscoplasticity. The corresponding IBVP is obtained by replacing inequality $(2.11\text{a})_2$ with (2.8), and the flows $t \in \mathbb{I} \mapsto \boldsymbol{\Sigma} \in \mathcal{T}$ and $t \in \mathbb{I} \mapsto \tilde{\boldsymbol{\Sigma}} \in \mathcal{T}$ associated with the two initial conditions (2.14) satisfy

$$\left.\begin{aligned} A(\dot{\boldsymbol{\Sigma}}, \boldsymbol{\Sigma} - \tilde{\boldsymbol{\Sigma}}) &\leq \ A(\dot{\boldsymbol{\Sigma}}^{\text{trial}}, \boldsymbol{\Sigma} - \tilde{\boldsymbol{\Sigma}}) + \frac{1}{\eta} \int_\Omega [g(J(\tilde{\boldsymbol{\Sigma}})) - g(J(\boldsymbol{\Sigma}))] \, d\Omega, \\ -A(\dot{\tilde{\boldsymbol{\Sigma}}}, \boldsymbol{\Sigma} - \tilde{\boldsymbol{\Sigma}}) &\leq -A(\dot{\tilde{\boldsymbol{\Sigma}}}^{\text{trial}}, \boldsymbol{\Sigma} - \tilde{\boldsymbol{\Sigma}}) - \frac{1}{\eta} \int_\Omega [g(J(\tilde{\boldsymbol{\Sigma}})) - g(J(\boldsymbol{\Sigma}))] \, d\Omega. \end{aligned}\right\} \tag{2.22}$$

Adding these inequalities yields again (2.18) and the contractivity result for viscoplasticity follows exactly as in the proof of Proposition 3.1. ■

The contractivity property of the elastoplastic problem relative to the natural norm $|||\cdot|||$ motivates the algorithmic definition of nonlinear stability employed subsequently.

3. Time Discretization. The Algorithmic Problem

Let $\mathbb{I} = \cup_{n=0}^{N} [t_n, t_{n+1}]$ be an arbitrary partition of the time interval $\mathbb{I} \subset \mathbb{R}_+$ of interest. To develop an algorithmic approximation to the IBVP (2.11) it suffices to consider a typical subinterval $[t_n, t_{n+1}] \subset \mathbb{I}$, and assume that the initial conditions

$\boldsymbol{\Sigma}_n \in \mathcal{T}$, $\boldsymbol{u}_n \in \mathcal{S}_n$ and $\boldsymbol{v}_n \in \mathcal{V}$ are given. The incremental algorithmic problem then reduces to finding $\boldsymbol{\Sigma}_{n+1} \in \mathcal{T}$, $\boldsymbol{u}_{n+1} \in \mathcal{S}_{n+1}$ and $\boldsymbol{v}_{n+1} \in \mathcal{V}$ for prescribed forcing function $\boldsymbol{f}(\cdot, t)$ and prescribed boundary conditions $\bar{\boldsymbol{u}}(\cdot, t)$, $\bar{\boldsymbol{t}}(\cdot, t)$ in $[t_n, t_{n+1}]$.

The construction of the algorithmic counterpart to the IBVP (2.11), which gives the approximations $\boldsymbol{\Sigma}_{n+1}$ and $\boldsymbol{u}_{n+1}$ to the exact values $\boldsymbol{\Sigma}(\cdot, t_{n+1})$ and $\boldsymbol{u}(\cdot, t_{n+1})$ is based on a classical one-step method adapted to the present infinite dimensional problem.

3.1. Time Discretization of the Variational Inequality

To motivate the time discretization of the system (2.11) of variational inequalities consider first the standard problem finite dimensional of evolution

$$\dot{\boldsymbol{z}}(t) = \boldsymbol{g}((\boldsymbol{z}(t), t), \quad \text{subject to} \quad \boldsymbol{z}(t)|_{t=0} = \boldsymbol{z}_0, \tag{3.1}$$

where $\boldsymbol{s}(t) \in \mathsf{R}^N$ for $t \in \mathsf{I}$. A generalized mid-point rule approximation to this system within a typical time interval $[t_n, t_{n+1}] \subset \mathsf{I}$ is defined by the algorithmic formulae

$$\boldsymbol{z}_{n+\alpha} - \boldsymbol{z}_n = \alpha \Delta t \boldsymbol{g}(\boldsymbol{z}_{n+\alpha}, t_{n+\alpha}) \quad \text{and} \quad \boldsymbol{z}_{n+1} = \tfrac{1}{\alpha}\, \boldsymbol{z}_{n+\alpha} - [1 - \tfrac{1}{\alpha}\,]\boldsymbol{z}_n, \tag{3.2}$$

where $\boldsymbol{z}_n \in \mathsf{R}^n$ is prescribed at time t_n and $\alpha \in (0, 1]$ is an algorithmic parameter. The properties of this approximation are well-known. Consistency holds for any $\alpha \in (0, 1]$ and second order accuracy is obtained for $\alpha = \frac{1}{2}$. In the linear regime, the scheme is the highest order possible linear multi-step method possessing the property of A-stability (DAHLQUIST [1963]). In the nonlinear regime the scheme is B-stable (BUTCHER [1975]), a property not inherited by the closely related trapezoidal rule method; see WANNER [1976]. Our goal is to extend this scheme to the system (2.11).

3.1.1. *Algorithmic approximation to the variational inequality.* Our first step is to construct the algorithmic counterpart of the variational inequality $(2.11)_3$ arising from the principle of maximum dissipation, via a generalized mid-point rule approximation. Motivated by (3.2), for a typical time interval $[t_n, t_{n+1}]$ one defines the generalized mid-point displacement and the generalized mid-point stress by the expressions

$$\left.\begin{aligned} \boldsymbol{u}_{n+\alpha} &:= \alpha \boldsymbol{u}_{n+1} + (1-\alpha)\boldsymbol{u}_n, \\ \boldsymbol{\Sigma}_{n+\alpha} &:= \alpha \boldsymbol{\Sigma}_{n+1} + (1-\alpha)\boldsymbol{\Sigma}_n \end{aligned}\right\}, \quad \alpha \in (0, 1]. \tag{3.3}$$

Let $\Delta \boldsymbol{u} := \boldsymbol{u}_{n+1} - \boldsymbol{u}_n$ be the incremental displacement field. Consistent with the generalized mid-point rule algorithm (3.2) and in view of (2.4) we set

$$\dot{\boldsymbol{\Sigma}}^{\text{trial}}_{n+\alpha} \Delta t = (\mathbf{C}\boldsymbol{\varepsilon}[\Delta \boldsymbol{u}], \mathbf{0}), \quad \text{and} \quad \dot{\boldsymbol{\Sigma}}_{n+\alpha} \Delta t = \boldsymbol{\Sigma}_{n+1} - \boldsymbol{\Sigma}_n. \tag{3.4}$$

The algorithmic counterpart of variational inequality $(2.11\text{a})_3$ then becomes

$$\begin{aligned} A(\dot{\boldsymbol{\Sigma}}^{\text{trial}}_{n+\alpha} - \dot{\boldsymbol{\Sigma}}_{n+\alpha}, \boldsymbol{T} - \boldsymbol{\Sigma}_{n+\alpha}) &= \tfrac{1}{\alpha \Delta t} A((\alpha \mathbf{C}\boldsymbol{\varepsilon}[\Delta \boldsymbol{u}], \mathbf{0}) - \alpha[\boldsymbol{\Sigma}_{n+1} - \boldsymbol{\Sigma}_n], \boldsymbol{T} - \boldsymbol{\Sigma}_{n+\alpha}) \\ &= \tfrac{1}{\alpha \Delta t} A((\boldsymbol{\sigma}_n + \alpha \mathbf{C}\boldsymbol{\varepsilon}[\Delta \boldsymbol{u}], \boldsymbol{q}_n) - \boldsymbol{\Sigma}_{n+\alpha}, \boldsymbol{T} - \boldsymbol{\Sigma}_{n+\alpha}). \end{aligned} \tag{3.5}$$

Thus, if one defines $\boldsymbol{\Sigma}_{n+\alpha}^{\text{trial}}$ by the expression

$$\boldsymbol{\Sigma}_{n+\alpha}^{\text{trial}} := (\boldsymbol{\sigma}_n + \alpha \mathbf{C}\boldsymbol{\varepsilon}[\Delta \boldsymbol{u}], \boldsymbol{q}_n), \tag{3.6}$$

then inequality $(2.11\text{a})_2$ along with approximation (3.5) leads to the variational inequality

$$A(\boldsymbol{\Sigma}_{n+\alpha}^{\text{trial}} - \boldsymbol{\Sigma}_{n+\alpha}, \boldsymbol{T} - \boldsymbol{\Sigma}_{n+\alpha}) \leq 0, \qquad \forall \boldsymbol{T} \in \mathsf{E} \cap \mathcal{T}. \tag{3.7}$$

An identical argument for the viscoplastic problem yields

$$A(\boldsymbol{\Sigma}_{n+\alpha}^{\text{trial}} - \boldsymbol{\Sigma}_{n+\alpha}, \boldsymbol{T} - \boldsymbol{\Sigma}_{n+\alpha}) \leq \frac{\alpha \Delta t}{\eta} \int_{\Omega} [g(J(\boldsymbol{T})) - g(J(\boldsymbol{\Sigma}_{n+\alpha}))] \, d\Omega, \qquad \forall \boldsymbol{T} \in \mathcal{T}. \tag{3.8}$$

which is the algorithmic counterpart of variational inequality (2.8).

3.2. The incremental (algorithmic) IBVP problem

As pointed out above, for the continuum viscoplastic problem the generalized stresses $\boldsymbol{\Sigma} \in \mathcal{T}$ need not lie within the elastic domain E; a fact also reflected in the algorithmic inequality (3.8). On the other hand, inequality (3.7) also shows that in the algorithmic version of rate–independent plasticity, with the exception of the initial condition $\boldsymbol{\Sigma}_0$, in general $\boldsymbol{\Sigma}_n$ and $\boldsymbol{\Sigma}_{n+1}$ need not be in the elastic domain. Only the algorithmic approximation $\boldsymbol{\Sigma}_{n+\alpha}$ at the generalized mid-time step is in $\mathsf{E} \cap \mathcal{T}$. This suggests the enforcement of the algorithmic counterpart of $(2.11\text{a})_1$ also at $t_{n+\alpha}$, and leads to the following algorithmic problem which is consistent with the generalized mid-point discretization (3.2):

Problem $W_{\Delta t}$: Find $\Delta \boldsymbol{u}$ and $\boldsymbol{\chi}_{n+\alpha} = (\boldsymbol{v}_{n+\alpha}, \boldsymbol{\Sigma}_{n+\alpha})$ such that:

$$\left.\begin{aligned} \tfrac{1}{\Delta t} \langle \rho_0 \Delta \boldsymbol{u}, \boldsymbol{\eta} \rangle - \langle \boldsymbol{v}_{n+\alpha}, \boldsymbol{\eta} \rangle &= 0, \\ \tfrac{1}{\alpha \Delta t} \langle \rho_0 (\boldsymbol{v}_{n+\alpha} - \boldsymbol{v}_n), \boldsymbol{\eta} \rangle + \langle \boldsymbol{\sigma}_{n+\alpha}, \boldsymbol{\varepsilon}[\boldsymbol{\eta}] \rangle &= \langle \boldsymbol{f}_{n+\alpha}, \boldsymbol{\eta} \rangle + \langle \bar{\boldsymbol{t}}_{n+\alpha}, \boldsymbol{\eta} \rangle_{\Gamma}, \\ A(\boldsymbol{\Sigma}_{n+\alpha}^{\text{trial}} - \boldsymbol{\Sigma}_{n+\alpha}, \boldsymbol{T} - \boldsymbol{\Sigma}_{n+\alpha}) &\leq 0, \end{aligned}\right\} \tag{3.9}$$

for all $(\boldsymbol{\eta}, \boldsymbol{T}) \in Z$. The variational equations (3.9) determine the velocity and generalized stress $\boldsymbol{\chi}_{n+\alpha}(\boldsymbol{v}_{n+\alpha}, \boldsymbol{\Sigma}_{n+\alpha}) \in Z$, along with the displacement increment $\Delta \boldsymbol{u} \in \mathcal{V}$. The initial condition $\boldsymbol{\chi}_n$ and the displacement field $\boldsymbol{u}_n$ are then updated by the formulae

$$\left.\begin{aligned} \boldsymbol{v}_{n+1} &= \tfrac{1}{\alpha} \boldsymbol{v}_{n+\alpha} + [1 - \tfrac{1}{\alpha}] \boldsymbol{v}_n, \\ \boldsymbol{\Sigma}_{n+1} &= \tfrac{1}{\alpha} \boldsymbol{\Sigma}_{n+\alpha} + [1 - \tfrac{1}{\alpha}] \boldsymbol{\Sigma}_n, \\ \boldsymbol{u}_{n+1} &= \boldsymbol{u}_n + \Delta \boldsymbol{u}. \end{aligned}\right\} \tag{3.10}$$

For viscoplasticity, equation $(3.9)_2$ is replaced by (3.8). For the quasi-static problem the preceding algorithm reduces to that proposed in SIMO & GOVINDJEE [1989]. The dynamic extension summarized above accounts for the inertia terms via a standard formulation of the generalized mid-point rule.

Remarks 3.1.

1. In general, the stress $\boldsymbol{\sigma}_{n+1}$ does not satisfy the equilibrium equation at t_{n+1} unless the forcing terms are linear in time. Hence, as in the nonlinear heat conduction, the equilibrium equation $(2.11)_1$ is enforced at the generalized mid–step $t_{n+\alpha}$.

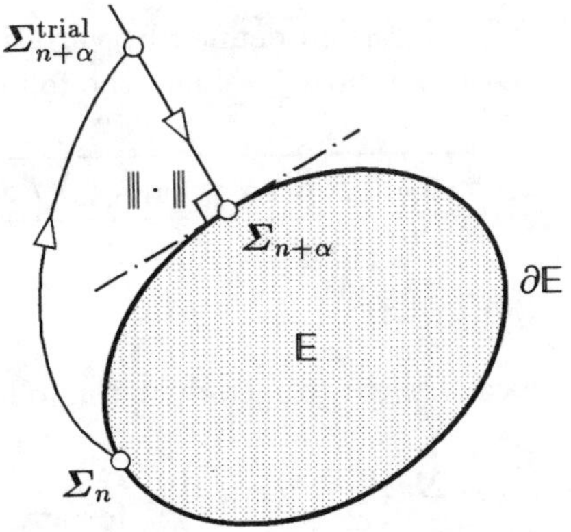

FIGURE 3.1. Geometric interpretation of the variational inequality $A\big(\boldsymbol{\Sigma}^{\text{trial}}_{n+\alpha} - \boldsymbol{\Sigma}_{n+\alpha}, \boldsymbol{T} - \boldsymbol{\Sigma}_{n+\alpha}\big) \leq 0$.

2. The geometric interpretation of variational inequality $(3.9)_2$ is illustrated in Figure 3.1. Using standard results in convex analysis one concludes that the point $\boldsymbol{\Sigma}_{n+\alpha} \in \mathsf{E} \cap \mathcal{T}$ is the closest point projection in the natural norm $|||\cdot|||$ of the trial elastic state $\boldsymbol{\Sigma}^{\text{trial}}_{n+\alpha}$ onto $\mathsf{E} \cap \mathcal{T}$. In the context of an isoparametric finite element approximation with numerical quadrature the stresses need to be evaluated only at the quadrature points. Consequently, the infinite dimensional constrained optimization problem defined by inequality $(3.9)_2$ reduces to a collection of *finite dimensional* constrained optimization problems at quadrature points. This is the crucial observation exploited in actual numerical implementations; see SIMO, KENNEDY & GOVINDJEE [1988], SIMO & HUGHES [1988] and references therein.

3. For the quasi-static problem, the value $\alpha = \frac{1}{2}$ results in second order accuracy and leads to an algorithm that exactly preserves the second law; see SIMO & GOVINDJEE [1989] for further details and numerical simulations. ■

The main objective of the discussion that follows is the study of the nonlinear stability properties of problem $W_{\Delta t}$. It is shown below that the algorithmic solutions generated by $W_{\Delta t}$ are nonlinearly stable for $\alpha \geq \frac{1}{2}$.

3.3. Nonlinear Stability Analysis

Let $\chi_n \in Z$ and $\widetilde{\chi}_n \in Z$ be two initial conditions at time t_n. The preceding algorithm then generates two sequences $\{\chi_n\}_{n\in\mathbb{N}}$ and $\{\widetilde{\chi}_n\}_{n\in\mathbb{N}}$. Nonlinear stability holds if the algorithm inherits the contractivity property of the continuum problem relative to the natural norm $|||\cdot|||$; a property also known as A-contractivity or B-stability. Equivalently, finite perturbations in the initial data are attenuated by the algorithm relative to the natural norm if the B-stability property holds.

Proposition 3.1. The algorithm defined by variational problem $W_{\Delta t}$ and the update formula (3.10) is B-stable for $\alpha \geq \frac{1}{2}$; i.e., the following inequality holds

$$\boxed{|||\chi_n - \widetilde{\chi}_n||| \leq |||\chi_0 - \widetilde{\chi}_0|||, \quad \forall n \in \mathbb{N},} \tag{3.11}$$

provided that $\alpha \geq \frac{1}{2}$.

Proof: It will prove convenient to introduce the following notation

$$\left.\begin{aligned} \Delta\boldsymbol{\Sigma}_{n+\alpha} &:= \boldsymbol{\Sigma}_{n+\alpha} - \widetilde{\boldsymbol{\Sigma}}_{n+\alpha}, \\ \Delta\boldsymbol{v}_{n+\alpha} &:= \boldsymbol{v}_{n+\alpha} - \widetilde{\boldsymbol{v}}_{n+\alpha}, \end{aligned}\right\} \quad \text{for any } \alpha \in [0,1]. \tag{3.12}$$

By hypothesis, the two sequences generated by the algorithm satisfy the variational inequality (3.9). Choosing successively $\boldsymbol{T} = \boldsymbol{\Sigma}_{n+\alpha}$ and $\boldsymbol{T} = \widetilde{\boldsymbol{\Sigma}}_{n+\alpha}$, using the definition of $\boldsymbol{\Sigma}^{\text{trial}}_{n+\alpha}$ and the notation introduced above yields

$$\left.\begin{aligned} a(\alpha\mathbf{C}\boldsymbol{\varepsilon}[\Delta\widetilde{\boldsymbol{u}}], \boldsymbol{\sigma}_{n+\alpha} - \widetilde{\boldsymbol{\sigma}}_{n+\alpha}) + A(\widetilde{\boldsymbol{\Sigma}}_n - \widetilde{\boldsymbol{\Sigma}}_{n+\alpha}, \Delta\boldsymbol{\Sigma}_{n+\alpha}) \leq 0, \\ -a(\alpha\mathbf{C}\boldsymbol{\varepsilon}[\Delta\boldsymbol{u}], \boldsymbol{\sigma}_{n+\alpha} - \widetilde{\boldsymbol{\sigma}}_{n+\alpha}) - A(\boldsymbol{\Sigma}_n - \boldsymbol{\Sigma}_{n+\alpha}, \Delta\boldsymbol{\Sigma}_{n+\alpha}) \leq 0. \end{aligned}\right\} \tag{3.13}$$

Adding these two inequalities and using the identity $\Delta\boldsymbol{\Sigma}_{n+\alpha} - \Delta\boldsymbol{\Sigma}_n = \alpha[\Delta\boldsymbol{\Sigma}_{n+1} - \Delta\boldsymbol{\Sigma}_n]$ yields the result

$$\begin{aligned} A(\Delta\boldsymbol{\Sigma}_{n+1} - \Delta\boldsymbol{\Sigma}_n, \Delta\boldsymbol{\Sigma}_{n+\alpha}) &= \tfrac{1}{\alpha}\, A(\Delta\boldsymbol{\Sigma}_{n+\alpha} - \Delta\boldsymbol{\Sigma}_n, \Delta\boldsymbol{\Sigma}_{n+\alpha}) \\ &\leq a(\mathbf{C}\boldsymbol{\varepsilon}[\Delta\boldsymbol{u} - \Delta\widetilde{\boldsymbol{u}}], \boldsymbol{\sigma}_{n+\alpha} - \widetilde{\boldsymbol{\sigma}}_{n+\alpha}). \end{aligned} \tag{3.14}$$

Since the difference $\Delta\boldsymbol{u} - \Delta\widetilde{\boldsymbol{u}}$ is in $\mathcal{V}$, use of definition $(2.2\text{a})_1$ along with the algorithmic weak form of the momentum equations yields

$$\begin{aligned} &a(\mathbf{C}\boldsymbol{\varepsilon}[\Delta\boldsymbol{u} - \Delta\widetilde{\boldsymbol{u}}], \boldsymbol{\sigma}_{n+\alpha} - \widetilde{\boldsymbol{\sigma}}_{n+\alpha}) \\ &\qquad = \langle \boldsymbol{\sigma}_{n+\alpha}, \boldsymbol{\varepsilon}[\Delta\boldsymbol{u} - \Delta\widetilde{\boldsymbol{u}}]\rangle - \langle \widetilde{\boldsymbol{\sigma}}_{n+\alpha}, \boldsymbol{\varepsilon}[\Delta\boldsymbol{u} - \Delta\widetilde{\boldsymbol{u}}]\rangle \\ &\qquad = -\tfrac{1}{\alpha\Delta t}\langle \rho_0(\Delta\boldsymbol{v}_{n+\alpha} - \Delta\boldsymbol{v}_n), \Delta\boldsymbol{u} - \Delta\widetilde{\boldsymbol{u}}\rangle \\ &\qquad = -\tfrac{1}{\Delta t}\langle \rho_0(\Delta\boldsymbol{v}_{n+1} - \Delta\boldsymbol{v}_n), \Delta\boldsymbol{u} - \Delta\widetilde{\boldsymbol{u}}\rangle, \end{aligned} \tag{3.15}$$

where we have used the identity $\alpha[\Delta \boldsymbol{v}_{n+\alpha} - \Delta \boldsymbol{v}_n] = \Delta \boldsymbol{v}_{n+1} - \Delta \boldsymbol{v}_n$. On the other hand, the variational equation $(3.9)_1$ along with a simple algebraic manipulation yield the local relation

$$\Delta \boldsymbol{u} - \Delta \tilde{\boldsymbol{u}} = \Delta t\, \Delta \boldsymbol{v}_{n+\alpha} = \Delta \boldsymbol{v}_{n+\frac{1}{2}} + (\alpha - \tfrac{1}{2})\, [\Delta \boldsymbol{v}_{n+1} - \Delta \boldsymbol{v}_n]. \tag{3.16}$$

Inserting this expression into (3.16), and using the identity

$$K(\Delta \boldsymbol{v}_{n+1}) - K(\Delta \boldsymbol{v}_n) = \langle \rho_0 \Delta \boldsymbol{v}_{n+\frac{1}{2}}, \Delta \boldsymbol{v}_{n+1} - \Delta \boldsymbol{v}_n \rangle \tag{3.17}$$

gives, after a straightforward manipulation the result

$$\begin{aligned} a(\mathbf{C}\boldsymbol{\varepsilon}[\Delta \boldsymbol{u} - \Delta \tilde{\boldsymbol{u}}], \boldsymbol{\sigma}_{n+\alpha} - \tilde{\boldsymbol{\sigma}}_{n+\alpha}) \\ = -[K(\Delta \boldsymbol{v}_{n+1}) - K(\Delta \boldsymbol{v}_n)] - (2\alpha - 1)[K(\Delta \boldsymbol{v}_{n+1} - \Delta \boldsymbol{v}_n)]. \end{aligned} \tag{3.18}$$

Combining (3.14) and (3.18) one obtains the following estimate

$$\begin{aligned} K(\Delta \boldsymbol{v}_{n+1}) - K(\Delta \boldsymbol{v}_n) + A(\Delta \boldsymbol{\Sigma}_{n+1} - \Delta \boldsymbol{\Sigma}_n, \Delta \boldsymbol{\Sigma}_{n+\alpha}) \\ \leq -(\alpha - \tfrac{1}{2})\, 2K(\Delta \boldsymbol{v}_{n+1} - \Delta \boldsymbol{v}_n). \end{aligned} \tag{3.19}$$

The proof is now concluded by exploiting once more the identity

$$\Delta \boldsymbol{\Sigma}_{n+\alpha} = \Delta \boldsymbol{\Sigma}_{n+\frac{1}{2}} + (\alpha - \tfrac{1}{2})[\Delta \boldsymbol{\Sigma}_{n+1} - \Delta \boldsymbol{\Sigma}_n]. \tag{3.20}$$

Combining (3.19) and (3.20) and using the bilinearity property of the bilinear form $A(\cdot,\cdot)$ along with the definition (2.2) of the natural norm $|||\cdot|||$ gives the estimate

$$\begin{aligned} |||\Delta \boldsymbol{\Sigma}_{n+1}|||^2 - |||\Delta \boldsymbol{\Sigma}_n|||^2 &\leq -(\alpha - \tfrac{1}{2})\, \left[|||\Delta \boldsymbol{\Sigma}_{n+1} - \Delta \boldsymbol{\Sigma}_n|||^2 + 2[K(\Delta \boldsymbol{v}_{n+1} - \Delta \boldsymbol{v}_n)]\right] \\ &= -(\alpha - \tfrac{1}{2})|||\boldsymbol{\chi}_{n+1} - \boldsymbol{\chi}_n|||^2 \leq 0 \iff \alpha \geq \tfrac{1}{2}, \end{aligned} \tag{3.21}$$

which in conjunction with a straightforward induction argument implies (3.11) and completes the proof of the contractivity result. ■

An argument entirely analogous to that given above also shows that the incremental algorithmic problem for viscoplasticity is A–contractive relative to the natural norm $|||\cdot|||$.

In summary, the analysis for dynamic plasticity and viscoplasticity proves described above proves nonlinear stability in the velocity and the stress field $\boldsymbol{\chi} = (\boldsymbol{v}, \boldsymbol{\Sigma},$ provided that $\alpha \geq \frac{1}{2}$, thus settling once and for all the question of nonlinear stability of widely used algorithms for infinitesimal plasticity. The accuracy of this class of return mapping algorithms is examined in SIMO & GOVINDJEE [1990]. As expected, the mid–point rule ($\alpha = \frac{1}{2}$) is second order accurate.

II. EXPONENTIAL ALGORITHMS FOR FINITE STRAIN PLASTICITY

In this Part II of the lectures we describe a possible extension of the classical models of infinitesimal plasticity characterized by the property that the structure return mapping algorithms is preserved exactly. To make matters as explicit as possible, suppose one is given:

i. A convex elastic domain $\mathbb{E}$ specified in stress space in terms of *true* stresses and internal variables defined on the current configuration of the body via a classical single-surface or multi-surface yield criterion. Typical examples include the Mises or Tresca yield conditions of metal plasticity with isotropic/kinematic hardening, and the Cambridge Cam-Clay models of soil mechanics.

ii. A poly-convex stored energy function relative to a local stress-free (unloaded) configuration which characterizes the elastic response of the material. For example the Ogden models; see OGDEN [1984].

iii. A finite element method suitable quasi-incompressible finite elasticity and a local convex optimization algorithm which defines the closest-point projection and the algorithmic moduli for the specific elastic domain $\mathbb{E}$.

The objective, then, is to construct a model of plasticity at finite strains and develop a numerical implementation possessing the following features:

1. The evolution equations of the continuum model defining plastic flow obey the classical principle of maximum dissipation leading to *symmetric* tangent elastoplastic moduli.

2. The local closest-point-projection algorithm of the infinitesimal theory remains unchanged. In particular, the algorithmic elastoplastic moduli are symmetric with the same form as in the infinitesimal theory.

3. For purely elastic response the formulation collapses to a classical model of finite strain hyperelasticity, with stored energy function possessing the correct asymptotic behavior for extreme strains.

4. For pressure insensitive yield criteria the algorithm inherits exactly the property of exact conservation of the plastic volume. In the general case, the algorithm ensures that the jacobian of the deformation remains positive for *any* step size.

5. For dynamic plasticity, the overall time-stepping algorithm inherits exactly the conservation laws of total linear and total angular momentum (i.e., for equilibrated external loads and for pure Neumann boundary data).

Below is is shown that all the preceding design conditions can be accomplished by considering a formulation of finite strain elastoplasticity based on a multiplicative decomposition of the deformation gradient, as suggested by LEE [1969], MANDEL

[1974] and others. Before outlining the main results, two observations will be made. First, by frame invariance, the specification of the elastic domain in terms of true stresses restricts the functional form the yield criterion to isotropic functions; i.e, functions of the joint invariants of the stress and the internal variables. Hence, the elastic response is also assumed to be isotropic. Second, even within the context of infinitesimal strains, conditions **i** and **ii** above lead to a formulation of the return mapping in strain space. The classical return mapping algorithms in stress space are recovered when the additional assumption of linear elastic response is introduced.

An overview of the topics covered is as follows. First, the local evolution equations of a model of elastoplasticity are described in which the associative flow rule arises as an optimality condition from the principle of maximum dissipation. Next, an algorithmic approximation to the flow rule is constructed leading to a hyperelastic trial step followed by a return mapping algorithm constructed via an exponential approximation to the flow rule at fixed (updated) configuration. Aspects related to the actual implementation of this approach are addressed subsequently. In particular, by introducing a spectral decomposition of the Kirchhoff stress and the left Cauchy-Green tensors one obtains a return mapping algorithm in principal logarithmic strains, with functional form identical to those of the infinitesimal theory, but now valid for arbitrary nonlinear elastic response. As an application, we consider the implementation of a classical model of J_2-flow theory incorporating kinematic hardening and viscoplastic response. Finally, we examine the extension of the preceding results to dynamic plasticity and presents a class of integration algorithms that inherit exactly the conservation law of total angular momentum. The actual accuracy, stability and convergence properties of the proposed algorithms are finally assessed in a number of representative large-scale numerical simulations.

1. Multiplicative Elastoplasticity at Finite Strains

We give a concise outline of a formulation of plasticity at finite strains in which the elastic domain E is specified by any classical model of infinitesimal plasticity, with the stress tensor now interpreted as the *true* (Kirchhoff) stress tensor acting on the current configuration of the body. As in the infinitesimal theory, in the present formulation the associative form of the evolution equations is uniquely defined via the principle of maximum plastic work (see HILL [1950]). For simplicity, attention is restricted to the purely mechanical theory, with no restrictions placed on the elastic domain, following the presentation given in SIMO & MESCHKE [1992]. A complete thermodynamical theory is found in SIMO & MIEHE [1991].

1.1. Motivation: The Infinitesimal Theory

To motivate our subsequent developments recall first the key assumptions underlying a classical model of infinitesimal elastoplasticity, as described in the preceding part of these lectures.

i. *Elastic domain.* A *convex* elastic domain $\mathbb{E}$ defined by a m–convex yield functions defined in stress space and intersecting (possibly) in a non–smooth fashion; i.e.,

$$\mathbb{E} = \{(\boldsymbol{\tau}, \boldsymbol{q}) \in \mathbb{S} \times \mathbb{R}^{m_{\text{int}}} : \hat{\phi}^{\mu}(\boldsymbol{\tau}, q) \leq 0, \quad \text{for} \quad \mu = 1, 2, \cdots, m\}. \tag{1.1}$$

where $\mathbb{S}$ is the vector space of symmetric rank-two tensors and $\boldsymbol{q}$ is a suitable vector of $m_{\text{int}} \geq 1$ (stress-like) internal variables characterizing the hardening response of the material.

ii. *Additive decomposition and free energy.* The total infinitesimal strain tensor is decomposed additively into elastic and plastic parts as

$$\boldsymbol{\varepsilon}(X,t) = \boldsymbol{\varepsilon}^{e}(X,t) + \boldsymbol{\varepsilon}^{p}(X,t) \tag{1.2}$$

. The elastic response of the material is assumed to depend only on the elastic part $\boldsymbol{\varepsilon}^e$ and is characterized by means of a free energy function $\psi(\boldsymbol{\varepsilon}^e, \boldsymbol{\xi})$. As in the preceding lectures, $\boldsymbol{\xi}$ is the vector of (strain-like) internal variables conjugate to $\boldsymbol{q}$; i.e., $\boldsymbol{q} := -\partial_{\boldsymbol{\xi}} \psi$.

iii. *Associative evolution of plastic flow.* Restricted to the purely mechanical theory the dissipation function $\mathcal{D}$ is the difference between the stress power and the rate of change of the internal energy. The second law then implies

$$\mathcal{D} = \boldsymbol{\tau} \cdot \dot{\boldsymbol{\varepsilon}} - \dot{\psi}(\boldsymbol{\varepsilon}^e, \boldsymbol{\xi}) > 0. \tag{1.3}$$

Clearly $\mathcal{D} \equiv 0$ for an elastic material. As pointed out in Part I, the simplest model that describes the evolution of the internal variables $\{\epsilon^p, \boldsymbol{\xi}\}$ corresponds to the assumption of *maximum dissipation.* There, it was shown that for the convex elastic domain $\mathbb{E}$ defined by (1.1), the local form of the optimality conditions associated with this postulate yield

$$\begin{gathered} \dot{\boldsymbol{\varepsilon}}^e = \dot{\boldsymbol{\varepsilon}} - \textstyle\sum_{\mu=1}^{m} \gamma_\mu \partial_{\boldsymbol{\tau}} \phi^\mu(\boldsymbol{\tau}, \boldsymbol{q}), \quad \dot{\boldsymbol{\xi}} = \textstyle\sum_{\mu=1}^{m} \gamma_\mu \partial_{\boldsymbol{q}} \phi^\mu(\boldsymbol{\tau}, \boldsymbol{q}), \\ \gamma_\mu \geq 0, \quad \phi^\mu(\boldsymbol{\tau}, \boldsymbol{q}) \leq 0, \quad \textstyle\sum_{\mu=1}^{m} \gamma_\mu \phi^\mu(\boldsymbol{\tau}, \boldsymbol{q}) = 0, \end{gathered} \tag{1.4}$$

which furnish the standard local form of the evolution equations for associative plasticity.

1.2. The Nonlinear Theory. Basic Assumptions

In our formulation of classical models of infinitesimal elastoplasticity in finite strain regime we retain the form of elastic domain $\mathbb{E}$ in the infinitesimal theory with

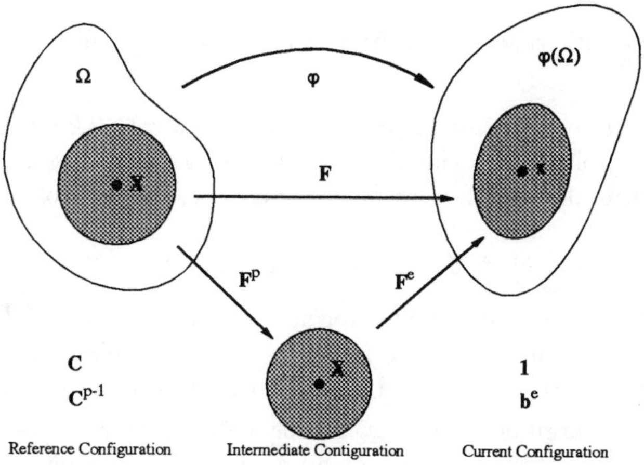

FIGURE 1.1. Illustration of the local multiplicative decomposition of the deformation gradient, and the basic strain tensors.

$\boldsymbol{\tau}$ now interpreted as the Kirchhoff stress tensor in the current configuration. This key design condition defines the structure of the nonlinear theory as follows.

i. *Elastic Domain.* For simplicity, attention is restricted in this section to a single scalar hardening variable; thus

$$\mathsf{E} := \{ (\boldsymbol{\tau}, q) \in \mathsf{S} \times \mathsf{R} : \quad \phi^{\mu}(\boldsymbol{\tau}, q) \leq 0 \quad \text{for} \quad \mu = 1, 2, \cdots, m \}. \tag{1.5}$$

We remark that the principle of objectivity restricts the possible forms of ϕ in (1.5) to isotropic functions. Consequently, a classical formulation of the yield criterion in terms of true stresses necessarily imposes the restriction to isotropy of the resulting theory.

ii. *Multiplicative decomposition and free energy.* The generalization to finite strains of the additive decomposition used in the infinitesimal theory is motivated by the structure of the single crystal models of metal plasticity, and takes the form of the local multiplicative decomposition

$$\boldsymbol{F}(X,t) = \boldsymbol{F}^{e}(X,t)\boldsymbol{F}^{p}(X,t) \tag{1.6}$$

where $\boldsymbol{F}(X,t) := D\boldsymbol{\varphi}(X,t)$ denotes the deformation gradient of the motion $\boldsymbol{\varphi}(\cdot,t)$ at $X \in \Omega$; From a phenomenological point of view $\boldsymbol{F}^{p}$ is an internal variable related to the amount of plastic flow while $\boldsymbol{F}^{e-1}$ defines the local, stress-free, unloaded configuration. Consistent with the restriction to isotropy it is further assumed that ψ is independent of the orientation of the intermediate configuration. Frame invariance then implies the functional form

$$\psi = \psi(\boldsymbol{b}^{e}, \xi), \quad \text{where} \quad \boldsymbol{b}^{e} := \boldsymbol{F}^{e}\boldsymbol{F}^{e\,T}. \tag{1.7}$$

The measure of elastic deformation $\boldsymbol{b}^e$ defined above is the elastic left Cauchy-Green tensor.

iii. *Associative evolution of plastic flow.* The local dissipation function $\mathcal{D}$ per unit of reference volume associated with a material point $X \in \mathcal{B}$ is defined as the difference between the local stress power and the local rate of change of the free energy; i.e,

$$\mathcal{D} := \boldsymbol{\tau} \cdot \boldsymbol{d} - \dot{\psi}(\boldsymbol{b}^e, \xi) \geq 0. \tag{1.8}$$

Here $\boldsymbol{d} = \text{sym}[\boldsymbol{l}]$ denotes the rate of deformation tensor and $\boldsymbol{l} := \dot{\boldsymbol{F}}\boldsymbol{F}^{-1}$ is the spatial velocity gradient. As in the infinitesimal theory, the simplest model describing the evolution of plastic flow is obtained by postulating *maximum dissipation* of the flow.

Following an argument entirely analogous to that used for the infinitesimal theory, we show below that the preceding assumptions uniquely define a model of associative plasticity at finite strains.

1.3. Constitutive Equations. Dissipation Inequality

Our first objective is to obtain an explicit expression for the dissipation. The rate of change $\dot{\psi}$ is computed by exploiting the relation

$$\boldsymbol{b}^e = \boldsymbol{F}\boldsymbol{C}^{p-1}\boldsymbol{F}^T \quad \text{with} \quad \boldsymbol{C}^p := \boldsymbol{F}^{pT}\boldsymbol{F}^p, \tag{1.9}$$

which follows from the multiplicative decomposition (1.6), and identifies $\boldsymbol{C}^p$ as the plastic right Cauchy-Green tensor. Time differentiation of this expression yields the identity

$$\dot{\boldsymbol{b}}^e = \boldsymbol{l}\boldsymbol{b}^e + \boldsymbol{b}^e\boldsymbol{l}^T + \boldsymbol{\pounds}_v\boldsymbol{b}^e \quad \text{with} \quad \boldsymbol{\pounds}_v\boldsymbol{b}^e := \boldsymbol{F}\frac{\partial}{\partial t}[\boldsymbol{C}^{p-1}]\boldsymbol{F}^T, \tag{1.10}$$

where $\boldsymbol{\pounds}_v\boldsymbol{b}^e$ is referred to as the *Lie derivative* of $\boldsymbol{b}^e$. Time differentiation of the free energy function and use of the relation $(1.10)_1$ yields the result

$$\frac{d\psi}{dt} = \partial_{\boldsymbol{b}^e}\psi \cdot \dot{\boldsymbol{b}}^e + \partial_\xi\psi\,\dot{\xi} = [\,2\partial_{\boldsymbol{b}^e}\psi\,\boldsymbol{b}^e\,] \cdot [\,\boldsymbol{l} + \tfrac{1}{2}(\boldsymbol{\pounds}_v\boldsymbol{b}^e)\,\boldsymbol{b}^{e-1}\,] + \partial_\xi\psi\,\dot{\xi}. \tag{1.11}$$

Since $\boldsymbol{b}^e$ commutes with $\partial_{\boldsymbol{b}^e}\psi$ as a result of the restriction to isotropy, combining (1.11) and (1.8) yields the following local form of the dissipation inequality:

$$\mathcal{D} = [\,\boldsymbol{\tau} - 2\partial_{\boldsymbol{b}^e}\psi\,\boldsymbol{b}^e\,] \cdot \boldsymbol{d} + [\,2\partial_{\boldsymbol{b}^e}\psi\,\boldsymbol{b}^e\,] \cdot [\,-\tfrac{1}{2}(\boldsymbol{\pounds}_v\boldsymbol{b}^e)\,\boldsymbol{b}^{e-1}\,] + [\,-\partial_\xi\psi\,]\,\dot{\xi} \geq 0. \tag{1.12}$$

Since (1.12) must hold for all admissible processes, the same standard argument used in the infinitesimal theory now gives the following constitutive equations and the reduced form of the dissipation inequality

$$\boldsymbol{\tau} = 2\,\partial_{\boldsymbol{b}^e}\psi\,\boldsymbol{b}^e, \quad \mathcal{D} = \boldsymbol{\tau} \cdot [\,-\tfrac{1}{2}\,(\boldsymbol{\pounds}_v\boldsymbol{b}^e)\,\boldsymbol{b}^{e-1}\,] + q\,\dot{\xi} \geq 0, \tag{1.13}$$

with $q := -\partial_\xi\psi$.

1.3.1. *Associative Evolution Equations. Maximum Dissipation.* According to the postulate of *maximum dissipation* the actual state $(\boldsymbol{\tau}, q) \in \mathsf{E}$ in the plastically deformed body at a given (fixed) configuration, with prescribed intermediate configuration and prescribed rates $\{\pounds_v \boldsymbol{b}^e, \dot{\xi}\}$, renders a maximum of the dissipation function $\mathcal{D}$. Equivalently, the following inequality holds:

$$[\boldsymbol{\tau} - \overset{*}{\boldsymbol{\tau}}] \cdot [-\tfrac{1}{2}(\pounds_v \boldsymbol{b}^e)\boldsymbol{b}^{e-1}] + [q - \overset{*}{q}]\,\dot{\xi} \geq 0, \quad (\overset{*}{\boldsymbol{\tau}}, \overset{*}{q}) \in \mathsf{E}. \tag{1.14}$$

This result is the counterpart of reduced dissipation inequality in the finite strain regime. As in the infinitesimal theory, inequality (1.14) holds if and only if the coefficients $\{-\frac{1}{2}(\pounds_v \boldsymbol{b}^e)\boldsymbol{b}^{e-1}, \dot{\xi}\}$ lie in the cone normal to $\partial\mathsf{E}$ at the point $(\boldsymbol{\tau}, q)$. In particular, if $\partial\mathsf{E}$ is defined by (1.5), the evolution equations read

$$\boxed{\begin{aligned} \left.\begin{aligned} -\tfrac{1}{2}\,\pounds_v \boldsymbol{b}^e &= [\,\textstyle\sum_{\mu=1}^{m} \gamma_\mu \partial_{\boldsymbol{\tau}} \phi^\mu(\boldsymbol{\tau}, q)\,]\,\boldsymbol{b}^e, \\ \dot{\xi} &= [\,\textstyle\sum_{\mu=1}^{m} \gamma_\mu \partial_q \phi^\mu(\boldsymbol{\tau}, q)\,], \end{aligned}\right\} \\ \gamma_\mu \geq 0, \quad \phi^\mu(\boldsymbol{\tau}, q) \leq 0 \quad \text{and} \quad \textstyle\sum_{\mu=1}^{m} \gamma_\mu \phi^\mu(\boldsymbol{\tau}, q) = 0. \end{aligned}} \tag{1.15}$$

Despite its unconventional appearance [compare with the corresponding expression $(1.4)_2$ in the infinitesimal theory] the flow rule (1.15) possesses a number of interesting properties. In particular, it gives the correct evolution of plastic volume changes as the following observations show.

i. The total and elastic volume changes are given by $J := \det[\boldsymbol{F}] > 0$ and $J^e := (\det[\boldsymbol{b}^e])^{\frac{1}{2}} > 0$, respectively.

ii. Let $J^p := \det[\boldsymbol{F}^p]$. Setting $\varepsilon_v^p := \log[J^p]$ the rate of plastic volume change predicted by the flow rule $(1.15)_1$ is given by the evolution equation

$$\dot{\varepsilon}_v^p = \textstyle\sum_{\mu=1}^{m} \gamma_\mu \mathrm{tr}\,[\,\partial_{\boldsymbol{\tau}} \phi^\mu\,], \tag{1.16}$$

which implies exact conservation of plastic volume for pressure insensitive yield conditions; i.e. if $\mathrm{tr}\,[\,\partial_{\boldsymbol{\tau}} \phi^\mu\,] = 0$.

The proof of the result in (1.16) follows easily from (1.10) along with $(1.15)_1$. The foregoing theory will be applied to a number of classical models in soil mechanics in the following sections.

1.3.2. *General Non-Associative Evolution Equations.* Expression (1.15) provides a crucial insight in the extension to finite strains of non–associative models of infinitesimal plasticity. For instance, for single–surface plasticity (i.e., $\mu = 1$) the non–associative flow rule takes the form

$$\dot{\boldsymbol{\varepsilon}}^e = \dot{\boldsymbol{\varepsilon}} - \gamma \partial_{\boldsymbol{\tau}} g(\boldsymbol{\tau}, q), \quad \dot{\xi} = \gamma \partial_q g(\boldsymbol{\tau}, q), \tag{1.17}$$

where $g: \mathsf{S} \times \mathsf{R} \to \mathsf{R}$ is the plastic flow potential; in general $g \neq \phi$. The extension of a model of this type to the finite strain regime then takes the form

$$-\tfrac{1}{2}\,\pounds_v \boldsymbol{b}^e = [\,\gamma \partial_{\boldsymbol{\tau}} g(\boldsymbol{\tau}, q)\,]\boldsymbol{b}^e, \quad \dot{\xi} = \gamma \partial_q g(\boldsymbol{\tau}, q), \tag{1.18}$$

where $g(\boldsymbol{\tau}, q)$ has identical form as in the infinitesimal theory.

2. Exponential Return Mapping Algorithms

Following SIMO [1991b] we consider an algorithmic approximation consistent with the associative flow rule derived above which is exact for incrementally elastic processes, independent of the specific form adopted for the stored energy function and exact plastic volume preserving for pressure insensitive plasticity models. Remarkably, the implementation of this algorithm takes a form essentially identical to the standard return maps of the infinitesimal theory. The case of multi-surface plasticity described below follows the presentation given in SIMO & MESCHKE [1992].

2.1. The Local Problem of Evolution. The Algorithmic Flow Rule

Consider a typical time sub-interval $[t_n, t_{n+1}]$, an arbitrary material point $X \in \mathcal{B}$ and assume that the deformation gradient $\boldsymbol{F}_n$ along with $\{\boldsymbol{b}_n^e, \xi_n\} \in \mathsf{S}_+ \times \mathsf{R}_+$ which define the constitutive response of the material, are prescribed initial data at $X \in \mathcal{B}$. The objective is to construct an algorithmic approximation to the plastic flow evolution equations for incremental displacement field $\boldsymbol{u}_t$ prescribed on $\varphi_n(\mathcal{B}) \times [t_n, t_{n+1}]$, where $\varphi_n(\cdot)$ is a given configuration at time t_n defined on $\mathcal{B}$. Denoting by $\boldsymbol{f}_t(\boldsymbol{x}_n) := \boldsymbol{I} + \nabla \boldsymbol{u}_t(\boldsymbol{x}_n)$ the relative deformation gradient at $\boldsymbol{x}_n = \varphi_n(X)$, the local evolution of plastic flow is governed by the first order constrained system

$$\left.\begin{aligned} \dot{\boldsymbol{b}}_t^e &= [\boldsymbol{l}_t \boldsymbol{b}_t^e + \boldsymbol{b}_t^e \boldsymbol{l}_t^T] - 2[\textstyle\sum_{\mu=1}^m \gamma_\mu \partial_{\boldsymbol{\tau}} \phi^\mu(\boldsymbol{\tau}_t, q_t)]\boldsymbol{b}_t^e, \\ \dot{\xi}_t &= \qquad\qquad\qquad [\textstyle\sum_{\mu=1}^m \gamma_\mu \partial_q \phi^\mu(\boldsymbol{\tau}_t, q_t)], \end{aligned}\right\} \tag{2.1}$$
$$\gamma_\mu \geq 0, \quad \phi^\mu(\boldsymbol{\tau}_t, q_t) \leq 0 \quad \text{and} \quad \textstyle\sum_{\mu=1}^m \gamma_\mu \phi^\mu(\boldsymbol{\tau}_t, q_t) = 0,$$

where $\boldsymbol{l}_t = \dot{\boldsymbol{f}}_t \boldsymbol{f}_t^{-1}$ is the spatial velocity gradient. The solution of (2.1) defines $\{\boldsymbol{b}_t^e, \xi_t\}$ at the point X within the interval of time $[t_n, t_{n+1}]$ for configuration prescribed by the relation $\varphi_t = \varphi_n + \boldsymbol{u}_t \circ \varphi_n$.

The key idea in the design of an integration algorithm for (2.1) is to recast this system as a standard problem of evolution, of the form $\dot{\boldsymbol{y}} = \boldsymbol{H}(\boldsymbol{y}, t)\boldsymbol{y}$, and then apply the the exponential approximation $\boldsymbol{y}_t = \exp[\Delta t \boldsymbol{H}(\boldsymbol{y}_t, t)]\boldsymbol{y}_t$. This idea is carried out in two steps:

Step **1.** Recast (2.1) in the form of a nonlinear problem of evolution of the form $\dot{\boldsymbol{y}} = \boldsymbol{H}(\boldsymbol{y}, t)\boldsymbol{y}$. In the present context this is accomplished by pull–back of (2.1) to the fixed initial configuration $\varphi_n(\mathcal{B})$, as follows. Define the tensor fields

$$\boldsymbol{b}_t^{e\,*} := \boldsymbol{f}_t^{-1} \boldsymbol{b}_t^e \boldsymbol{f}_t^{-T}, \quad \boldsymbol{c}_t := \boldsymbol{f}_t^T \boldsymbol{f}_t \quad \text{and} \quad \boldsymbol{n}_t^\mu := \boldsymbol{f}_t^T \partial_{\boldsymbol{\tau}} \phi^\mu(\boldsymbol{\tau}_t, q_t) \boldsymbol{f}_t. \tag{2.2}$$

In geometric terms $\boldsymbol{n}_t^\mu$, $\mu = 1, 2, \cdots, m$ are interpreted as the pull-back of the normal $\partial_{\boldsymbol{\tau}} \phi^\mu$ to the configuration $\varphi_n(\mathcal{B})$, with the tensor $\boldsymbol{c}_t = \boldsymbol{f}_t^T \boldsymbol{f}_t$ viewed as a Riemannian metric on $\varphi_n(\mathcal{B})$. A straightforward manipulation shows that the system $(2.1)_1$ is equivalent to

$$\frac{\partial}{\partial t}\boldsymbol{b}_t^{e\,*} = [-2\sum_{\mu=1}^{m}\gamma_\mu \boldsymbol{c}_t^{-1}\boldsymbol{n}_t^\mu]\,\boldsymbol{b}_t^{e\,*}, \qquad \text{with} \quad \boldsymbol{b}_t^{e\,*}|_{t=t_n} = \boldsymbol{b}_n^e, \tag{2.3}$$

which is now written in standard form.

Step **2.** A exponential approximation to equation (2.3) within $[t_n, t] \subset [t_n, t_{n+1}]$ yields the first order accurate scheme $\boldsymbol{b}_t^{e\,*} \approx \exp[-2\sum_{\mu=1}^{m}\Delta\gamma_\mu \boldsymbol{c}_t^{-1}\boldsymbol{n}_t^\mu]\boldsymbol{b}_n^e$, where $\Delta\gamma_\mu \geq 0$ is a Lagrange multiplier to be determined by enforcing consistency as described below. Using $(2.2)_1$ this relation becomes

$$\boldsymbol{b}_t^e = (\boldsymbol{f}_t \exp[-2\sum_{\mu=1}^{m}\Delta\gamma_\mu \boldsymbol{c}_t^{-1}\boldsymbol{n}_t^\mu]\boldsymbol{f}_t^{-1})\boldsymbol{b}_t^{e\,\mathrm{tr}} \quad \text{where} \quad \boldsymbol{b}_t^{e\,\mathrm{tr}} = \boldsymbol{f}_t\boldsymbol{b}_n^e\boldsymbol{f}_t^T, \tag{2.4}$$

where the symbol $\approx$ has been replaced by the equal sign. The algorithmic approximation is completed by exploiting well-known property $\boldsymbol{F}\exp[\boldsymbol{A}]\boldsymbol{F}^{-1} = \exp[\boldsymbol{F}\boldsymbol{A}\boldsymbol{F}^{-1}]$, which yields

$$\boxed{\begin{aligned} &\left.\begin{aligned} \boldsymbol{b}_t^e &= \exp[-2\sum_{\mu=1}^{m}\Delta\gamma_\mu \partial_{\boldsymbol{\tau}}\phi^\mu(\boldsymbol{\tau}_t, q_t)]\boldsymbol{b}_t^{e\,\mathrm{tr}}, \\ \xi_t &= \xi_n + \sum_{\mu=1}^{m}\Delta\gamma_\mu \partial_q\phi^\mu(\boldsymbol{\tau}_t, q_t), \end{aligned}\right\} \\ &\Delta\gamma_\mu \geq 0, \quad \phi^\mu(\boldsymbol{\tau}_t, q_t) \leq 0, \quad \sum_{\mu=1}^{m}\Delta\gamma_\mu\phi^\mu(\boldsymbol{\tau}_t, q_t) = 0, \end{aligned}} \tag{2.5}$$

where $(\boldsymbol{\tau}_t, q_t)$ are defined by the elastic stress-strain relations evaluated at $(\boldsymbol{b}_t^e, \xi_t)$. A standard argument using a Taylor series expansion confirms that the product formula algorithm (2.5) is first order accurate. Unconditional stability follows from the properties of the backward Euler scheme and the exponential map. A geometric illustration of (2.5) is given in Figure 2.1.

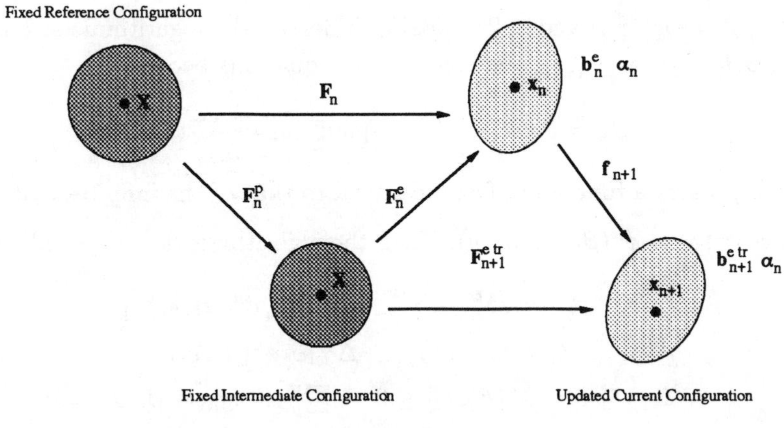

FIGURE 2.1. Interpretation of the update procedure defined by the trial elastic state. The (local) reference and intermediate configurations remain fixed, while the current configuration is updated by the prescribed relative deformation gradient.

The derivation of the algorithmic flow rule (2.5) is identical to the algorithmic treatment in SIMO [1988], with the backward–Euler scheme replaced here by an exponential approximation. Although both algorithms possess identical accuracy properties, the backward Euler method does not inherit the following conservation properties which are identically satisfied by (2.5).

i. *Preservation of the constraint* $\det[\boldsymbol{b}^e_t] > 0$; a property which follows from the identity $\det[\exp[\boldsymbol{A}]] = \exp[\mathrm{tr}[\boldsymbol{A}]]$.

ii. *Exact conservation of the plastic volume.* For pressure insensitive yield criteria the plastic volume preserving property of the exact flow rule is preserved by the algorithmic approximation, since (2.5) implies $\mathrm{tr}[\partial_{\boldsymbol{\tau}} \phi^\mu(\boldsymbol{\tau}_t, q_t)] \equiv 0 \iff J^p_t = J^p_n$.

A modification of the backward Euler method designed to preserve the property **ii** above is given in SIMO & MIEHE [1991].

2.2. Spectral Decomposition: Formulation in Principal Axes

By isotropy the principal directions $\{\boldsymbol{n}^{(A)}_t\}_{A=1,2,3}$ of the Kirchhoff stress tensor $\boldsymbol{\tau}_t$ coincide with the principal directions of the elastic left Cauchy-Green tensor $\boldsymbol{b}^e_t$. Let $\{\beta_A\}$ and $\{\lambda^{e\,2}_A\}$ denote the respective set of principal values. The key to the implementation of the algorithm lies in the following observations:

i. Let $\boldsymbol{n}^{\mathrm{tr}\,(A)}_t$ and $\lambda^{e\,\mathrm{tr}}_A > 0$ denote the principal directions and the square root of the associated principal values of $\boldsymbol{b}^{e\,\mathrm{tr}}_t$. respectively. Then the final principal directions *coincide* with the principal directions; i.e, $\boldsymbol{n}^{(A)}_t \equiv \boldsymbol{n}^{\mathrm{tr}\,(A)}_t$ for $A = 1,2,3$.

ii. Let $\varepsilon^e_{t\,A} := \log[\lambda^e_{t\,A}]$ and $\varepsilon^{e\,\mathrm{tr}}_{t\,A} := \log[\lambda^{e\,\mathrm{tr}}_{t\,A}]$ denote the logarithmic stretches. Setting $\psi(\boldsymbol{b}^e, \xi) = \hat{\psi}(\varepsilon^e_A, \xi)$, the constitutive equations become

$$\beta_A = \partial_{\varepsilon^e_A} \hat{\psi}(\varepsilon^e_A, \xi_t), \quad \text{quad} \quad q_t = -\partial_\xi \hat{\psi}(\varepsilon^e_t, \xi_t). \tag{2.6}$$

which possess a functional form identical to that of the infinitesimal theory.

iii. Set $\phi^\mu(\boldsymbol{\tau}, q) = \hat{\phi}^\mu(\beta_1, \beta_2, \beta_3, q)$. Then the algorithmic flow rule collapses to

$$\left.\begin{aligned} \varepsilon^e_{t\,A} &= \varepsilon^{e\,\mathrm{tr}}_{t\,A} - \textstyle\sum_{\mu=1}^m \Delta\gamma_\mu \partial_{\beta_A} \hat{\phi}^\mu(\boldsymbol{\beta}_t, q_t), \\ \xi_t &= \xi_n \ + \textstyle\sum_{\mu=1}^m \Delta\gamma_\mu \partial_q \hat{\phi}^\mu(\boldsymbol{\beta}_t, q_t), \end{aligned}\right\} \tag{2.7}$$
$$\Delta\gamma_\mu \geq 0, \quad \hat{\phi}^\mu(\boldsymbol{\beta}_t, q_t) \leq 0, \quad \textstyle\sum_{\mu=1}^m \Delta\gamma_\mu \hat{\phi}^\mu(\boldsymbol{\beta}_t, q_t) = 0.$$

which along with (2.6) define a return mapping algorithm at *fixed principal axis* (defined by the trial state) identical to that of the infinitesimal theory.

The advantage of the preceding scheme is clear: It allows the extension to the finite strain regime of the classical return mapping algorithms of the infinitesimal theory without modification but with the following additional simplification. The return map is now formulated in the principal (Eulerian) axis defined by the trial state, which can be computed in *closed form* directly from $\boldsymbol{b}_t^{e\,\mathrm{tr}}$, as explained next.

2.3. Closed–Form Implementation

The step-by-step procedure outlined below summarizes the actual implementation of the updating scheme defined by preceding algorithm within the interval $[t_n, t_{n+1}]$. In a finite element context, this update procedure takes place at each quadrature point of a typical element.

Step **1.** *Trial state.* Given $\{\boldsymbol{b}_n^e, \xi_n\}$ [at a specific quadrature point $\boldsymbol{x}_n \in \varphi_n(\mathcal{B})$] compute the trial elastic left Cauchy-Green tensor and the relative deformation gradient $\boldsymbol{f}_{n+1}$ for prescribed deformation $\boldsymbol{u}_{n+1}: \varphi_n(\mathcal{B}) \to \mathbf{R}^3$ as

$$\boldsymbol{f}_{n+1}(\boldsymbol{x}_n) := \boldsymbol{I} + \nabla \boldsymbol{u}_{n+1}(\boldsymbol{x}_n), \quad \boldsymbol{b}_{n+1}^e = \boldsymbol{f}_{n+1} \boldsymbol{b}_n^e \boldsymbol{f}_{n+1}^T. \tag{2.8}$$

Step **2.** *Spectral decomposition of* $\boldsymbol{b}_{n+1}^{e\,\mathrm{tr}}$. Find the principal stretches $\{\lambda_{n+1\,A}^{e\,\mathrm{tr}}\}$ by solving the cubic characteristic equation in closed form via Cardano's formulae. Compute the principal directions $\{\boldsymbol{n}_{n+1}^{\mathrm{tr}\,(A)}\}$ via the closed form formula

$$\boldsymbol{n}_{n+1}^{\mathrm{tr}\,(A)} \otimes \boldsymbol{n}_{n+1}^{\mathrm{tr}\,(A)} = \left[\frac{\boldsymbol{b}_{n+1}^{e\,\mathrm{tr}} - (\lambda_{n+1\,B}^{e\,\mathrm{tr}})^2 \boldsymbol{I}}{(\lambda_{n+1\,B}^{e\,\mathrm{tr}})^2 - (\lambda_{n+1\,A}^{e\,\mathrm{tr}})^2}\right] \left[\frac{\boldsymbol{b}_{n+1}^{e\,\mathrm{tr}} - (\lambda_{n+1\,C}^{e\,\mathrm{tr}})^2 \boldsymbol{I}}{(\lambda_{n+1\,C}^{e\,\mathrm{tr}})^2 - (\lambda_{n+1\,A}^{e\,\mathrm{tr}})^2}\right], \tag{2.9}$$

for $A = 1, 2, 3$, with $B = 1 + \mathrm{mod}(3, A)$ and $C = 1 + \mathrm{mod}(3, B)$.

Step **3.** *Return mapping in principal (elastic trial state) axes.* Compute the logarithmic stretches and the principal trial stresses:

$$\left.\begin{aligned} \boldsymbol{\varepsilon}_{n+1}^{e\,\mathrm{tr}} &= [\log(\lambda_{n+1\,1}^{e\,\mathrm{tr}}) \quad \log(\lambda_{n+1\,2}^{e\,\mathrm{tr}}) \quad \log(\lambda_{n+1\,3}^{e\,\mathrm{tr}})]^T, \\ \boldsymbol{\beta}_{n+1}^{\mathrm{tr}} &= \partial_{\boldsymbol{\varepsilon}^e} \hat{\psi}(\boldsymbol{\varepsilon}_{n+1}^{e\,\mathrm{tr}}, \xi_n), \quad q_{n+1}^{\mathrm{tr}} = -\partial_\xi \hat{\psi}(\boldsymbol{\varepsilon}_{n+1}^{e\,\mathrm{tr}}, \xi_n). \end{aligned}\right\} \tag{2.10}$$

If the trial stress (in principal axis) lies outside of the elastic domain; i.e., $\boldsymbol{\beta}_{n+1}^{\mathrm{tr}} \notin \mathbb{E}$, then perform the return mapping algorithm (see remarks below) to compute $\{\boldsymbol{\varepsilon}_{n+1}^e, \xi_{n+1}\}$. Compute the Kirchhoff stress tensor via the explicit formula

$$\boldsymbol{\tau}_{n+1} = \sum_{A=1}^{3} \beta_{n+1\,A}\, \boldsymbol{n}_{n+1}^{\mathrm{tr}\,(A)} \otimes \boldsymbol{n}_{n+1}^{\mathrm{tr}\,(A)}, \tag{2.11}$$

where $\beta_{n+1\,A} = \partial_{\varepsilon_A^e} \hat{\psi}(\boldsymbol{\varepsilon}_{n+1}^e, \xi_{n+1})$ are the principal Kirchhoff stresses.

Step **4.** *Update of the intermediate configuration.* Compute the updated left Cauchy-Green tensor via the spectral decomposition

$$\boldsymbol{b}_{n+1}^e = \sum_{A=1}^{3} \exp[2\,\varepsilon_{n+1\,A}^e] \boldsymbol{n}_{n+1}^{\mathrm{tr}\,(A)} \otimes \boldsymbol{n}_{n+1}^{\mathrm{tr}\,(A)}, \tag{2.12}$$

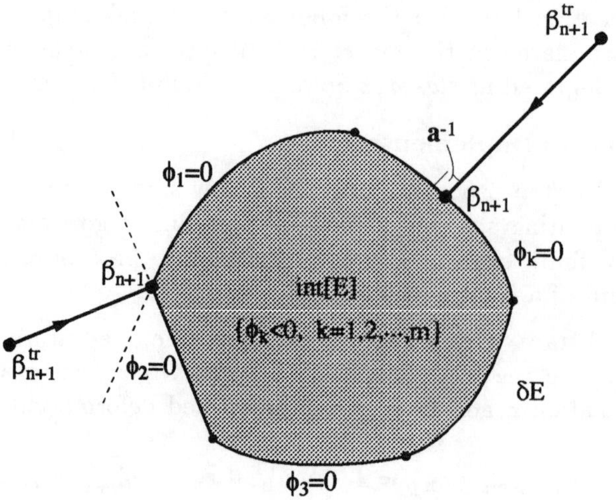

FIGURE 2.2. Geometric interpretation of the return mapping algorithm for multisurface plasticity in principal stress space for two possible trial elastic states.

where $\boldsymbol{n}_{n+1}^{\mathrm{tr}\,(A)}$ $(A = 1, 2, 3)$ are defined in Step **2.** ∎

Remarks 2.1.

1. The explicit formula for the principal directions given in (2.9) above assumes three different principal stretches, similar expressions can be derived for repeated principal stretches; see the Appendix. However, as first pointed out in SIMO & TAYLOR [1991], it is more convenient to alway employ the expression for three different roots and perform a numerical perturbation if some (or all) the roots are repeated.

2. The classical return mapping algorithms in *stress space* of the infinitesimal theory, described in the first part of these lectures, are exactly recovered in the present finite strain context, for a free energy function quadratic in the principal logarithmic stretches, of the form

$$\hat{\psi}(\varepsilon_A^e, \xi) = \tfrac{1}{2}\lambda[\varepsilon_1^e + \varepsilon_2^e + \varepsilon_3^e]^2 + \mu[(\varepsilon_1^e)^2 + (\varepsilon_2^e)^2 + (\varepsilon_3^e)^2] + \hat{K}(\xi), \tag{2.13}$$

where $\lambda > 0$ and $\mu > 0$ are the Lamé constants, $\hat{K}(\cdot)$ is a function which characterizes isotropic hardening in the material and $\varepsilon_A^e := \log(\lambda_A^e)$, for $A = 1, 2, 3$. Using vector notation, the linear elastic stress–logarithmic strain relations emanating from (2.13) read

$$\boldsymbol{\beta} = \boldsymbol{a}\boldsymbol{\varepsilon}^e, \quad \text{where} \quad \boldsymbol{a} := \kappa \mathbf{1} \otimes \mathbf{1} + 2\mu[\boldsymbol{I}_3 - \tfrac{1}{3}\mathbf{1} \otimes \mathbf{1}], \tag{2.14}$$

where $\boldsymbol{I}_3$ denotes the 3×3 identity matrix, $\kappa := \lambda + \frac{2}{3}\mu$ is the bulk-modulus. and $\boldsymbol{a}$ is the (3×3) matrix of constant elastic moduli in principal stress space.

3. The geometric interpretation of the stress–space return mapping algorithms of the infinitesimal theory, now phrased in principal stress space, is illustrated in Figure 2.2 for the general case of multi–surface plasticity. Assuming for simplicity linear isotropic hardening, as in the infinitesimal theory one can show that (SIMO [1991b])

$$\boxed{[\overset{*}{\boldsymbol{\beta}} - \boldsymbol{\beta}_{n+1}] \cdot \boldsymbol{a}^{-1}[\boldsymbol{\beta}^{\mathrm{tr}}_{n+1} - \boldsymbol{\beta}_{n+1}] + [\overset{*}{q} - q_{n+1}]H^{-1}[q_n - q_{n+1}] \leq 0 \quad \forall(\overset{*}{\boldsymbol{\beta}}, \overset{*}{q}) \in \mathsf{E}.} \tag{2.15}$$

Standard results in convex analysis yield the following interpretation of inequality (2.15): *The final state $(\boldsymbol{\beta}_{n+1}, q_{n+1}) \in \mathsf{E}$ in principal stress space is the closest point projection of the trial state $(\boldsymbol{\beta}^{\mathrm{tr}}_{n+1}, q_n)$ onto the elastic domain in the metric defined by the inverse moduli $(\boldsymbol{a}^{-1}, H^{-1})$.* This result reduces the algorithmic problem in principal Kirchhoff stress space to the standard problem in convex optimization: Find the distance of a point (the trial state) to a convex set (the elastic domain) in a given metric (the inverse elastic moduli). The standard algorithm for this problem is a variant of the gradient-projection method proposed in SIMO, KENNEDY & GOVINDJEE [1988]. ■

Form a practical standpoint, the preceding developments translate into the following prescription for the implementation in the finite strain regime of *any* infinitesimal model of classical plasticity.

i. Express the model in principal axes. Note that the classical linear infinitesimal elastic stress–strain relations translate into a linear relation between principal Kirchhoff stresses and principal logarithmic strains.

ii. Construct the algorithmic counterpart of the model (in principal axes) by applying the standard return mapping algorithms, along with the algorithmic elastoplastic moduli $\boldsymbol{a}^{ep}$ (in principal axes).

This defines *Step* **3** in the step–by–step outline of the algorithm given above. The other steps in this algorithm are independent of the specific elastoplastic constitutive model and remain, therefore, unchanged.

2.4. The Exact Algorithmic Tangent Moduli

It only remains to compute the linearization of the algorithm summarized above to obtain, in the general case, the closed–form expression of the algorithmic tangent moduli. The final expression is given by the result derived in SIMO & TAYLOR [1991] for finite elasticity formulated in principal stretches, with the second derivatives of the stored energy function expressed in principal stretches now replaced by the linearization of the return mapping algorithm. Remarkably, the linearization of the algorithm in principal stretches yields a 3×3 symmetric matrix with functional form identical to the algorithmic elastoplastic tangent moduli of the infinitesimal theory.

By exploiting in a crucial manner the property that the principal directions of $\boldsymbol{\tau}_t$ and and $\boldsymbol{b}_t^e$ coincide with those of $\boldsymbol{b}_t^{e\,\mathrm{tr}}$ defined in the trial state phase, the spectral decomposition of the Kirchhoff stress tensor $\boldsymbol{\tau}_t$ takes the following form

$$\boldsymbol{\tau}_t = \sum_{A=1}^{3} \beta_{t\,A} \boldsymbol{m}_t^{\mathrm{tr}\,(A)}, \quad \text{where} \quad \boldsymbol{m}_t^{\mathrm{tr}\,(A)} := \boldsymbol{n}_t^{\mathrm{tr}\,(A)} \otimes \boldsymbol{n}_t^{\mathrm{tr}\,(A)}. \tag{2.16}$$

Let $\boldsymbol{S}_t := \boldsymbol{F}_t^{-1} \boldsymbol{\tau}_t \boldsymbol{F}_t^{-T}$ denote the (symmetric) second Piola-Kirchhoff stress tensor. The objective is to compute the *spatial* moduli $\mathbf{c}_t$ defined by the relations

$$c_t^{ijkl} := 2 F_{t\,I}^{i} F_{t\,J}^{j} F_{t\,K}^{k} F_{t\,L}^{l} \frac{\partial S_t^{IJ}}{\partial C_{t\,KL}}, \tag{2.17}$$

where $C_{t\,KL}$ denote the components of the *total* right Cauchy-Green deformation tensor at current time $t \in [t_n, t_{n+1}]$. The explicit computation of (2.17) relies crucially on the following facts:

i. The algorithm (2.6)-(2.7) defines the vector of principal Kirchhoff stress $\boldsymbol{\beta}_t$ as an implicit function of the vector of trial elastic logarithmic strain $\boldsymbol{\epsilon}_t^e$. The exact linearization of this relation defines the 3×3 matrix of algorithmic moduli:

$$\boldsymbol{a}_t^{\mathrm{ep}} := \frac{\partial \boldsymbol{\beta}_t}{\partial \boldsymbol{\varepsilon}_t^{e\,\mathrm{tr}}}, \quad \text{i.e.,} \quad a_{t\,AB}^{\mathrm{ep}} := \frac{\partial \beta_{t\,A}}{\partial \varepsilon_{t\,B}^{e\,\mathrm{tr}}}. \tag{2.18}$$

ii. Since the intermediate configuration is held fixed in the trial state (modulo a rigid rotation), the logarithmic stretches $\boldsymbol{\varepsilon}_t^{e\,\mathrm{tr}}$ are functions of the total deformation and independent of the plastic flow within the time step $[t_n, t]$. The following crucial relation is proved in the Appendix

$$2 \boldsymbol{F}_t \frac{\partial \varepsilon_{t\,A}^{e\,\mathrm{tr}}}{\partial \boldsymbol{C}_t} \boldsymbol{F}_t^T = \boldsymbol{m}_t^{\mathrm{tr}\,(A)}, \quad A = 1, 2, 3. \tag{2.19}$$

iii. The rank-two tensor $\boldsymbol{m}_t^{\mathrm{tr}\,(A)} = \boldsymbol{n}_t^{\mathrm{tr}\,(A)} \otimes \boldsymbol{n}_t^{\mathrm{tr}\,(A)}$ of principal directions is also a function of the total deformation independent of the plastic flow within the step $[t_n, t]$. The linearization of $\boldsymbol{m}_t^{\mathrm{tr}\,(A)}$ can be computed once and for all leading is a rank-four tensor, denoted by $\mathbf{c}_t^{\mathrm{tr}\,(A)}$, with explicit expression given in the Appendix.

Combining the preceding results, a straight forward application of the chain rule yields the following expression for the algorithmic moduli defined by (2.17):

$$\boxed{\mathbf{c}_t = \sum_{A=1}^{3} \sum_{B=1}^{3} a_{t\,AB}^{\mathrm{ep}} \boldsymbol{m}_t^{\mathrm{tr}\,(A)} \otimes \boldsymbol{m}_t^{\mathrm{tr}\,(B)} + 2 \sum_{A=1}^{3} \beta_{t\,A}\, \mathbf{c}_t^{\mathrm{tr}\,(A)}.} \tag{2.20}$$

The moduli $\mathbf{c}_t^{\mathrm{tr}\,(A)}$ in (2.20) are independent of the specific model of plasticity under consideration and depend only on the form of the stored energy function in principal stretches. Similarly, the tensor products $\boldsymbol{m}_t^{\mathrm{tr}\,(A)}$ are independent of the specific model of plasticity and, therefore, the algorithmic moduli defined by (2.20) are completely specified once the coefficients $a_{t\,AB}^{\mathrm{ep}}$ are determined. These coefficients depend solely on the specific model of plasticity and the structure of the return mapping algorithm in principal stretches. In fact, for the case of single-surface plasticity, a direct calculation gives the following expression for the the moduli $\boldsymbol{a}_t^{\mathrm{ep}}$ during plastic loading:

$$a_{AB}^{\mathrm{ep}} = h_{AB} - \sum_{C,B=1}^{3} \frac{[h_{AC}\frac{\partial\hat{\phi}}{\beta_C}]\,[h_{BD}\frac{\partial\hat{\phi}}{\beta_D}]}{\frac{\partial\hat{\phi}}{\beta_C}\,h_{AB}\,\frac{\partial\hat{\phi}}{\beta_D}}, \tag{2.21}$$

where $h_{AB} := [(\frac{\partial^2\hat{\psi}}{\partial\varepsilon_A^e\partial\varepsilon_B^e})^{-1} + \Delta\gamma\,\frac{\partial^2\hat{\phi}}{\partial\beta_A\partial\beta_B}]^{-1}$. The preceding result is identical to that given in SIMO [1985] within the context of the infinitesimal theory.

3. Application: J_2–Flow Theory

The preceding theory and numerical analysis is applied in this Section to the canonical example of J_2–flow theory. Remarkably, the classical radial return method is exactly recovered if one assumes an uncoupled free energy function, *quadratic* in the principal elastic logarithmic stretches.

3.1. Quadratic stored energy in principal logarithmic stretches

Let $\bar{\boldsymbol{\varepsilon}}^e$ denote the vector of deviatoric principal logarithmic stretches, defined by the relation

$$\bar{\boldsymbol{\varepsilon}}^e := \boldsymbol{\varepsilon}^e - \tfrac{1}{3}\mathrm{tr}[\boldsymbol{\varepsilon}^e]\mathbf{1}, \quad \text{where} \quad \mathrm{tr}[\boldsymbol{\varepsilon}^e] := \boldsymbol{\varepsilon}^e\cdot\mathbf{1}. \tag{3.1}$$

From the definition of the vector of principal logarithmic stretches it follows that

$$\mathrm{tr}[\boldsymbol{\varepsilon}^e] = \sum_{A=1}^{3}\varepsilon_A^e = \sum_{A=1}^{3}\log(\lambda_A^e) = \log(\lambda_1^e\lambda_2^e\lambda_3^e) = \log(J^e). \tag{3.2}$$

Combining (3.2) and (3.1) one concludes that $\bar{\boldsymbol{\varepsilon}}_e$ is in fact the vector of principal volume-preserving logarithmic stretches; i.e.,

$$\bar{\boldsymbol{\varepsilon}}^e = [\log(\bar{\lambda}_1^e) \quad \log(\bar{\lambda}_2^e) \quad \log(\bar{\lambda}_3^e)]^T \quad \text{where} \quad \bar{\lambda}_A^e := J^{e\,-\frac{1}{3}}\lambda_A^e, \; A=1,2,3. \tag{3.3}$$

Denoting by $\kappa = \lambda + \frac{2}{3}\mu$ the bulk modulus, the preceding results yield the following equivalent form of the free energy function (2.13):

$$\hat{\psi}(\boldsymbol{b}^e,\xi) = \tfrac{1}{2}\kappa[\log(J^e)]^2 + \mu\bar{\boldsymbol{\varepsilon}}^e\cdot\bar{\boldsymbol{\varepsilon}}^e + \hat{K}(\xi). \tag{3.4}$$

Consequently, $\hat{\psi}$ has the correct behavior for extreme strains in the sense that $\hat{\psi} \to \infty$ as $J^e \to 0$ and, likewise, $\hat{\psi} \to \infty$ as $J^e \to \infty$. Unfortunately, $\hat{\psi}$ is *not* a convex function of J^e and hence $\hat{\psi}$ cannot be a poly-convex function of the deformation gradient; see e.g., CIARLET [1988] for an explanation of this terminology. Therefore, the stored energy function defined by (2.13) cannot be accepted as a correct model of elasticity for *extreme strains*. Despite this shortcoming, the model provides an excellent approximation for moderately large elastic strains, vastly superior to the usual Saint Venant-Kirchhoff model of finite elasticity. Furthermore, this limitation has negligible practical implications in realistic models of classical plasticity, which are typically restricted to small elastic strains, and is more than offset by the simplicity of the return mapping algorithm in stress space, which takes a format identical to that of the infinitesimal theory. This property is illustrated below in the concrete setting of classical J_2–flow theory.

3.2. Application to J_2–Flow Theory

The elastic domain E of this classical single surface model of metal plasticity is defined by the Mises yield criterion expressed in true (Kirchhoff) stresses as

$$\phi(\boldsymbol{\tau}, \xi) := \|\operatorname{dev}[\boldsymbol{\tau}]\| - \sqrt{\tfrac{2}{3}}[\sigma_Y + \hat{K}'(\xi)] \leq 0, \tag{3.5}$$

where $\sigma_Y > 0$ is the flow stress, ξ is the equivalent plastic strain and $\|\operatorname{dev}[\boldsymbol{\tau}]\|$ is the square root of (twice) the J_2 invariant of the Kirchhoff stress deviator. Denoting by $\bar{\boldsymbol{\beta}} := \boldsymbol{\beta} - \frac{1}{3}[\boldsymbol{\beta} \cdot \mathbf{1}]\mathbf{1}$ the principal deviatoric Kirchhoff stress, the following relations are easily verified

$$\|\operatorname{dev}[\boldsymbol{\tau}]\| = \|\bar{\boldsymbol{\beta}}\| := \sqrt{\bar{\boldsymbol{\beta}} \cdot \bar{\boldsymbol{\beta}}} \quad \text{and} \quad \partial_{\boldsymbol{\beta}}[\|\operatorname{dev}[\boldsymbol{\tau}]\|] = \frac{\bar{\boldsymbol{\beta}}}{\|\bar{\boldsymbol{\beta}}\|} =: \bar{\boldsymbol{\nu}}, \tag{3.6}$$

where $\bar{\boldsymbol{\nu}}$ is the unit normal to the Mises cylinder in principal stress space with $\bar{\boldsymbol{\nu}} \cdot \bar{\boldsymbol{\nu}} = 1$. Since $\bar{\boldsymbol{\nu}} \cdot \mathbf{1} = 0$, the return mapping algorithm reduces to

$$\left.\begin{aligned} \boldsymbol{\beta}^{\text{tr}}_{n+1} &= \boldsymbol{a}\,\boldsymbol{\varepsilon}^{e\,\text{tr}}_{n+1}, \\ \boldsymbol{\beta}_{n+1} &= \boldsymbol{\beta}^{\text{tr}}_{n+1} - 2\mu\Delta\gamma\bar{\boldsymbol{\nu}}_{n+1}, \\ \xi_{n+1} &= \xi_n + \sqrt{\tfrac{2}{3}}\Delta\gamma, \end{aligned}\right\} \tag{3.7}$$
$$\Delta\gamma \geq 0, \quad \hat{\phi}_{n+1} \leq 0, \quad \Delta\gamma\hat{\phi}_{n+1} = 0,$$

where $\bar{\boldsymbol{\nu}}_{n+1} := \bar{\boldsymbol{\beta}}_{n+1}/\|\bar{\boldsymbol{\beta}}_{n+1}\|$. The constraint $\bar{\boldsymbol{\nu}}_{n+1} \cdot \mathbf{1} = 0$ yields $\boldsymbol{\beta}_{n+1} \cdot \mathbf{1} = \boldsymbol{\beta}^{\text{tr}}_{n+1} \cdot \mathbf{1}$, which implies $J^e_{n+1} = J^{e\,\text{tr}}_{n+1} = J_{n+1}$. Equation $(3.7)_2$ then determines $\bar{\boldsymbol{\nu}}_{n+1}$ solely in terms of the trial state. The resulting expression along with the yield condition evaluated at the trial state take the form:

$$\left.\begin{aligned} \bar{\boldsymbol{\nu}}_{n+1} &= \bar{\boldsymbol{\beta}}^{\text{tr}}_{n+1}/\|\bar{\boldsymbol{\beta}}^{\text{tr}}_{n+1}\|, \\ \hat{\phi}^{\text{tr}}_{n+1} &= \|\bar{\boldsymbol{\beta}}^{\text{tr}}_{n+1}\| - \sqrt{\tfrac{2}{3}}[\sigma_Y + \hat{K}'(\xi_n)], \end{aligned}\right\} \tag{3.8}$$

where $\|\bar{\boldsymbol{\beta}}^{\rm tr}_{n+1}\| := \sqrt{\bar{\boldsymbol{\beta}}^{\rm tr}_{n+1} \cdot \bar{\boldsymbol{\beta}}^{\rm tr}_{n+1}}$. The algorithmic loading condition $\hat{\phi}^{\rm tr}_{n+1} > 0$ implies that $\Delta\gamma > 0$, and the consistency condition $\hat{\phi}_{n+1} = 0$ yields the following scalar equation which defines $\Delta\gamma$ during plastic loading:

$$\hat{\phi}_{n+1} := \hat{\phi}^{\rm tr}_{n+1} - 2\mu\Delta\gamma - \sqrt{\tfrac{2}{3}}[\hat{K}'(\xi_n + \sqrt{\tfrac{2}{3}}\Delta\gamma) - \hat{K}'(\xi_n)] = 0. \qquad (3.9)$$

Due to the convexity of $\hat{K}(\cdot)$, equation (3.9) is ideally suited for an iterative solution by Newton's method for fixed $\hat{\phi}^{\rm tr}_{n+1}$. Of course, the resulting algorithm is identical to the radial return method with nonlinear isotropic hardening, now formulated in principal stresses.

3.2.1. *Extension to viscoplasticity.* An extension of the preceding algorithm to accommodate viscoplastic response is constructed as follows. Consider the standard viscoplastic regularization in which the Kuhn-Tucker relations are replaced by the constitutive equation $\gamma = \langle g(\phi(\boldsymbol{\tau}, q)\rangle/\eta \geq 0$, where $\langle x\rangle := [x + H(x)]/2$ with $H(\cdot)$ denoting the Heaviside function (see the first part of these lectures). To circumvent the characteristic ill-conditioning exhibited by the standard viscoplastic algorithms in the rate independent limit as the viscosity $\eta \to 0$, the viscoplastic regularization is rewritten in terms of the inverse function $g^{-1}(\cdot)$ as

$$\hat{\phi}_{n+1} = g^{-1}(\frac{\eta}{\Delta t}\Delta\gamma) \quad \text{for} \quad \hat{\phi}^{\rm tr}_{n+1} > 0 \iff \Delta\gamma > 0. \qquad (3.10)$$

Observe that $g^{-1}(\cdot)$ is given in *closed form* for practically all rate-dependent models of interest, which typically characterize $g(\cdot)$ via exponential functions or power laws. Combining (3.10) with the consistency condition (3.9) results in the modified scalar equation

$$-g^{-1}(\frac{\eta}{\Delta t}\Delta\gamma) + \hat{\phi}^{\rm tr}_{n+1} - 2\mu\Delta\gamma - \sqrt{\tfrac{2}{3}}[\hat{K}'(\xi_n + \sqrt{\tfrac{2}{3}}\Delta\gamma) - \hat{K}'(\xi_n)] = 0, \qquad (3.11)$$

which defines $\Delta\gamma > 0$ and can be easily solved by Newton's method. Observe that (3.11) is *well-conditioned for any values of the viscosity coefficient* $\eta \in [0, \infty)$. In particular, it is possible to set $\eta \equiv 0$ in (3.11) to recover exactly the inviscid limit defined by equation (3.9).

3.3. Unified Scheme for Kinematic Hardening and Viscoplasticity

The extension of the preceding model of J_2–flow theory at finite strains to incorporate kinematic hardening relies on the following observation. By frame invariance, a general yield criterion of the form $\phi(\boldsymbol{\tau}, \boldsymbol{q}) \leq 0$ must be an isotropic function of its arguments and can only depend on the principal values of $\boldsymbol{\tau}$ and $\boldsymbol{q}$. Accordingly, in the presence of kinematic hardening, the Mises yield criterion (3.5) takes the form

$$\hat{\phi}(\bar{\boldsymbol{\beta}} - \bar{\boldsymbol{\xi}}, \xi) = \|\bar{\boldsymbol{\beta}} - \bar{\boldsymbol{\xi}}\| - \sqrt{\tfrac{2}{3}}[\sigma_Y + \hat{K}'(\xi)] \leq 0, \qquad (3.12)$$

where $\bar{\boldsymbol{\xi}} \in \mathbf{R}^3$ denotes the vector of principal values of the back-stress tensor in principal deviatoric (Kirchhoff) stress space; i.e., $\bar{\boldsymbol{\xi}} \cdot \mathbf{1} = 0$. Equations (3.7) defining the return mapping algorithm for J_2–flow theory in principal stresses are then replaced by the following relations

$$\left.\begin{aligned}
\boldsymbol{\beta}^{\mathrm{tr}}_{n+1} &= \boldsymbol{a}\,\boldsymbol{\varepsilon}^{e\,\mathrm{tr}}_{n+1}, \\
\boldsymbol{\beta}_{n+1} &= \boldsymbol{\beta}^{\mathrm{tr}}_{n+1} - 2\mu\Delta\gamma\bar{\boldsymbol{\nu}}_{n+1}, \\
\bar{\boldsymbol{\xi}}_{n+1} &= \bar{\boldsymbol{\xi}}_n + \tfrac{2}{3}\Delta\gamma\bar{H}\bar{\boldsymbol{\nu}}_{n+1}, \\
\xi_{n+1} &= \xi_n + \sqrt{\tfrac{2}{3}}\Delta\gamma,
\end{aligned}\right\} \tag{3.13}$$
$$\Delta\gamma = \tfrac{\Delta t}{\eta}\langle g(\hat{\phi}(\boldsymbol{\beta} - \bar{\boldsymbol{\xi}}, \xi))\rangle \geq 0,$$

where $\bar{H}$ is the kinematic hardening modulus and $\bar{\boldsymbol{\nu}}_{n+1}$ is the normal to Mises yield condition now defined by

$$\bar{\boldsymbol{\nu}}_{n+1} = \partial_{\boldsymbol{\beta}}\hat{\phi}_{n+1} = \frac{\bar{\boldsymbol{\zeta}}_{n+1}}{\|\bar{\boldsymbol{\zeta}}_{n+1}\|}, \quad \text{where} \quad \bar{\boldsymbol{\zeta}}_{n+1} := \boldsymbol{\beta}_{n+1} - \bar{\boldsymbol{\xi}}_{n+1}. \tag{3.14}$$

Relations (3.13) define a return mapping algorithm in principal stress space which incorporates a form of kinematic hardening consistent with the Prager-Ziegler evolution law of the infinitesimal theory. Proceeding exactly as in the derivation leading to (3.11) one concludes that the trial state completely determines the unit normal the the Mises yield criterion as $\bar{\boldsymbol{\nu}}_{n+1} = \bar{\boldsymbol{\zeta}}^{\mathrm{tr}}_{n+1}/\|\bar{\boldsymbol{\zeta}}^{\mathrm{tr}}_{n+1}\|$. Similarly, one easily sees that $\Delta\gamma = 0$ if and only if $\hat{\phi}^{\mathrm{tr}}_{n+1} \leq 0$, while $\Delta\gamma > 0$ is determined for prescribed $\hat{\phi}^{\mathrm{tr}}_{n+1} > 0$ by solving the scalar equation

$$\hat{\phi}^{\mathrm{tr}}_{n+1} - g^{-1}(\frac{\eta}{\Delta t}\Delta\gamma) - 2\mu\Delta\gamma[1 + \frac{\bar{H}}{3\mu}] - \sqrt{\tfrac{2}{3}}[\hat{K}'(\xi_n + \sqrt{\tfrac{2}{3}}\Delta\gamma) - \hat{K}'(\xi_n)] = 0. \tag{3.15}$$

Since $g(x) = 0 \Leftrightarrow x = 0$, equations (3.8)–(3.15) collapse to the standard radial return method of KRIEG & KEY [1977] for combined linear isotropic/kinematic hardening by setting $\eta = 0$ and assuming that $\hat{K}'(\cdot)$ is linear. See Figure 3.1.

Observe that the form of (3.15) precludes ill-conditioning for any values of the viscosity parameter $\eta \in [0, \infty)$ To complete the algorithmic treatment of J_2–flow theory it only remains to compute the algorithmic elastoplastic moduli $\boldsymbol{a}^{\mathrm{ep}}_{n+1}$ associated with the return map (3.8)–(3.15), and defined by the general expression (2.21) in the rate independent case. A direct claculation identical to that described in SIMO & TAYLOR [1984] yields the result

$$\boxed{\boldsymbol{a}^{\mathrm{ep}}_{n+1} = \kappa\mathbf{1}\otimes\mathbf{1} + 2\mu[\, s_{n+1}(\boldsymbol{I}_3 - \tfrac{1}{3}\mathbf{1}\otimes\mathbf{1}) - \delta_{n+1}(\bar{\boldsymbol{\nu}}_{n+1}\otimes\bar{\boldsymbol{\nu}}_{n+1})\,],} \tag{3.16}$$

where $\bar{\boldsymbol{\nu}}_{n+1}$ is defined by (3.8), while the scaling factor s_{n+1} and the coefficient δ_{n+1} are given by the explicit formulae

$$s_{n+1} := 1 - \frac{2\mu\Delta\gamma}{\|\bar{\boldsymbol{\zeta}}^{\mathrm{tr}}_{n+1}\|}, \quad \delta_{n+1} := \frac{1}{1 + \frac{K'' + \bar{H}}{3\mu} + \frac{\eta}{\Delta t}\frac{\bar{g}'}{2\mu}} - \frac{2\mu\Delta\gamma}{\|\bar{\boldsymbol{\zeta}}^{\mathrm{tr}}_{n+1}\|}. \tag{3.17}$$

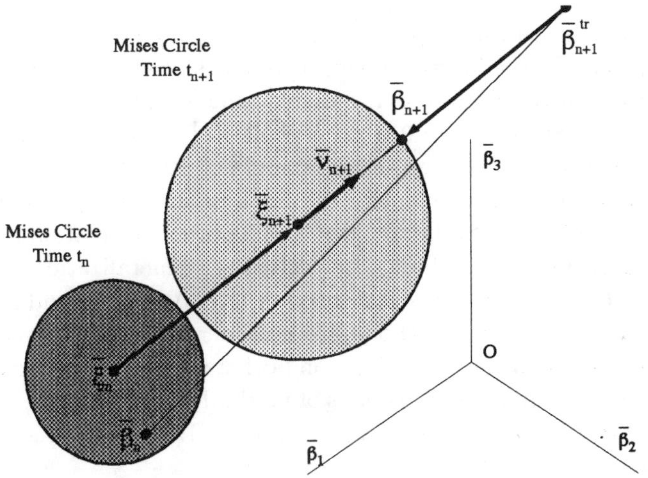

FIGURE 3.1. Geometric Illustration of the standard radial return method for combined isotropic-kinematic hardening obtained here for zero viscosity ($\eta \equiv 0$). In the present approach, the algorithm takes place in principal Kirchhoff stresses and is the same as in the infinitesimal theory.

Observe that $\boldsymbol{a}^{\text{ep}}_{n+1}$ as defined by (3.16) is symmetric. For $\eta \equiv 0$ one recovers from (3.16) the algorithmic moduli of the rate–independent infinitesimal theory.

4. Time-Stepping Algorithms for Dynamic Plasticity

The finite element implementation of the initial boundary value problem defined by the formulation of elastoplasticity described above involves the following steps:

i. A finite element discretization in space of the dynamic weak form of the momentum equations leading to a nonlinear coupled system of ordinary differential equations (ODE's) which describe the time evolution of the nodal degrees of freedom in the time interval of interest $\mathbf{I} = \cup_{n=0}^{N}[t_n, t_{n+1}]$.

ii. A discretization in time of the preceding system of ODE's leading to an incremental nonlinear algebraic problem for the update of the nodal position vectors $\{\boldsymbol{d}_I : I = 1, 2, \cdots n_{\text{node}}\}$ within a typical time subinterval $[t_n, t_{n+1}]$.

The main difficulty in step **i** arises when the plastic flow is constrained by the volume–preserving condition $J^p_t = 1$ for $\forall t \in \mathbf{I}$, a condition induced by pressure–insensitive yield criteria such as the Mises yield condition of classical metal plasticity. This problem is currently well-understood and, since first pointed out in the pioneering work NAGTEGAAL,. PARK & RICE [1974], it has been extensive addressed in the literature. In particular, use of low order Galerkin finite element methods is well-known to result in an over-constrained pressure field that may render overly stiff (locking) numerical solutions. In the linear regime, techniques designed to circumvent this difficulty include classical LBB-satisfying displacement/pressure mixed finite element methods, see the recent review in BREZZI & FORTIN [1990], high order Galerkin finite element methods, such as the high-order triangle of SCOTT & VOGELIUS [1985], and special three field mixed finite element methods as in SIMO & RIFAI [1990], among others. Extensions of these methodologies to the finite deformation regime have been considered by a number of authors. In particular, generalizations of the approach in NAGTEGAAL, PARK & RICE are presented in HUGHES [1979] and extensions of the displacement/pressure mixed methods are considered in SIMO, TAYLOR & PISTER [1985] and SUSSMAN & BATHE [1987] among others. Enhanced mixed finite element methods which include nonlinear versions of method of incompatible modes (see TAYLOR, BERESFORD & WILSON [1976] and CIARLET [1978]) are presented in SIMO & ARMERO [1991].

The discussion that follows will address aspects concerned only with the implementation of step **ii** above. Details pertaining to the actual implementation of step **i** can be found in the aforementioned references.

4.1. The Semi-discrete Initial Boundary Value Problem

Conceptually, the application of any of the finite element strategies outlined above yields a coupled system of ODE's written in conservation form as:

$$\left.\begin{aligned} \frac{d}{dt}\boldsymbol{p}_t &= \boldsymbol{F}_t^{\text{ext}} - \boldsymbol{F}^{\text{int}}(\boldsymbol{d}_t, \boldsymbol{\tau}_t) \\ \frac{d}{dt}\boldsymbol{d}_t &= \boldsymbol{M}^{-1}\boldsymbol{p}_t =: \boldsymbol{v}_t \end{aligned}\right\} \quad \text{for } t \in \mathsf{I}, \tag{4.1}$$

and subject to the initial condition $(\boldsymbol{d}_t, \boldsymbol{p}_t)|_{t=0} = (\boldsymbol{d}_0, \boldsymbol{p}_0)$. Here $\boldsymbol{M}$ is the mass matrix and $t \in \mathsf{I} \mapsto (\boldsymbol{d}_t, \boldsymbol{p}_t) \in P$ is the flow in the finite dimensional phase space $P = \mathsf{R}^{n_{\text{dof}}} \times \mathsf{R}^{n_{\text{dof}}}$, with $n_{\text{dof}} = 3 \times n_{\text{node}}$. In addition, $t \in \mathsf{I} \mapsto \boldsymbol{F}_t^{\text{ext}}$ denotes the prescribed nodal forcing vector and $\boldsymbol{F}^{\text{int}}(\boldsymbol{d}_t, \boldsymbol{\tau}_t)$ is the finite element counterpart of the stress-divergence term in the momentum equation, as defined by

$$\boldsymbol{F}^{\text{int}}(\boldsymbol{\tau}_t, \boldsymbol{d}_t) := \mathop{\mathbf{A}}_{e=1}^{n_{\text{elem}}} \int_{\mathcal{B}_e} [\mathsf{B}(\boldsymbol{d}_t)]^T \boldsymbol{\tau}_t \, d\Omega. \tag{4.2}$$

In this expression $\mathcal{B}_e \subset \mathcal{B}$ denotes the domain of a typical finite element, $\mathsf{B}(\boldsymbol{d}_t)$ is the configuration dependent discrete gradient operator defined by the specific finite element method and $\mathbf{A}_{e=1}^{n_{\text{elem}}}[\cdot]$ is the standard assembly operator over the total number of elements n_{elem} in the discretization. Since the integral in (4.2) is evaluated by numerical quadrature knowledge of the stress field $\boldsymbol{\tau}_t$ is required only at discrete points $X_l \in \mathcal{B}_e$, $l = 1, 2, \cdots, n_{\text{int}}^e$, where $n_{\text{int}}^e \geq 1$ is the total number of points adopted in the quadrature formula.

Within a typical time step $[t_n, t_{n+1}] \subset \mathbb{I}$, the return mapping algorithms of plasticity and viscoplasticity define the stress field $\boldsymbol{\tau}_t$ (at the quadrature points $X_l \in \mathcal{B}_e$) in terms of the nodal displacement field $\boldsymbol{d}_t : [t_n, t_{n+1}] \to \mathbb{R}^{n_{\text{dof}}}$ and the initial data $\{\boldsymbol{b}_n^e, \boldsymbol{\alpha}_n^e\}$ (associated with the quadrature points $X_l \in \mathcal{B}_e$). The specification of a global time-stepping algorithm then defines the solution vector $(\boldsymbol{d}_{n+1}, \boldsymbol{p}_{n+1}) \in P$ as an implicit nonlinear function of the initial data $(\boldsymbol{d}_n, \boldsymbol{p}_n) \in P$, and reduces the semidiscrete problem (4.1) to a nonlinear algebraic system of equations.

4.1.1. *Global balance laws.* Assuming a 'consistent mass' approximation, the dynamics described by the evolution equations (4.1) inherits two conservation laws present in the continuum problem. Let $t \in \mathbb{I} \mapsto \boldsymbol{L}_t \in \mathbb{R}^3$ and $t \in \mathbb{I} \mapsto \boldsymbol{J}_t \in \mathbb{R}^3$ be the total linear and total angular momentum maps defined by

$$\boldsymbol{L}_t = \sum_{I=1}^{n_{\text{node}}} \boldsymbol{p}_t^I \quad \text{and} \quad \boldsymbol{J}_t = \sum_{I=1}^{n_{\text{node}}} \boldsymbol{d}_{t\,I} \times \boldsymbol{p}_t^I, \tag{4.3}$$

respectively. Then, the relations

$$\boldsymbol{L}_t = \sum_{I=1}^{n_{\text{node}}} \boldsymbol{F}_t^{\text{ext}\,I} \quad \text{and} \quad \boldsymbol{J}_t = \sum_{I=1}^{n_{\text{node}}} \boldsymbol{d}_{t\,I} \times \boldsymbol{F}_t^{\text{ext}\,I} \tag{4.4}$$

furnish the discrete version of the classical Euler laws of motion which, in the present context, are a direct consequence of the translational and rotational invariance properties of the divergence term in the momentum equations. These invariance properties should be exactly preserved by any reasonable spatial finite element discretization since they are a manifestation of the frame indifference property of the constitutive equations. For equilibrated external loading balance of moments and balance of forces holds and relations (4.4) yield the conservation laws $\frac{d}{dt}\boldsymbol{L}_t = \boldsymbol{0}$ and $\frac{d}{dt}\boldsymbol{J}_t = \boldsymbol{0}$. We shall require that these conservation laws be also exactly preserved by the temporal discretization.

4.2. Momentum Conserving Time–Stepping Algorithms

Two possible time-discretization schemes of the semidiscrete initial boundary value problem (4.1) are considered which exactly preserve the conservation laws of total linear and total angular momentum. For a proof of the results quoted below see SIMO, TARNOW & WONG [1991].

4.2.1. *The classical Newmark family of algorithms.* This time stepping algorithm is defined by the semi-discrete momentum equation $(4.1)_1$ enforced at t_{n+1} along with the standard Newmark formulae; i.e.,

$$\left.\begin{aligned} \boldsymbol{M}\boldsymbol{a}_{n+1} &= \boldsymbol{F}_{n+1}^{\text{ext}} - \boldsymbol{F}^{\text{int}}(\boldsymbol{\tau}_{n+1}, \boldsymbol{d}_{n+1}), \\ \boldsymbol{d}_{n+1} &= \boldsymbol{d}_n + \Delta t \boldsymbol{v}_n + \Delta t^2[(\tfrac{1}{2} - \beta)\boldsymbol{a}_n + \beta \boldsymbol{a}_{n+1}], \\ \boldsymbol{v}_{n+1} &= \boldsymbol{v}_n + \Delta t[(1-\gamma)\boldsymbol{a}_n + \gamma \boldsymbol{a}_{n+1}]. \end{aligned}\right\} \tag{4.5}$$

Two noteworthy properties of this class of schemes are:

i. The only member of the classical Newmark family that preserves exactly conservation of total angular momentum (for equilibrated loading) is the *explicit central difference method* obtained for $\beta = 0$ and $\gamma = \frac{1}{2}$.

ii. The Newmark–*trapezoidal* rule, corresponding to the values $\gamma = \frac{1}{2}$ and $\beta = \frac{1}{4}$, does not inherit the conservation property of total angular momentum.

Observe that the Newmark-trapezoidal rule defines an acceleration dependent time stepping algorithm. Furthermore, in the nonlinear regime, the trapezoidal rule *does not* define an A-contractive algorithm (see the counter-example of WANNER [1976]). Since the implementation of the preceding class of algorithms is fairly standard further details will be omitted.

4.2.2. *Time-stepping algorithms in conservation form.* As an illustration if this class of methods, consider the algorithm defined by the semi-discrete momentum equation $(4.1)_1$ enforced in conservation form at $t_{n+\alpha}$ along with the standard Newmark formulae; i.e.,

$$\left.\begin{aligned} \boldsymbol{M}[\boldsymbol{v}_{n+1} - \boldsymbol{v}_n] &= \boldsymbol{F}^{\text{ext}}_{n+\alpha} - \boldsymbol{F}^{\text{int}}(\boldsymbol{\tau}_{n+\alpha}, \boldsymbol{d}_{n+\alpha}), \\ \boldsymbol{d}_{n+1} &= \boldsymbol{d}_n + \Delta t[(1-\beta/\gamma)\boldsymbol{v}_n + \beta/\gamma \boldsymbol{v}_{n+1}] + \Delta t^2(\tfrac{1}{2} - \beta/\gamma)\boldsymbol{a}_n, \\ \boldsymbol{a}_{n+1} &= [\boldsymbol{v}_{n+1} - \boldsymbol{v}_n]/\gamma\Delta t + (1 - 1/\gamma)\boldsymbol{a}_n, \end{aligned}\right\} \tag{4.6}$$

where $\boldsymbol{d}_{n+\alpha} := \alpha \boldsymbol{d}_{n+1} + (1-\alpha)\boldsymbol{d}_n$ and $\boldsymbol{F}^{\text{ext}}_{n+\alpha} := \alpha \boldsymbol{F}^{\text{ext}}_{n+1} + (1-\alpha)\boldsymbol{F}^{\text{ext}}_n$. In the preceding expressions $\boldsymbol{\tau}_{n+\alpha}$ denotes the Kirchhoff stress field evaluated at the generalized mid-point configuration $\varphi_{n+\alpha}: \mathcal{B} \to \mathsf{R}^3$, with nodal position vector $\boldsymbol{d}_{n+\alpha}$. Noteworthy properties of this algorithm are:

i. Exact conservation of the total angular momentum (for equilibrated loading) is achieved for the values $\alpha = \frac{\beta}{\gamma} = \frac{1}{2}$ corresponding to the *conservation form of the mid-point rule.* These values define an *acceleration independent* time-stepping algorithm.

ii. In general, formulae (4.6) define a one-leg multistep method. Second order accuracy is obtained if and only if $\alpha = \frac{1}{2}$. For the parameter values $\alpha = \frac{\beta}{\gamma} = \frac{1}{2}$ a linear analysis shows that the spurious root at zero sampling frequency vanishes if and only if $\gamma = 1$. This value yields the post-processing formula $\boldsymbol{a}_{n+1} = (\boldsymbol{v}_{n+1} - \boldsymbol{v}_n)/\Delta t$.

In sharp contrast with the trapezoidal rule, the implicit mid-point rule *does* define an A-contractive algorithm in the nonlinear regime. In addition to the property of exact momentum conservation, both the explicit central difference method and mid-point rule possess the strong property of defining symplectic schemes for nonlinear elastodynamics; see SIMO & TARNOW [1992] for a detailed discussion of these issues.

4.2.3. *Algorithms incorporating high frequency dissipation.* Extensions to the nonlinear regime of a number of algorithms exhibiting numerical dissipation in the high frequency range often require the evaluation of the stress field at a generalized mid-point configuration. As an example, consider the algorithm defined by a modified momentum equation and the standard Newmark formulae as:

$$\left.\begin{aligned} \boldsymbol{M}\boldsymbol{a}_{n+1} &= \boldsymbol{F}^{\text{ext}}_{n+\alpha} - \boldsymbol{F}^{\text{int}}(\boldsymbol{\tau}_{n+\alpha}, \boldsymbol{d}_{n+\alpha}), \\ \boldsymbol{d}_{n+1} &= \boldsymbol{d}_n + \Delta t \boldsymbol{v}_n + \Delta t^2[(\tfrac{1}{2} - \beta)\boldsymbol{a}_n + \beta \boldsymbol{a}_{n+1}], \\ \boldsymbol{v}_{n+1} &= \boldsymbol{v}_n + \Delta t[(1-\gamma)\boldsymbol{a}_n + \gamma \boldsymbol{a}_{n+1}]. \end{aligned}\right\} \tag{4.7}$$

For linear elastodynamics, this scheme reduces to the α-method of HILBER, HUGHES & TAYLOR [1977]. However, in the nonlinear regime the preceding scheme uses a generalized mid-point rule evaluation of the stress divergence term in place of the commonly used trapezoidal rule evaluation (see MIRANDA, FERENCZ & HUGHES [1989]. The former preserves A-contractivity and leads to nonlinearly stable schemes for infinitesimal plasticity (SIMO [1990]) whereas the latter does not. Noteworthy properties of this algorithm are

i. The values $\alpha = \frac{1}{2}$, $\beta = \frac{1}{2}$ $\gamma = 1$ the algorithm reduces to (4.6) and yields the acceleration-independent, exact momentum–preserving, *mid-point rule* in conservation form. As pointed out above, this algorithm possesses zero spurious root (compare with HILBER [1976]).

ii. The values $\alpha = 1$, $\beta = \frac{1}{4}$ and $\gamma = \frac{1}{2}$ give the acceleration dependent Newmark-*trapezoidal rule* in (4.5). The acceleration field predicted by this algorithm is notoriously noisy.

For $\alpha \in (\frac{1}{2}, 1)$ the algorithm exhibits numerical dissipation in the high frequency range but the property of exact conservation of total angular momentum no longer holds.

4.3. The Return Map in the Generalized Mid-Point Configuration

The implementation of the last two algorithms outlined above requires the interpolation of the external load vector, a task easily accomplished, and the evaluation of the stress field $\boldsymbol{\tau}_{n+\alpha}$ at the mid-point configuration. Let $\boldsymbol{\varphi}_{n+\alpha}: \mathcal{B} \to \mathbb{R}^3$ be the generalized mid-point configuration defined by the convex combination

$$\boldsymbol{\varphi}_{n+\alpha} := \alpha \boldsymbol{\varphi}_{n+1} + (1-\alpha)\boldsymbol{\varphi}_n, \quad \alpha \in (0, 1]. \tag{4.8}$$

Setting $\boldsymbol{F}_n := D\boldsymbol{\varphi}_n$ and $\boldsymbol{F}_{n+1} := D\boldsymbol{\varphi}_{n+1}$, the associated deformation gradient and elastic left Cauchy-Green tensor are given by

$$\boldsymbol{F}_{n+\alpha} := \alpha \boldsymbol{F}_{n+1} + (1-\alpha)\boldsymbol{F}_n, \quad \boldsymbol{b}^e_{n+\alpha} := \boldsymbol{F}_{n+\alpha}\boldsymbol{C}^{p-1}_{n+\alpha}\boldsymbol{F}^T_{n+\alpha}. \tag{4.9}$$

The evaluation of the Kirchhoff stress tensor can be accomplished via a simple extension of the preceding return mapping algorithms. The following alternative schemes are discussed in detail in SIMO [1992]:

i. *Shifted backward-Euler scheme.* This update is defined by setting $\{\boldsymbol{b}^e_{n+1}, \xi_{n+1}\} = \{\boldsymbol{b}^{e\,\mathrm{tr}}_{n+1}, \xi_{n+\alpha}\}$. It can be shown that the resulting scheme yields a backward Euler return mapping algorithm identical to that presented in Section 4 but within time steps shifted by α; i.e., $[t_{n+\alpha}, t_{n+1+\alpha}]$, for $n = 0, 1, \cdots$.

ii. *Generalized mid-point rule.* This update is defined via an explicit step with initial data $\{\boldsymbol{b}^{e\,\mathrm{tr}}_{n+1}, \xi_{n+\alpha}\}$, inspired in the following two-step formulation of the generalized mid-point rule for the standard problem $\dot{\boldsymbol{y}} = \boldsymbol{f}(\boldsymbol{y}, t)$; i.e.,

$$\left.\begin{aligned} \boldsymbol{y}_{n+\alpha} &= \boldsymbol{y}_n + \alpha\Delta t \boldsymbol{f}(\boldsymbol{y}_{n+\alpha}, t_{n+\alpha}), \\ \boldsymbol{y}_{n+1} &= \boldsymbol{y}_{n+\alpha} + (1-\alpha)\Delta t \boldsymbol{f}(\boldsymbol{y}_{n+\alpha}, t_{n+\alpha}), \end{aligned}\right\} \tag{4.10}$$

with $\boldsymbol{y}_{n+\alpha} = \alpha \boldsymbol{y}_{n+1} + (1-\alpha)\boldsymbol{y}_n$. Different variants of this scheme have been considered by a number of authors in the context of the infinitesimal theory; in particular, RICE & TRACEY [1973], HUGHES & TAYLOR [1978], ORTIZ & POPOV [1985] and SIMO & TAYLOR [1986], among others.

iii. *Product formula algorithm.* This update procedure defines $\{\boldsymbol{b}_{n+1}, \xi_{n+1}\}$ via a return mapping algorithm essentially identical described in Part I with the trial state now defined by $\{\boldsymbol{b}^{e\,\mathrm{tr}}_{n+1}, \xi_{n+\alpha}\}$.

In the first two algorithms only the stresses at the generalized mid-point configuration remain in the elastic domain; i.e, $(\boldsymbol{\tau}_{n+\alpha}, q_{n+\alpha}) \in \mathsf{E}$. In general, however, the stresses $(\boldsymbol{\tau}_{n+1}, q_{n+1})$ will not lie in the elastic domain E; a feature often regarded as a disadvantage from a practical standpoint.

5. Representative Numerical Simulations

The formulation presented in the preceding Sections is illustrated below in a number of full three dimensional numerical simulations taken from SIMO [1992]. The goals are to provide a practical accuracy assessment of the different return mapping algorithms in actual calculations and to demonstrate the robustness of the overall finite element formulation in both static and dynamic analyses. The calculations are performed with an enhanced version of the finite element program FEAP developed by R.L.TAYLOR and the author from the version documented in ZIENKIEWICZ & TAYLOR [1989].

5.1. Three Dimensional Necking of a Circular Bar

This simulation is the three dimensional version presented in SIMO & ARMERO [1991] of a well-known two dimensional problem considered by a number of authors.

A bar possessing a circular cross section with radius $R_0 = 6.413$ and total length $L = 53.334$ is subjected to a displacement–controlled pure tension test with simply supported boundary conditions corresponding to perfectly lubricated end grips. For a perfect specimen these boundary conditions lead to a bifurcation problem from an initially homogeneous uniaxial state of stress. The bifurcation problem is transformed into a limit point problem via a geometric imperfection induced by a linear reduction in the radius along the length of the bar to a maximum value $R_{\text{sym}} = 0.982 \times R_0$ at the cross section in the symmetry plane. By obvious symmetry considerations only one eighth of the specimen is discretized in the analysis. The constitutive response of the material is characterized by the model of J_2–flow theory describe above, with the logarithmic free energy function (3.4) and the Mises yield criterion (3.5). The material properties are

$$\kappa = 164.21, \quad \mu = 80.1938, \quad \text{and} \quad \sigma_Y = 0.45. \tag{5.1}$$

Hardening in the material is characterized by an isotropic hardening function of the saturation type, with functional form

$$\hat{K}(\xi) := H\xi + [\sigma_Y^\infty - \sigma_Y](1 - \exp[-\delta\xi]), \tag{5.2a}$$

where $\sigma_Y^\infty \geq \sigma_Y > 0$, $H \geq 0$ and $\delta \geq 0$ are material constants chosen here as

$$\sigma_Y^\infty = 0.715, \quad \delta = 16.93, \quad H = 0.12924. \tag{5.2b}$$

The specimen is subjected to a total increase in length of $\Delta L/L \times 100 = 26.25\%$. The finite element discretization used consists of three dimensional Q1/P0 mixed finite elements, as described in SIMO, TAYLOR & PISTER [1985]. This simulation is used in SIMO [1992] as a practical accuracy assessment of a number of alternative return mapping algorithms in different finite element meshes. The results presented there (see Figure 5.1 for a representative sample) support the remakably good performance exhibited by the exponential return mapping algorithms described in the preceding sections. In particular, the lack of significant degradation in accuracy exhibited by these algorithms suggests that the choice of time step should be dictated by the convergence of the global solution scheme.

The simulations described above are performed with a global Newton iterative solution procedure together with a linear line-search algorithm which renders the scheme globally convergent. Remarkably, successful computations are completed in only *10 load increments*; a feature which demonstrates the robustness of the formulation. In spite of the increase in the condition number of the Hessian with finer meshes, a successful solution for the 960 finite element mesh can also be accomplished in 10 load steps scheme employing 6 BFGS updates, a line search with factor 0.6 and periodic refactorizations. Simulations with 20, 50 and 100 load steps were successfully completed with only two refactorizations per load step.

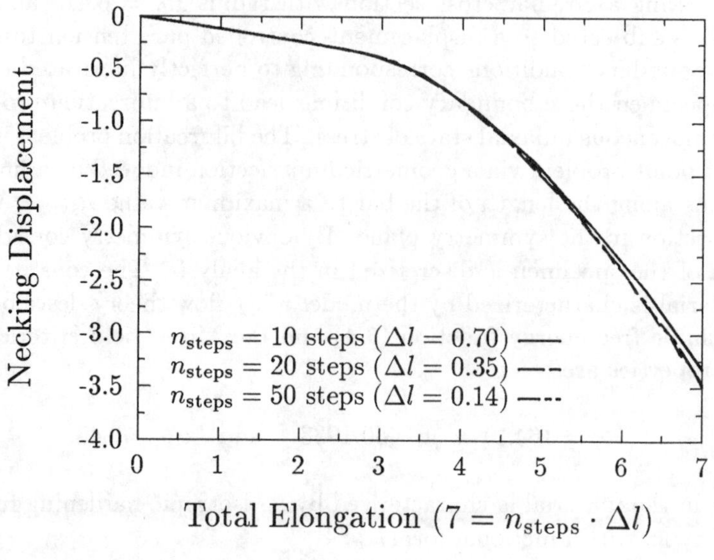

FIGURE 5.1. Displacement at the section of extreme necking versus total elongation. Exponential return mapping algorithm. 960-Finite element mesh.

5.2. Three Dimensional Dynamic Impact of a Circular Bar

The goal of this simulation is to illustrate the performance of the time-stepping algorithms for dynamic plasticity described in the preceding Section; in particular, the the conservation form of the mid-point rule. These results are compared with with those obtained via the standard form of the Newmark time-stepping algorithm. The model problem selected is the dynamic impact on a rigid frictionless wall of a three-dimensional bar, with length $L = 32.4mm$ and circular cross section with radius $R_0 = 3.2mm$, presented in HALLQUIST & BENSON [1988]. The initial velocity is $\boldsymbol{v}_0 = 0.227mm/\mu s$. The material response is again characterized by the same model of J_2 flow theory as in the preceding example, with the following values of the material constants

$$\kappa = 0.4444GPa, \quad \sigma_Y = \sigma_Y^\infty = 0.40Gpa \quad H = 0.1GPa. \tag{5.3}$$

The density in the reference configuration is taken as $\rho_0 = 8930.00Kg/m^3$.
The sequence of deformed configurations corresponding to times $t = 20s$ $t = 40s$ and $t = 80$, computed with the conservation form of the mid-point rule, is shown in Figure 5.3 (for the 144–finite element mesh) and Figure 5.3 (for the 972-element mesh). These simulations are successfully completed in 16 and 32 time steps, respectively. Despite the extremely large time steps used in the simulation, the final shape of the specimen at time $t = 80s$ agrees well with the results reported in HALLQUIST & BENSON [1988]. The time histories of the maximum lateral displacement at the impact

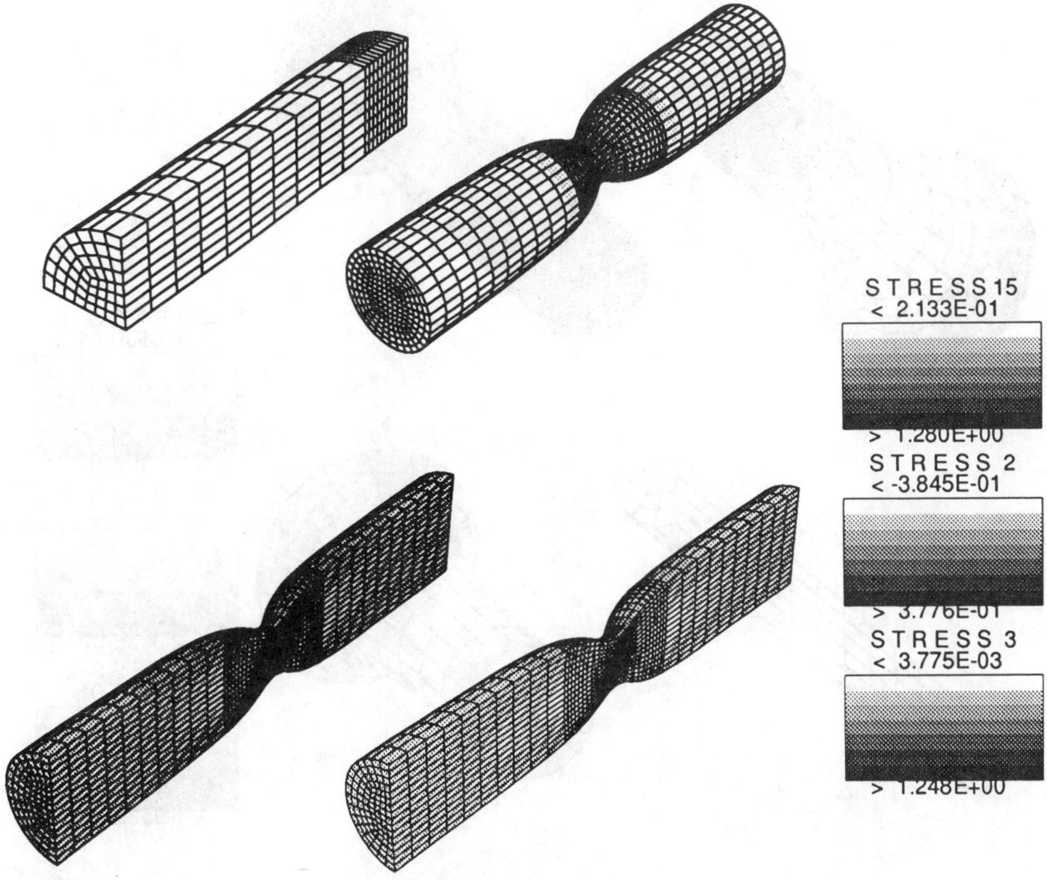

FIGURE 5.2. Fine finite element mesh consisting of 960 Q1/P0 elements. Contours of the equivalent plastic strain and stress distribution for the three dimension bar. Final solution obtained in 20 time steps with the product formula (PF) algorithm.

section, computed with different time steps, are shown in Figure 5.5 (144-element mesh) and in the following figure for the 972-element mesh. The same simulations were attempted for the same two meshes with the standard Newmark algorithm and the one-step exponential backward-Euler return map. Successful computations were completed only when the time steps where reduced to 2.5 and 1.25, respectively. In all these simulations a quadratic rate of asymptotic convergence was attained in a Newton iterative solution strategy.

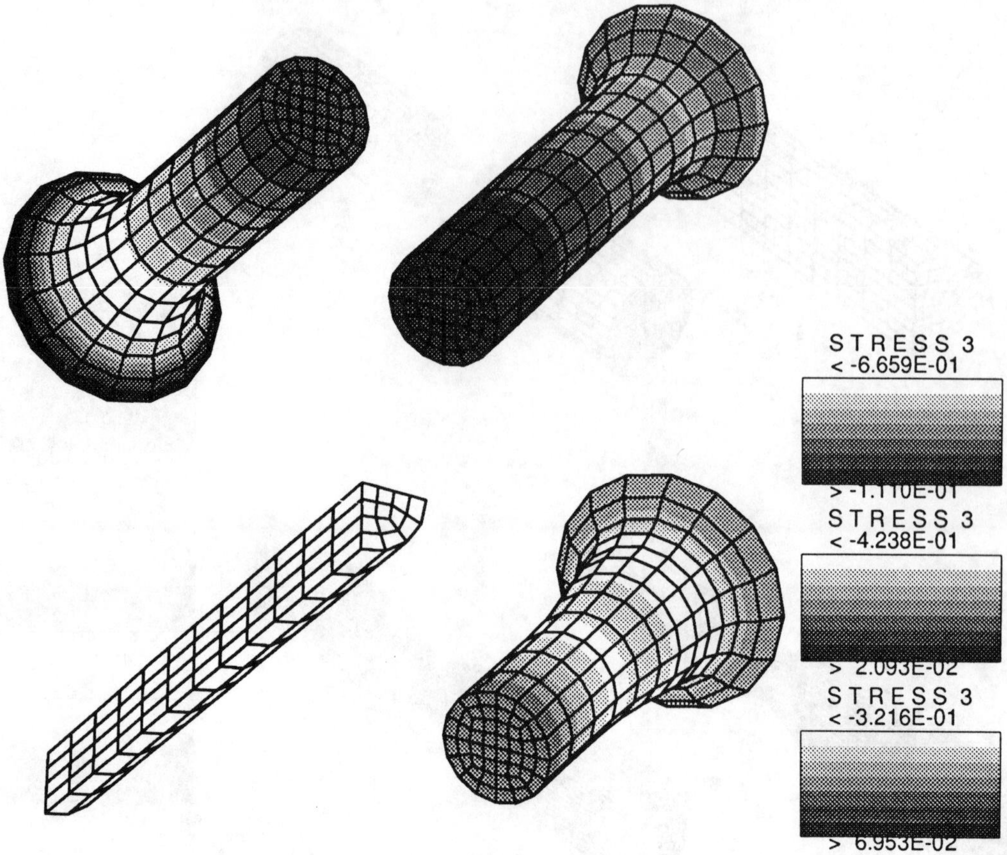

FIGURE 5.3. Three dimensional impact of a circular bar. Small mesh consisting of 144 Q1/P0 mixed finite elements. Sequence of deformed shapes corresponding to times $t = 20$, $t = 40$ and $t = 80$, obtained with the conservation form of the mid-point rule in 16 time steps.

5.3. Simple Shear of a Block: Pure Kinematic Hardening Law

This final simulation is concerned with the simple shear of a rectangular three dimensional block, with constitutive response characterized by J_2–flow theory, in the ideal case of pure kinematic hardening. As first noted in NAGTEGAAL & DE JONG [1981], this situation gives rise to spurious oscillations in the stress-strain response when the kinematic hardening law is described by the Jaumann derivative. The goal of this simulation is to assess the response exhibited by the extension to the finite strain regime of the classical Prager-Ziegler kinematic hardening law described above.

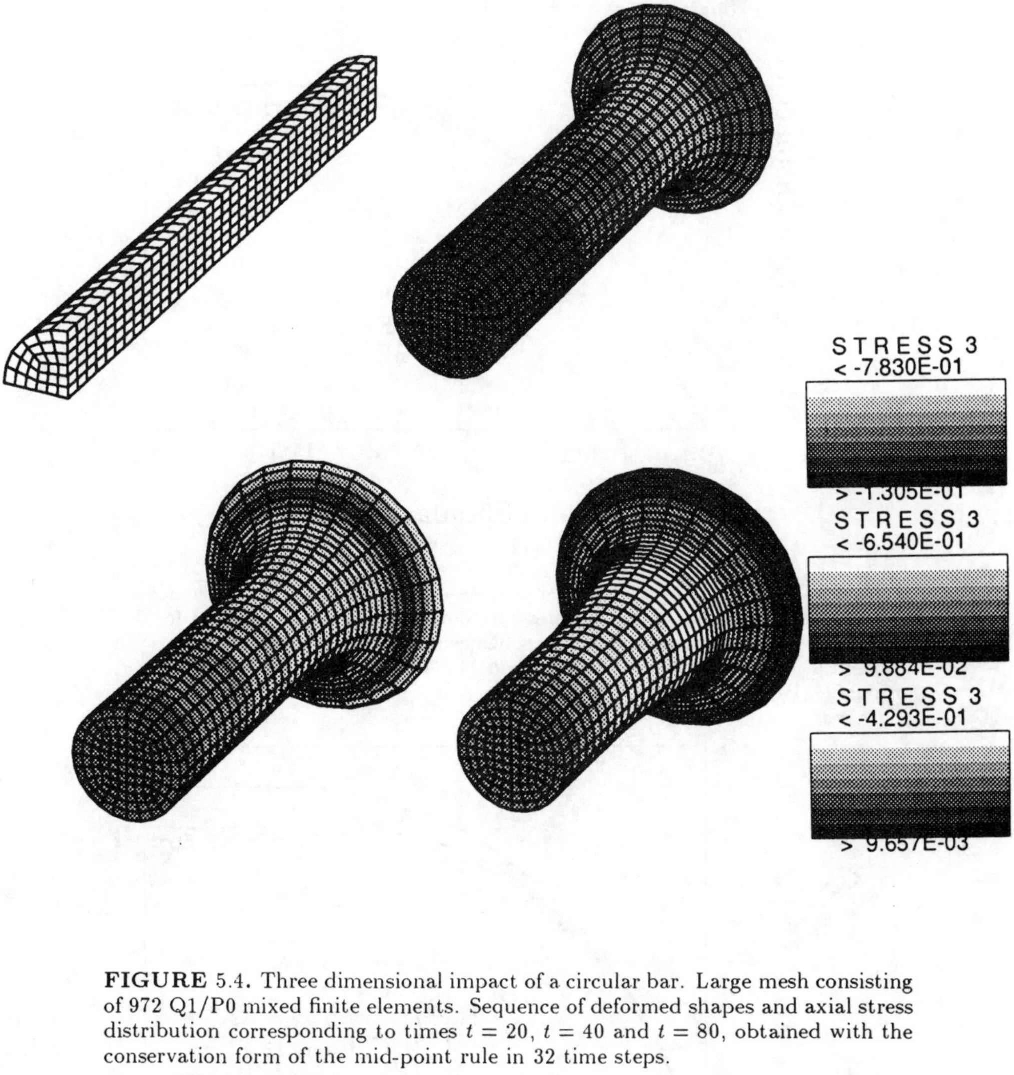

FIGURE 5.4. Three dimensional impact of a circular bar. Large mesh consisting of 972 Q1/P0 mixed finite elements. Sequence of deformed shapes and axial stress distribution corresponding to times $t = 20$, $t = 40$ and $t = 80$, obtained with the conservation form of the mid-point rule in 32 time steps.

The finite element mesh consists of 27 three dimensional Q1/P0 elements as shown in Figure 5.7. The elastic properties are the same as in Example 7.1., while the model of J_2–flow is now characterized by a flow stress $\sigma_Y = 0.45$ and a kinematic hardening modulus $\bar{H} = 0.1$. The final configuration is also shown in Figure 5.7 and the plot of the Kirchhoff shear stress τ_{12} versus the amount of shear is recorded in the last figure. It is apparent from these results that the stress response is monotonically increasing; in fact, essentially linear, and exhibits no spurious oscillations in the entire range of deformations.

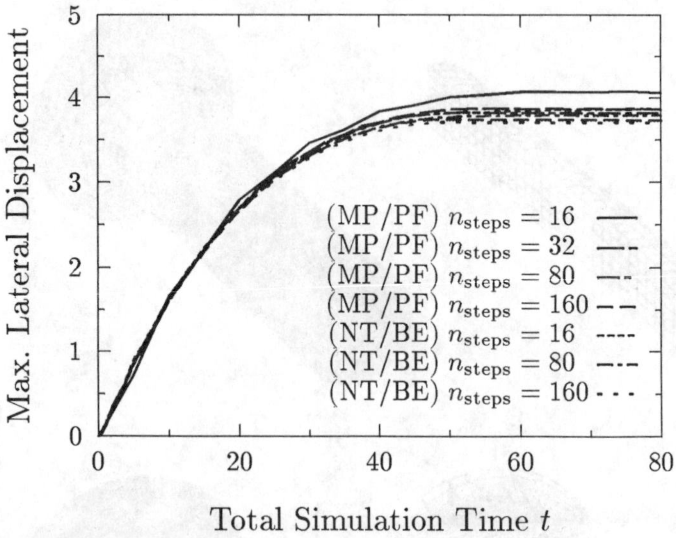

FIGURE 5.5. Impact of a circular bar: 144-element mesh (Q1/P0 elements). Time history of the maximum lateral displacement at the impact section for different time steps, computed with two schemes: Newmark-Trapezoidal rule with exponential Backward Euler (NT/BE) and Mid-Point with exponential Product Formula (MP/PF).

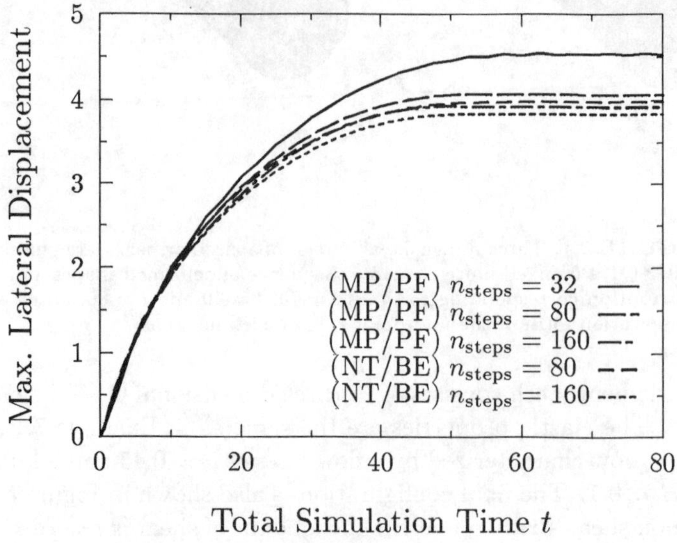

FIGURE 5.6. Impact of a circular bar: 972-element mesh (Q1/P0 elements). Time history of the maximum lateral displacement at the impact section for different time steps, computed with two schemes: Newmark-Trapezoidal rule with exponential Backward Euler (NT/BE) and Mid-Point with exponential Product Formula (MP/PF).

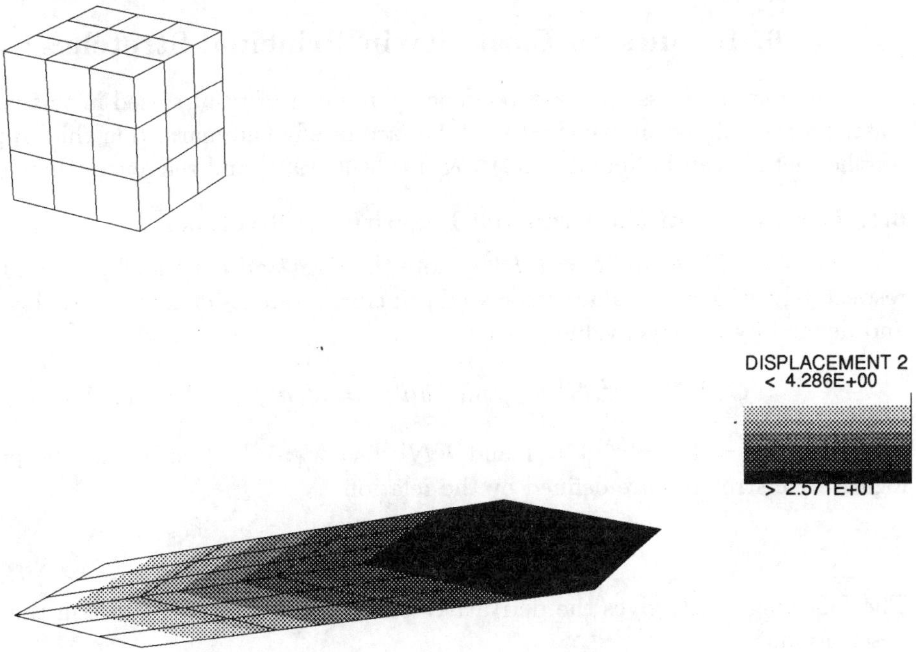

FIGURE 5.7. Simple shear of a three dimensional block. J_2–flow theory with pure kinematic hardening. Initial mesh and final meshes consisting of 27 Q1/P0 mixed finite elements.

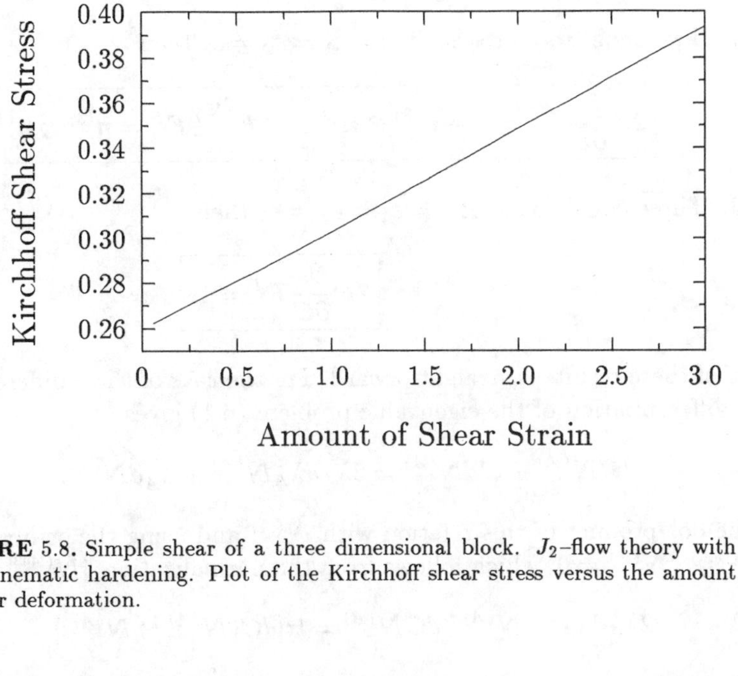

FIGURE 5.8. Simple shear of a three dimensional block. J_2–flow theory with pure kinematic hardening. Plot of the Kirchhoff shear stress versus the amount of shear deformation.

6. Results on Elasticity in Principal Stretches

For completeness and easy reference, a number of results used in this paper on elasticity formulated in principal stretches are briefly summarized in this Appendix. Further details can be found in SIMO & TAYLOR [1991] and references therein.

6.1. Derivatives of the Principal Logarithmic Stretches

Let $\boldsymbol{C} := \boldsymbol{F}^T\boldsymbol{F}$ and $\boldsymbol{b} := \boldsymbol{F}\boldsymbol{F}^T$ denote the right and left Cauchy-Green tensors, respectively, with eigenvalues associated principal stretches denoted $\{\lambda_A,\ A = 1,2,3\}$ and defined by the eigenvalue problems:

$$\boldsymbol{C}\boldsymbol{N}^{(A)} = \lambda_A^2\boldsymbol{N}^{(A)} \quad \text{and} \quad \boldsymbol{b}\boldsymbol{n}^{(A)} = \lambda_A^2\boldsymbol{n}^{(A)}, \quad A = 1,2,3, \tag{6.1}$$

where $\|\boldsymbol{n}^{(A)}\| = 1$, $\|\boldsymbol{N}^{(A)}\| = 1$ and $\boldsymbol{F}\boldsymbol{N}^{(A)} = \lambda_A\boldsymbol{n}^{(A)}$. Recall that the principal logarithmic stretches are defined by the relation

$$\varepsilon_A := \log(\lambda_A), \quad A = 1,2,3. \tag{6.2}$$

The following result gives the derivatives of the logarithmic stretches in the spatial description:

Case **1.** Three different roots. If $\varepsilon_1 \neq \varepsilon_2 \neq \varepsilon_3$ then

$$\boxed{2\boldsymbol{F}\frac{\partial\varepsilon_A}{\partial\boldsymbol{C}}\boldsymbol{F}^T = \boldsymbol{n}^{(A)} \otimes \boldsymbol{n}^{(A)}, \quad A = 1,2,3.} \tag{6.3a}$$

Case **2.** Two equal roots roots. If $\varepsilon := \varepsilon_1 = \varepsilon_2 \neq \varepsilon_3$ then

$$\boxed{2\boldsymbol{F}\frac{\partial\varepsilon}{\partial\boldsymbol{C}}\boldsymbol{F}^T = \boldsymbol{1} - \boldsymbol{n}^{(3)} \otimes \boldsymbol{n}^{(3)}, \quad 2\boldsymbol{F}\frac{\partial\varepsilon_3}{\partial\boldsymbol{C}}\boldsymbol{F}^T = \boldsymbol{n}^{(3)} \otimes \boldsymbol{n}^{(3)}.} \tag{6.3b}$$

Case **3.** Three equal roots. If $\varepsilon = \varepsilon_1 = \varepsilon_2 = \varepsilon_3$ then

$$\boxed{2\boldsymbol{F}\frac{\partial\varepsilon}{\partial\boldsymbol{C}}\boldsymbol{F}^T = \boldsymbol{1}.} \tag{6.3c}$$

The proof of these results is straightforward. For the case of three different roots, for instance, differentiation of the eigenvalue problem (6.1) gives

$$d\boldsymbol{C}\boldsymbol{N}^{(A)} + \boldsymbol{C}d\boldsymbol{N}^{(A)} = 2\lambda_A d\lambda_A\boldsymbol{N}^{(A)} + \lambda_A^2 d\boldsymbol{N}^{(A)}. \tag{6.4}$$

Taking the dot product of this relation with $\boldsymbol{N}^{(A)}$ and using the orthogonality condition $d\boldsymbol{N}^{(A)} \cdot \boldsymbol{N}^{(A)} = 0$, which follows from the normalization $\|\boldsymbol{N}^{(A)}\| = 1$, gives

$$2\lambda_A d\lambda_A = \boldsymbol{N}^{(A)} \cdot d\boldsymbol{C}\boldsymbol{N}^{(A)} = \mathrm{tr}[d\boldsymbol{C}(\boldsymbol{N}^{(A)} \otimes \boldsymbol{N}^{(A)})]. \tag{6.5}$$

Since tr[·, ·] defines an inner product in the linear space $\mathsf{L}(3)$ of 3×3 matrices, the preceding relation along with the directional derivative formula implies

$$\frac{\partial \varepsilon_A}{\partial \boldsymbol{C}} = \lambda_A^{-2} \boldsymbol{N}^{(A)} \otimes \boldsymbol{N}^{(A)}. \tag{6.6}$$

The push-forward of this result along with the relation $\lambda_A \boldsymbol{n}^{(A)} = \boldsymbol{F}\boldsymbol{N}^{(A)}$ gives (6.3a).

Let $\hat{\omega}(\varepsilon_1, \varepsilon_2, \varepsilon_3)$ be the stored energy function of an elastic material formulated in principal logarithmic stretches. The chain rule along with relations (6.3) then gives the following expression for the Kirchhoff stress tensor

$$\boldsymbol{\tau} = \sum_{A=1}^{3} \frac{\partial \hat{\omega}(\varepsilon_A)}{\partial \varepsilon_A} 2\boldsymbol{F} \frac{\partial \varepsilon}{\partial \boldsymbol{C}} \boldsymbol{F}^T = \sum_{A=1}^{3} \frac{\partial \hat{\omega}(\varepsilon_A)}{\partial \varepsilon_A} \boldsymbol{n}^{(A)} \otimes \boldsymbol{n}^{(A)}. \tag{6.7}$$

Comparing this result with the spectral decomposition $\boldsymbol{\tau} = \sum_{A=1} \beta_A \boldsymbol{n}^{(A)} \otimes \boldsymbol{n}^{(A)}$ yields:

$$\boxed{\beta_A = \frac{\partial}{\partial \varepsilon_A} \hat{w}(\varepsilon_1, \varepsilon_2, \varepsilon_3), \quad A = 1, 2, 3.} \tag{6.8}$$

This expression defines the constitutive equations for the principal Kirchhoff stresses in terms of the principal logarithmic stretches.

6.2. Closed-Form Expression for the Principal Stretches and Directions

The squares of the principal stretches are the roots of the characteristic polynomial

$$p(\lambda^2) := -\lambda^6 + I_1 \lambda^4 - I_2 \lambda^2 + I_3 = 0, \tag{6.9}$$

where $\{I_A : A = 1, 2, 3\}$ are the principal invariants of $\boldsymbol{C}$ (or $\boldsymbol{b}$) defined by the standard relations

$$I_1 := \mathrm{tr}[\boldsymbol{C}], \quad I_2 = \tfrac{1}{2}[I_1^2 - \mathrm{tr}[\boldsymbol{C}^2]], \quad I_3 := \det[\boldsymbol{C}]. \tag{6.10}$$

The roots of the characteristic polynomial are computed via the well-known closed-form solution for a cubic equation. The rank-one matrices of principal directions are also computed in closed form as follows.

Case **1.** Three different roots, $\lambda_1 \neq \lambda_2 \neq \lambda_3$. The spectral decomposition of $\boldsymbol{b}$ gives

$$[\boldsymbol{b} - \lambda_A \mathbf{1}] = \sum_{B=1,\, B\neq A}^{3} (\lambda_B^2 - \lambda_A^2) \boldsymbol{n}^{(B)} \otimes \boldsymbol{n}^{(B)}. \tag{6.11}$$

Since $\boldsymbol{n}^{(A)} \cdot \boldsymbol{n}^{(B)} = \delta_{AB}$, it follows that

$$\boxed{\boldsymbol{n}^{(A)} \otimes \boldsymbol{n}^{(A)} = \left[\frac{\boldsymbol{b} - \lambda_B^2 \mathbf{1}}{\lambda_B^2 - \lambda_A^2}\right] \left[\frac{\boldsymbol{b} - \lambda_C^2 \mathbf{1}}{\lambda_C^2 - \lambda_A^2}\right], \quad A = 1, 2, 3,} \tag{6.12}$$

where $B = 1 + \mathrm{mod}(3, A)$ and $C = 1 + \mathrm{mod}(3, B)$.

Case **2.** Two equal roots, $\lambda := \lambda_1 = \lambda_2 \neq \lambda_3$. The spectral decomposition of $\boldsymbol{b}$ now gives

$$[\boldsymbol{b} - \lambda \mathbf{1}] = (\lambda_3^2 - \lambda^2)\boldsymbol{n}^{(3)} \otimes \boldsymbol{n}^{(3)}, \quad [\boldsymbol{b} - \lambda_3 \mathbf{1}] = (\lambda^2 - \lambda_3^2)[\mathbf{1} - \boldsymbol{n}^{(3)} \otimes \boldsymbol{n}^{(3)}]. \tag{6.13}$$

Therefore, the rank-one matrices of principal directions are now given by

$$\boxed{\boldsymbol{n}^{(3)} \otimes \boldsymbol{n}^{(3)} = \left[\frac{\boldsymbol{b} - \lambda^2 \mathbf{1}}{\lambda_3^2 - \lambda^2}\right], \quad [\mathbf{1} - \boldsymbol{n}^{(3)} \otimes \boldsymbol{n}^{(3)}] = \left[\frac{\boldsymbol{b} - \lambda_3^2 \mathbf{1}}{\lambda^2 - \lambda_3^2}\right].} \tag{6.14}$$

Note that the formula for the principal direction associated with a single root remains uniformly valid in the two cases above since squaring $(6.14)_1$ gives (6.12). The case of three equal roots is trivial since the tensor $\boldsymbol{b}$ is speherical.

6.3. Derivatives of the Rank-one Matrices of Principal Directions

Consider the case of three different roots $\lambda_1 \neq \lambda_2 \neq \lambda_3$. Associated with each rank-one matrix of principal directions $\boldsymbol{n}^{(A)} \otimes \boldsymbol{n}^{(A)}$ one has a spatial tensor $\mathbf{c}^{(A)}$ of order four defined in components by the push-forward relation

$$c^{ijkl(A)} := F^i_I F^j_J F^k_K F^l_L \, 4 \frac{\partial [F_a^{-1\,I} (n^{a\,(A)} n^{b\,(A)}) F_b^{-1\,J}]}{\partial C_{KL}}. \tag{6.15}$$

Implicit in this definition are the following steps: Pull-back $[\boldsymbol{n}^{(A)} \otimes \boldsymbol{n}^{(A)}]$ to the reference configuration to obtain $\boldsymbol{F}^{-1}[\boldsymbol{n}^{(A)} \otimes \boldsymbol{n}^{(A)}]\boldsymbol{F}^{-T}$, compute in the reference configuration the derivative of this material tensor relative to $\boldsymbol{C}$ and push-forward the result to the current configuration to obtain $\mathbf{c}^{(A)}$. [1] The final result takes the following form

$$\boxed{\begin{aligned} \mathbf{c}^{(A)} &= \frac{1}{d_A}[\boldsymbol{I}_b - \boldsymbol{b} \otimes \boldsymbol{b} - I_3 \lambda_A^{-1}[\boldsymbol{I} - (\mathbf{1} - \boldsymbol{m}^{(A)}) \otimes (\mathbf{1} - \boldsymbol{m}^{(A)})]] \\ &+ \frac{\lambda_A^2}{d_A}[[\boldsymbol{b} \otimes \boldsymbol{m}^{(A)} + \boldsymbol{m}^{(A)} \otimes \boldsymbol{b}] + (I_1 - 4\lambda_A^2)\boldsymbol{m}^{(A)} \otimes \boldsymbol{m}^{(A)}], \end{aligned}} \tag{6.16}$$

where $I_b^{ijkl} := \frac{1}{2}[b^{ik}b^{jl} + b^{il}b^{jk}]$, $d_A := (\lambda_A^2 - \lambda_B^2)(\lambda_A^2 - \lambda_C^2)$ and $\boldsymbol{m}^{(A)} := \boldsymbol{n}^{(A)} \otimes \boldsymbol{n}^{(A)}$. This expression is strictly valid only if the principal roots are different since $d_A = 0$ otherwise. Although it is possible to derive a similar result for the case of repeated

[1] By covariance this computation boils down to evaluating the derivative of $\boldsymbol{n}^{(A)} \otimes \boldsymbol{n}^{(A)}$ relative to an aribitrary spatial metric tensor $\boldsymbol{g}$ and particularizing the result to $\boldsymbol{g} = \mathbf{1}$.

roots, from a computational standpoint it is preferable to reduce this situation to the general case of different roots by introducing a perturbation of the repeated roots.

Acknowledgements. I am indebted to F. Armero, S. Govindjee, T. Laursen, G. Meschke, C. Miehe, P. Pimenta, R.L. Taylor and K. Wong for numerous discussion on the subject of this paper. Support for this research was provided by IBM Contract No. 2-DJA-991, NSF Grant No. 2DJA-491 with Stanford University and CENTRIC Engineering Systems. This support is gratefully acknowledged.

REFERENCES

J.H. Argyris and J. St. Doltsinis, [1979], "On the Large Strain Inelastic Analysis in Natural Formulation. Part I. Quasistatic Problems", *Computer Methods in Applied Mechanics and Engineering* **20**, 213–252.

J.H. Argyris J.St. Doltsinis, P.M. Pimenta and H. Wüstenberg, [1982], "Thermomechanical Response of Solids at High Strains — Natural Approach", *Computer Methods in Applied Mechanics and Engineering* **32**, 3–57.

R. Asaro, [1983], "Micromechanics of Crystals and Polycrystals" In: *Advances in Applied Mechanics* (Ed.: T.Y. Wu, J.W. Hutchinson), **23**, 1–115.

K. Burrage, and J.C. Butcher, [1979], "Stability Criteria for Implicit Runge-Kutta Methods," *SIAM J. Num. Math.*, **16**, 46-57.

K. Burrage, and J.C. Butcher, [1980], "Nonlinear Stability of a General Class of Differential Equation Methods," *BIT*, **20**, 185-203.

J.C. Butcher, [1975], "A Stability Property of Implicit Runge-Kutta Methods," *BIT*, **15**, 358-361.

F. Brezzi and M. Fortin, [1990], *Mixed and Hibrid Finite Element Methods,* Springer-Verlag, Berlin (to appear).

P.G. Ciarlet, [1978], *The Finite Element Method for Elliptic Problems,* North Holland, Amsterdam.

P.G. Ciarlet, [1981], *Three-dimensional Mathematical Elasticity*, North-Holland, Amsterdam.

B.D. Coleman, and M.E. Gurtin, [1967], "Thermodynamics with Internal State Variables", *J. Chemistry and Physics* **47**, 597–613.

A. Corigliano and U. Perego, [1991], "Convergent and unconditionally stable finite-step dynamic analysis of elastoplastic structures," *Proceedings of the European Conferemce on New Advanced in Computational Structural Mechanics,* pp.577-584.

Commi, C. and G. Maier [1989], "On the Convergence of a Backward Difference Iterative Procedure in Elastoplasticity with Nonlinear Kinematic and Isotropic Hardening," *Computational Plasticity, Models, Software and Applications*, D.R.J. Owen, E. Hinton and E. Onate, Editors. Pineridge Press **1**, 3230334.

Dahlquist, G., [1963], "A Special Stability Problem for Linear Multistep Methods," *BIT*, **3**, 27-43.

G. Dahlquist and R, Jeltsch, [1979], "Generalized Disks of Contractivity and Implicit Runge–Kutta Methods," *Inst for Numerisk Analys*, Report No. TRITA–NA–7906.

F. Demengel [1989] "Compactness Theorems for Spaces of Functions with Bounded Derivatives and Applications to Limit Analysis Problems in Plasticity," *Archi ve for Rational Mechanics and Analysis,* **105**, No.2, pp.123-161.

G. Duvaut, and J.L. Lions, [1972], *Les Inequations en Mecanique et en Physique,* Dunot, Paris.

A.L. Eterovich and K.J. Bathe, [1990], "A hyperelastic-based large strain elastoplastic constitutive formulation with combined isotropic-kinematic hardening using logarithmic stresses and strain measures," *International Journal for Numerical Methods in Engineering,* **30**, No.6, 1099-1115.

R.A. Eve, B.D. Reddy & R.T. Rocckafellar, [1991], "An internal variable theory of elastoplasticity based on the maximum work inequality," *Quarterly J. Applied Mathematics,* to appear.

J.O. Hallquist and D.J. Benson, [1987], *DYNA3D User's Manual*, Report No. UCID-19592, Rev.3, Lawrence Livermore National Laboratory.

H.M. Hilber [1976], *Analysis and design of numerical integration methods in structural dynamics,* Report No. E.E.R.C. 76-29. Earthquake Eng. Research Center. University of California, Berkeley, CA.

H.M. Hilber, T.J.R. Hughes & R.L. Taylor, [1977], "Improved numerical dissipation for time integration algorithms in structural dynamics," *Earthquake Engng. & Struct. Dyn.,* **5**, 283–292.

R. Hill, [1950], *The Mathematical Theory of Plasticity*, Clarendon, Oxford.

T.J.R. Hughes, [1980], "Generalization of Selective Integration Procedures to Aniso tropic and Nonlinear Media," *International J. Numerical Methods in Engineering,* **15**, 1413-1418.

T.J.R. Hughes, [1984], "Numerical Implementation of Constitutive Models: Rate Independent Deviatoric Plasticity," in *Theoretical Foundations for Large Scale Computations of Nonlinear Material Behavior,* Editors S. Nemat-Nasser, R. Asaro and G. Hegemier, Martinus Nijhoff Publishers, The Netherlands.

T.J.R. Hughes, [1983], "Analysis of Transient Algorithms with Special Reference to Stability Behavior," in *Computational Methods for Transient Analysis, Volume 1,* T. Belytschko & T.J.R. Hugues Editors, Noth–Holland, Amsterdam.

T.J.R. Hughes and R.L. Taylor, [1978], "Unconditionally Stable Algorithms for Quasi Static Elasto/Viscoplastic Finite Element Analysis," *Computers and Structures*, **8**, 169-173.

C. Johnson, [1976a], "Existency Theorems for Plasticity Problems," *Journal De Mathematiques Pures et Appliques,* **55**, 431-444.

C. Johnson, [1978], "On Plasticity with Hardening," *Journal of Applied Mathematical Analysis,* **62**, 325-336.

T. Kato, [1974], "On the Trotter-Lie Product Formula," *Proceedings Japan Academy,* **50**, 694-698.

W.T. Koiter, [1960], "General Theorems for Elastic–Plastic Solids, Chapter IV in *Progress in Solid Mechanics*, **1**, 167–221.

R.D. Krieg and S.W. Key, [1976], "Implementation of a Time Dependent Plasticity Theory into Structural Computer Programs," *Constitutive Equations in Viscoplasticity: Computational and Engineering Aspects,* Editors J.A. Stricklin and K.J. Saczlski, AMD-20, ASME, New York.

R.D. Krieg and D.B. Krieg, [1977], "Accuracies of Numerical Solution Methods for the Elastic-Perfectly Plastic Model," *Journal of Pressure Vessel Technology,* ASME, Volume **99**.

S.J. Kim and J.T. Oden. [1990], "Finite element analysis of a class of problems in finite strain elastoplasticity based on the thermodynamical theory of materials of type N," *Computer Methods in Applied Mechanics and Engineering,* **53**, 277-302.

E.H. Lee, [1969], "Elastic-plastic Deformation at Finite Strains", *Journal of Applied Mechanics*, 1–6.

D.G. Luenberger, [1972], *Optimization By Vector Space Methods*, John Wiley and Sons Inc., New York.

G. Maier, [1970], "A Matrix Structural Theory of Piecewise Linear Elastoplasticity with Interacting Yield Planes," *Meccanica*, pp. 54-66.

G. Maenchen and S. Sacks, [1964], "The tensor code," in *Methods of Computational Physics*, Volume **3**, 181-210. B. Alder, S. Fernback and M. Rotenberg Editors. Academic Press, New York.

J.B. Martin, [1988], "Convergence and Shakedwon for Discrete Load Steps in Statically Loaded Elastic–Plastic Bodies," *Mech., Struct. & Mach.*, **16**(1), 1–16.

J.B. Martin & Caddemi [1990], "Suffient Conditions for the Convergence of the Newton–Raphson Iterative Algorithm in Incremental Elastic Plastic Analysis," (preprint).

J. Mandel, [1974], "Thermodynamics and Plasticity", In: *Foundations of Continuum Thermodynamics* (Ed.: J.J. Delgado Domingers, N.R. Nina, J.H. Whitelaw), Macmillan, London, 283–304.

H. Matthies, G. Strang, and E. Christiansen, [1979], "The saddle point of a differential," in *Energy Methods in Finite Element Analysis,* Glowinski, Rodin & Zienkiewicz Edts., J. Wiley & Sons.

H. Matthies, [1978], "Problems in Plasticity and Their Finite Element Approximation", Ph.D. Thesis, Department of Mathematics, Massachusetts Institute of Technology, Cambridge Massachusetts.

Matthies, H. [1979], "Existence theorems in thermoplasticity," *J. de Mechanique,* **18**, No.4, pp.695-711.

I. Miranda, R.M. Ferencz and T.J.R. Hughes, [1989], "An improved Implicit-Explicit Time Integration Method for Structural Dynamics," *Earthquake Engineering and Structural Dynamics,* **18**, 643-653.

B. Moran, M. Ortiz and F. Shi [1990], "Formulation of implicit finite element methods for multiplicative plasticity," *International J. Numerical Methods in Engineering,* **29**, 483-514.

Moreau, J.J., [1976], "Applications of Convex Analysis ot the Treatment of Elastoplastic Systems," in *Applications of Functional Analysis to Problems in Mechanics,* Lecture Notes in Mathematics, Vol. **503**, 56-89, Springer–Verlag, Berlin.

Moreau, J.J., [1977], "Evolution Problem Associated with a Moving Convex Set in a Hilbert Space", *Journal of Differential Equations,* **26**, 347.

J.C. Nagtegaal, D.M. Parks and J.R. Rice, [1974], "On numerically accurate finite element solutions in the fully plastic range," *Computer Methods Applied Mechanics Engineering,* **4**, 153-177.

J.C. Nagtegaal and J.E. de Jong, [1981], "Some aspects of non-isotropic work hardening in in finite strain plasticity," in *Plasticity of of Metals at Finite Strains,* Proceedings Research Workshop, Stanford University, E.H. Lee and R.L. Mallet Editors, pp. 65-102.

A. Needleman and V. Tvergaard, [1984], "Finite Element Analysis of Localization Plasticity," in *Finite elements, Vol V: Special problems in solid mechanics,* Editors J.T. Oden and G.F. Carey, Prentice-Hall, Englewood Cliffs, New Jersey.

Nguyen, Q.S., [1977], "On the elastic-plastic initial boundary value problem and its numerical integration," *International Journal for Numerical Methods in Engineering* **11**, 817.

R.W. Ogden, [1984], *Nonlinear Elastic Deformations,* Ellis Hardwood Limited Chichester, U.K.

M. Ortiz and E.P. Popov, [1985], "Accuracy and stability of integration algorithms for elastoplastic constitutive relations," *International J. Numerical Methods in Engineering,* **21**, 1561-1576.

Fj. Peric, D.R.J. Owen and M.E. Honnor, [1989], "A model for finite strain elastoplasticity," in *Proceedings of the 2nd International Conference on Computational Plasticity,* R. Owen, E. Hinton and E. Onate Editors, 111-126, Pineridge Press, U.K.

R.D. Richtmyer and K.W. Morton [1967], *Difference Methods for Initial Value Problems,* 2nd Edition, Interscience, New York

J.R. Rice and D.M. Tracey, [1973], "Computational fracture mechanics, " in *Proceedings of the Symposium on Numerical Methods in Structural Mechanics,* Editor S.J. Fenves, Urbana Illinois, Academic Press.

W.D. Rolph III and K.J. Bathe, [1984], "On a large strain finite element formulation for elastoplastic analysis," in *Constitutive Equations: Macro and Computational Aspects,* K.J. Willan editor, Winter Anual Meeting, 131-147, ASME.

L.R. Scott and M. Vogelius, [1985], "Conforming Finite Element Methods for Incompressible and Nearly Incompressible Continua," *Lectures in Applied Mathematics,* **22**, 221-244.

J.C. Simo, [1985], "On the computational significance of the intermediate configuration and hyperelastic stress relations in finite deformation elastoplasticity," *Mechanics of Materials*, **4**, 439-451.

J.C. Simo and M. Ortiz, [1985], "A unified Approach to Finite Deformation Elastoplastic Analysis based on the use of Hyperelastic Constitutive Equations", *Computer Methods in Applied Mechanics and Engineering* **49**, 221–245.

J.C. Simo R.L. Taylor, and K.S. Pister, [1985], "Variational and Projection Methods for the Volume Constraint in Finite Deformation Elasto-plasticity", *Computer Methods in Applied Mechanics and Engineering* **51**, 177–208.

J.C. Simo and R.L. Taylor, [1985], "Consistent Tangent Operators for Rate Independent Elastoplasticity", *Computer Methods in Applied Mechanics and Engineering* **48**, 101–118.

J.C. Simo and R.L. Taylor, [1986], "A return mapping algorithm for plane stress elastoplasticity," *Internation J. Numerical Methods in Engineering,* **22**, 649-670.

J.C. Simo, [1988a,b], "A Framework for Finite Strain Elastoplasticity Based on Maximum Plastic Dissipation and Multiplicative Decomposition: Part I. Continuum Formulation; Part II.; Computational Aspects", *Computer Methods in Applied Mechanics and Engineering* **66**, 199–219 and **68**, 1-31.

J.C. Simo, J.G.Kennedy and S.Govindjee, [1988], "General Return Mapping Algorithms for Multisurface Plasticity and Viscoplasticity," *International Journal of Numerical Methods in Engineering,* **26**, No. 2, 2161-2185.

Simo, J.C, and S. Govindjee, [1989], "B–Stability and Symmetry Preserving Return Mapping Algorithms for Plasticity and Viscoplasticity," *International Journal for Numerical Methods in Engineering,* in press

J.C. Simo, [1990], "Nonlinear Stability of the Time Discrete Variational Problem in Nonlinear Heat Conduction and Elastoplasticity," *Computer Methods in Applied Mechanics and Engineering,* **88**, p.111-121.

J.C. Simo and R.L. Taylor, [1991], "Finite Elasticity in Principal stretches; Formulation and Finite element implementation," *Computer Methods in Applied Mechanics and Engineering,* **85**, 273-310.

J.C. Simo and C.Miehe [1990], "Coupled associative thermoplasticity at finite strains. Formulation, numerical analysis and implementation, *Computer Methods in Applied Mechanics and Engineering,* (in press)

J.C. Simo & F. Armero [1992], "Geometrically nonlinear enhanced strain mixed methods and the method of incompatible modes," *International J. Numerical Methods in Engineering,* **33**, 1413-1449.

J.C. Simo, [1992], "Algorithms for Multiplicative Plasticity that Preserve the Form of the Return Mappings of the Infinitesimal Theory," *Computer Methods in Applied Mechanics and Engineering,* in press.

J.C. Simo and G. Meschke, [1992], "A New Class of Algorithms for Classical Plasticity Extended to Finite Strains. Application to Geomaterials," *International J. Computational Mechanics,* in press.

J.C. Simo, N. Tarnow & K. Wong [1991], "Exact energy–momentum conserving algorithms and symplectic schemes for nonlinear dynamics," *Computer Methods in Applied Mechanics and Engineering,* in press.

J.C. Simo and N. Tarnow [1992], "Conserving Algoritms for Nonlinear Elastodynamics: The Energy-Momentum Method," *ZAMM,* in press.

Strang, H., H. Matthies, and R. Temam, [1980], "Mathematical and Computational Methods in Plasticity," in *Variational Methods in the Mechanics of Solids,* S. Nemat–Nasser Edt. Vold **3**, Pergamon Press, Oxford.

Suquet, P., [1979], "Sur les Eq́uations de la Plasticite;" *Ann. Fac. Sciences Tolouse,* **1**, pp.77-87.

Temam, R., and G. Strang [1980], "Functions of bounded deformation," *Archive for Rational Mechanics and Analysis,,* **75**, pp.7-21.

T. Sussman, and K.J. Bathe, [1987], "A finite element formulation for nonlinear incompressible elastic and inelastic analysis," *Computers & Structures,* **26**, No1/2, 357–109.

R.L.Taylor, P.J. Beresford and E.L. Wilson, [1976], "A non-conforming element for stress analysis," *International Journal for Numerical Methods in Engineering,* **10**, No.6, 1211-1219.

Temam, R., and G. Strang [1980], "Functions of bounded deformation," *Archive for Rational Mechanics and Analysis,,* **75**, pp.7-21.

C. Truesdell and W. Noll, [1965], "The Nonlinear Field Theories of Mechanics", In: *Handbuch der Physik Bd. III/3* (Ed.: S. Fluegge), Springer–Verlag, Berlin.

G. Wanner, [1976], "A short Proof of Nonlinear A–Stability," *BIT*, **16**, 226-227.

G. Weber and L. Anand, [1990], "Finite Deformation Constitutive Equations and Time Integration Procedure for Isotropic Hiperelastic-Viscoelastic Solids," *Computer Methods in Applied Mechanics and Engineering,* **79**, 173-202.

M.L. Wilkins, [1964], "Calculation of Elastic-plastic Flow, in *Methods of Computational Physics*, Volume **3**, 211-272. B. Alder, S. Fernback and M. Rotenberg Editors. Academic Press, New York.

Zeidler, E. [1985], *Nonlinear Functional Analysis and its Applications III: Variational Methods and Optimization*, Springer–Verlag, Berlin.

O.C. Zienkiewicz and R.L.Taylor, [1989], *The Finite Element Method, Volume 1*, McGraw-Hill, London.

SHAKE-DOWN ANALYSIS FOR PERFECTLY PLASTIC AND KINEMATIC HARDENING MATERIALS

E. Stein, G. Zhang and R. Mahnken
University of Hannover, Hannover, Germany

SUMMARY AND SCOPE

This course will give an introduction to theoretical and numerical shake-down analysis of perfectly plastic and kinematic hardening materials in the framework of geometrical linear theory.

The static shake-down theorem for perfectly plastic and linear unlimited kinematic hardening material is due to MELAN. In order to extend the shake-down theory to non-linear limited kinematic hardening material, an overlay-model is presented which is a generalization of the one-dimensional MASING model and a corresponding static shake-down theorem is formulated. The numerical treatment is based on the FE-method. In order to solve the resulting non-linear optimization problem, different methods (a special SQP-method, a reduced basis technique and a local optimization technique) are presented. The relationship between shake-down behavior of perfectly plastic and kinematic hardening materials is discussed by showing the relationship between the optimization problems for different materials. The numerical examples given at the end of this course will illustrate the influence of kinematic hardening on shake-down behavior.

---

†Dedicated to Prof. Dr. Jan A. König on the occasion of his sudden death last year

1 Introduction

Many metallic materials show a plastic behavior if their yield limits are exceeded. If the material behaves elastic, perfectly plastic, yielding of an uniaxial rod makes it incapable of supporting any further increase of the external load. However, this must not be true for a hyperstatic system. Yielding at some points or parts of the system due to monotonously increasing loading does not always imply that a collapse mechanism occurs. Instead, a redistribution of stresses takes place. Since in general the stresses are not homogeneous, those points which yield firstly are embedded in an elastic region. Unrestricted yielding of these points as in the uniaxial case is avoided first of all by the surrounding elastic region. The capability of the system is reached, if the load has been increased, such that due to the redistribution of stresses a collapse mechanism occurs. For this collapse mechanism the plastic strains can grow indefinitely under a constant load. A further increase of the external load is not possible. The load which leads to a collapse mechanism is called the *ultimate load.*

Apart from plastifications of the material the ultimate load is dependent on different factors, e.g. instability of the system due to large deformations and strains. However, this shall not be considered in this paper.

The fact that a hyperstatic system can support loads which are higher than the elastic limit load has been recognized by engineers some decades ago. In general, in comparison to a design based on the elasticity theory, the design according to the plastic ultimate load theory implies a saving on material.

However, the simplicity of the static and kinematic formulations of the plastic ultimate load theory has led many engineers to apply it also to systems subjected to a variable loading. The usual way is to design the system according to the ultimate load theory by choosing the most "unfavorable" combination of the possible loads. But, in many cases this may lead to an unsafe construction due to the fact that during the change of loading a part of the system can yield continually, although all possible load combinations remain below the corresponding ultimate load. The consequence is that failure can occur either in form of incremental collapse or alternating plasticity. In the first case the plastic strain increments in the continually flowing part are of the same sign and the remaining deformation accumulates during the change of loads which finally leads to a loss of the initial geometry of the system and makes it unserviceable. In the second case the plastic strain increments change their sign alternately during the loading which will lead to a local material failure due to *low cycle fatigue*.

The incremental collapse is accompanied by a collapse mechanism arising during the loading, and therefore it has global (system) character, while alternating plasticity is of local character. Here, it may happen, that only one point is responsible for failure. Very often alternating plasticity is connected with local stress concentrations. In a system, which is subjected to a variable loading, incremental collapse and alternating plasticity can occur simultaneously.

Due to incompatible plastic strains residual stresses appear. It is possible that after initial yielding eventually the system behavior becomes elastic by forming a suitable residual stress field. This phenomenon is called *shake-down*.

The *shake-down* theory can be used to investigate the elastic, plastic behavior of a system subjected to variable loading, such as

- a reactor which is subjected simultaneously to a variable internal temperature and a variable internal pressure,
- a bridge subjected to moving loads,
- a container which is subjected to a variable internal pressure due to reloading and unloading.

In the comprehensive book by KÖNIG [30] one can find details about the history of the shake-down theory. There extensive references can also be found.

2 Basics of continuum mechanics and plasticity

2.1 Strains, stresses, equilibrium and principles of work

The object of our investigation is a body $\mathcal{B}$ embedded in a 3-dimensional Euclidian point space $\mathcal{E}^3$. To describe its position in the initial configuration at time $t = 0$, any material point $P \in \mathcal{B}$ is assigned to a vector $\mathbf{x}(P) \in \mathbb{E}^3$, where $\mathbb{E}^3$ denotes the 3-dimensional Euclidian vector space. Using an orthonormal Cartesian basis $\mathbf{e}_i$ for $\mathbb{E}^3$, the position vector $\mathbf{x}$ can be represented by means of its components $x_i\,(i = 1, 2, 3)$ in the following way

$$\mathbf{x} = x_i\,\mathbf{e}_i\,. \tag{2.1}$$

Here, the summation rule due to EINSTEIN is employed. Since the relation between P and its position vector $\mathbf{x}(P)$ in the initial configuration is bijective, any material point P is uniquely identified by $\mathbf{x}(P)$. For this reason we do not distinguish between P and $\mathbf{x}(P)$ in the following considerations.

The set of all position vectors shall be denoted by Ω and is defined by

$$\Omega := \{\mathbf{x}(P) \mid \forall\, P \in \mathcal{B}\}\,. \tag{2.2}$$

The motion of a material point $\mathbf{x}$ during the time t is described by a displacement vector $\mathbf{u}(\mathbf{x}, t)$. With respect to the basis $\mathbf{e}_i$ it can be given the form

$$\mathbf{u}(\mathbf{x}, t) = u_i(\mathbf{x}, t)\,\mathbf{e}_i\,. \tag{2.3}$$

For the following presentations vectors and tensors are written in terms of their components. Since we restrict ourselves to an orthonormal Cartesian coordinate system, only lower indices will be needed.

The classical shake-down theory is based on a geometrical linear theory, i.e. we assume that only small strains ε_{ij} occur. Thus, strains and displacements are related by the linear kinematic relation

$$\varepsilon_{ij} = \frac{1}{2}(u_{i,j} + u_{j,i}), \tag{2.4}$$

where $(.)_{,i}$ means $\partial(.)/\partial x_i$. The displacements u_i must satisfy the geometrical boundary conditions

$$u_i = u_i^\circ \qquad \text{on} \qquad \partial\Omega_u \tag{2.5}$$

with $\partial\Omega_u$ as that part of the surface $\partial\Omega$ of Ω, where the values for u_i are prescribed. Analogously to eq. (2.4) the rates of ε_{ij} and u_i, denoted by $\dot{\varepsilon}_{ij}$ and $\dot{u}_i$, respectively, are related by

$$\dot{\varepsilon}_{ij} = \frac{1}{2}(\dot{u}_{i,j} + \dot{u}_{j,i}). \tag{2.6}$$

In the context of a geometrical linear theory the static equilibrium conditions for the stresses σ_{ij} and the body forces per unit volume b_i can be formulated in the initial configuration as

$$\sigma_{ij,j} + b_i = 0 \qquad \text{in} \qquad \Omega\,. \tag{2.7}$$

Additionally the stresses σ_{ij} have to satisfy the static boundary conditions

$$\sigma_{ij}\, n_j = p_i \qquad \text{on} \qquad \partial\Omega_\sigma\,. \tag{2.8}$$

In eq. (2.8) $\partial\Omega_\sigma$ denotes the remaining part of the surface $\partial\Omega$ of Ω, where surface tractions p_i are prescribed. For simplicity of notation it is assumed that $\partial\Omega_u \bigcap \partial\Omega_\sigma = \emptyset$ and $\partial\Omega_u \bigcup \partial\Omega_\sigma = \partial\Omega$ hold. The unit outward vector normal to the surface $\partial\Omega_\sigma$ is denoted by n_i.

The principle of virtual work states that the stresses σ_{ij} and the external loads b_i and p_i are in equilibrium if

$$\int_\Omega \sigma_{ij}\, \bar{\varepsilon}_{ij}\, d\Omega = \int_\Omega b_i\, \bar{u}_i d\Omega + \int_{\partial\Omega_u + \partial\Omega_\sigma} p_i\, \bar{u}_i\, da \tag{2.9}$$

holds for any displacements $\bar{u}_i$ and strains $\bar{\varepsilon}_{ij}$ satisfying eq. (2.4). It is also valid for the rates $\dot{\bar{u}}_i$ and $\dot{\bar{\varepsilon}}_{ij}$, i.e.

$$\int_\Omega \sigma_{ij}\, \dot{\bar{\varepsilon}}_{ij}\, d\Omega = \int_\Omega b_i\, \dot{\bar{u}}_i\, d\Omega + \int_{\partial\Omega_u + \partial\Omega_\sigma} p_i\, \dot{\bar{u}}_i\, da\,. \tag{2.10}$$

2.2 Constitutive equations

For further considerations it is assumed that the actual strains ε_{ij} can be decomposed additively into an elastic part ε_{ij}^E, a plastic part ε_{ij}^P and a thermal part ε_{ij}^T according to

$$\varepsilon_{ij} = \varepsilon_{ij}^E + \varepsilon_{ij}^P + \varepsilon_{ij}^T\,. \tag{2.11}$$

2.2.1 Elasticity and thermal strains

For modelling of the elastic material behavior it is assumed that the stresses σ_{ij} and the elastic strains ε_{ij}^E are related linearly by HOOKE's law as

$$\sigma_{ij} = E_{ijkl}\,\varepsilon_{kl}^E\,. \tag{2.12}$$

Here, E_{ijkl} denotes the fourth order, positive-definite symmetric elasticity tensor. For an isotropic material E_{ijkl} can be written as

$$E_{ijkl} = \lambda\,\delta_{ij}\,\delta_{kl} + \mu\,(\delta_{ik}\,\delta_{jl} + \delta_{il}\,\delta_{jk}) \tag{2.13}$$

by use of the KRONECKER symbol δ_{ij}. The LAMÉ constants λ and μ are related to YOUNG's modulus E, the shear modulus G and POISSON's ratio ν by

$$\lambda = \frac{\nu\,E}{(1+\nu)(1-2\nu)} \quad \text{and} \quad \mu = \mathrm{G} = \frac{E}{2(1+\nu)}\,. \tag{2.14}$$

If temperature loading T is considered, we assume that the thermal strains ε_{ij}^T can be derived from

$$\varepsilon_{ij}^T = \alpha_\vartheta\,T\,\delta_{ij}\,, \tag{2.15}$$

where α_ϑ denotes a linear thermal strain coefficient. Note that the relation (2.15) holds only for an isotropic material.

2.2.2 Perfect plasticity

For an elastic, perfectly plastic material plastic strains occur if a yield function $f(\sigma_{ij})$ of stresses σ_{ij} reaches a critical value σ_0, i.e.

$$f(\sigma_{ij}) = \sigma_0\,. \tag{2.16}$$

In eq. (2.16) σ_0 is the initial yield stress of the material in an uniaxial tension or pressure test.

The choice of the yield function $f(\sigma_{ij})$ is dependent on the considered material. For most metallic materials the HUBER-V. MISES or the TRESCA yield functions are used. However, while the V. MISES yield function is continuously differentiable, the TRESCA yield function possesses singularities in the first derivative. Thus, from the numerical point of view, the V. MISES yield function is preferable, although most theorems of the plasticity theory hold for both yield functions.

Next, some basic relations of the plasticity theory shall be given. Let the 6-dimensional stress space be denoted by $\mathcal{S}$. Then the elastic domain

$$\mathcal{E} := \{\,\sigma_{ij}\,|\,f(\sigma_{ij}) < \sigma_0\,\} \quad \subset \mathcal{S} \tag{2.17}$$

of a material point is a subset of this stress space. The set

$$\mathcal{F} := \{\sigma_{ij}\,|\,f(\sigma_{ij}) = \sigma_0\} \quad \subset \mathcal{S} \tag{2.18}$$

denotes the yield surface of the material point. Stresses σ_{ij} which are part of $\mathcal{E}$ cause no plastic deformations and, therefore, shall be denoted by $\sigma_{ij}^{(s)}$, where the upper index s corresponds to "safe". Plastic strains can only occur or grow, respectively, if the stresses σ_{ij} are on the yield surface, i.e. if $\sigma_{ij} \in \mathcal{F}$. In this case the stresses are called "allowable" and shall be denoted by $\sigma_{ij}^{(a)}$. Note that stresses $\sigma_{ij} \notin \mathcal{E} \cup \mathcal{F}$ are physically impossible.

For further considerations we restrict ourselves to materials which are stable in the sense of DRUCKER's postulate [15, 16]. Then the following well known consequences can be drawn:

1. If stresses σ_{ij} are on the yield surface $\mathcal{F}$, and if $\dot{\varepsilon}_{ij}^{P}$ are the corresponding plastic strain rates, then the inequalities

$$(\sigma_{ij} - \sigma_{ij}^{(s)})\,\dot{\varepsilon}_{ij}^{P} > 0 \qquad \forall\ \sigma_{ij}^{(s)} \in \mathcal{E} \tag{2.19}$$

and

$$(\sigma_{ij} - \sigma_{ij}^{(a)})\,\dot{\varepsilon}_{ij}^{P} \geq 0 \qquad \forall\ \sigma_{ij}^{(a)} \in \mathcal{F} \tag{2.20}$$

hold.

2. If $\dot{\sigma}_{ij}$ are stress rates corresponding to the plastic strain rates $\dot{\varepsilon}_{ij}^{P}$, it follows that

$$\dot{\sigma}_{ij}\,\dot{\varepsilon}_{ij}^{P} \geq 0. \tag{2.21}$$

3. The yield surface $\mathcal{F}$ is convex.

The third consequence is satisfied a priori for both the v. MISES yield function and the TRESCA yield function. The inequalities (2.19) and (2.20) imply that the strain rates $\dot{\varepsilon}_{ij}^{P}$ are normal to the yield surface $\mathcal{F}$, i.e.

$$\dot{\varepsilon}_{ij}^{P} = \dot{\lambda}\,\frac{\partial f}{\partial \sigma_{ij}}, \tag{2.22}$$

where $\dot{\lambda}$ is a multiplier satisfying

$$\dot{\lambda} = 0, \text{ if } f < \sigma_0 \text{ or } \dot{f} = \frac{\partial f}{\partial \sigma_{ij}}\,\dot{\sigma}_{ij} < 0 \tag{2.23}$$

and

$$\dot{\lambda} \geq 0, \text{ if } f = \sigma_0 \text{ and } \dot{f} = 0, \tag{2.24}$$

respectively. Eq. (2.22) means that the yield function $f(\sigma_{ij})$ can be considered as a potential function for the plastic strain rates. The relation (2.22) is called *associated flow rule.*

As mentioned before, for numerical reasons, the v. MISES yield function is preferable to other yield functions, and therefore we shall restrict ourselves to it. Denoting the v. MISES yield function by $\Phi(\cdot)$ it is redefined as

$$\begin{aligned}\Phi(\sigma_{ij}) : &= f^2(\sigma_{ij}) \\ &= \frac{1}{2}[(\sigma_{11}-\sigma_{22})^2 + (\sigma_{22}-\sigma_{33})^2 + (\sigma_{11}-\sigma_{33})^2 \\ &+ 3\,\sigma_{12}^2 + 3\,\sigma_{13}^2 + 3\,\sigma_{23}^2] \\ &= 3\,J_2\,, \end{aligned} \tag{2.25}$$

where J_2 denotes the second invariant of the stress deviator. The corresponding yield condition is given by

$$\Phi(\sigma_{ij}) \leq \sigma_0^2 \,. \tag{2.26}$$

Equation (2.25) shows, that $\Phi(\sigma_{ij})$ is a quadratic function. Thus, by introducing the stress vector

$$\boldsymbol{\sigma}^T := [\,\sigma_{11}, \sigma_{22}, \sigma_{33}, \sigma_{12}, \sigma_{13}, \sigma_{23}\,]\,, \tag{2.27}$$

it can also be written as

$$\Phi(\boldsymbol{\sigma}) = \boldsymbol{\sigma}^T \cdot \mathbf{Q} \cdot \boldsymbol{\sigma}\,, \tag{2.28}$$

where the matrix $\mathbf{Q}$ is given by

$$\mathbf{Q} = \begin{bmatrix} 1 & -\frac{1}{2} & -\frac{1}{2} & 0 & 0 & 0 \\ -\frac{1}{2} & 1 & -\frac{1}{2} & 0 & 0 & 0 \\ -\frac{1}{2} & -\frac{1}{2} & 1 & 0 & 0 & 0 \\ 0 & 0 & 0 & 3 & 0 & 0 \\ 0 & 0 & 0 & 0 & 3 & 0 \\ 0 & 0 & 0 & 0 & 0 & 3 \end{bmatrix}. \tag{2.29}$$

Note that for any vectors $\boldsymbol{\sigma}_1$ and $\boldsymbol{\sigma}_2$ the inequality

$$\sqrt{\Phi(\boldsymbol{\sigma}_1 + \boldsymbol{\sigma}_2)} \leq \sqrt{\Phi(\boldsymbol{\sigma}_1)} + \sqrt{\Phi(\boldsymbol{\sigma}_2)} \tag{2.30}$$

can easily be shown.

2.3 Residual stresses

The total stresses σ_{ij} occurring in an elastic-plastic system can be decomposed into two parts as

$$\sigma_{ij} = \sigma_{ij}^E + \rho_{ij}\,. \tag{2.31}$$

In eq. (2.31) σ_{ij}^E are stresses which would appear if the system would behave purely elastic. Therefore the part ρ_{ij} results from plastic deformations. Since σ_{ij}^E and σ_{ij} are in equilibrium with the same external loads, it follows that ρ_{ij} must satisfy the homogeneous static equilibrium conditions

$$\rho_{ij,j} = 0 \quad \text{in} \quad \Omega \tag{2.32}$$

and the homogeneous static boundary conditions

$$\rho_{ij}\, n_j = 0 \quad \text{on} \quad \partial\Omega_\sigma . \tag{2.33}$$

For this reason ρ_{ij} are called *residual stresses.*

By use of eq. (2.31) the total strains ε_{ij} can be decomposed in the following way:

$$\begin{aligned} \varepsilon_{ij} &= \varepsilon^E_{ij} + \varepsilon^P_{ij} + \varepsilon^T_{ij} \\ &= E^{-1}_{ijkl}\, \sigma_{kl} + \varepsilon^P_{ij} + \varepsilon^T_{ij} \\ &= E^{-1}_{ijkl}\, \sigma^E_{kl} + E^{-1}_{ijkl}\, \rho_{kl} + \varepsilon^P_{ij} + \varepsilon^T_{ij} . \end{aligned} \tag{2.34}$$

In general the elastic strains ε^E_{ij} and the plastic strains ε^P_{ij} are incompatible, i.e. they are no integrable fields. However, using eq. (2.34) the total strains ε_{ij} can be decomposed into the compatible strain fields

$$\varepsilon^*_{ij} = E^{-1}_{ijkl}\, \sigma^E_{kl} + \varepsilon^T_{ij} \tag{2.35}$$

and

$$\varepsilon^{**}_{ij} = E^{-1}_{ijkl}\, \rho_{kl} + \varepsilon^P_{ij} . \tag{2.36}$$

Accordingly, the total displacements u_i can be decomposed into an elastic part u^E_i and a residual part u^R_i as

$$u_i = u^E_i + u^R_i . \tag{2.37}$$

Here, u^R_i are displacements, which result from the plastic deformation. Note that

$$u^R_i = 0 \quad \text{on} \quad \partial\Omega_u \tag{2.38}$$

holds. Using the above decompositions for the strains and the displacements we can derive the relations

$$\varepsilon^*_{ij} = \frac{1}{2}\,(u^E_{i,j} + u^E_{j,i}) \tag{2.39}$$

and

$$\varepsilon^{**}_{ij} = \frac{1}{2}\,(u^R_{i,j} + u^R_{j,i}) . \tag{2.40}$$

Applying eq. (2.9) to the residual stresses ρ_{ij} and the compatible strains ε^{**}_{ij} one obtains

$$\int_\Omega \rho_{ij}\,(E^{-1}_{ijkl}\, \rho_{kl} + \varepsilon^P_{ij})\, d\Omega = 0 . \tag{2.41}$$

Instead of ε^{**}_{ij} one can also use its rate form $\dot{\varepsilon}^{**}_{ij}$ in eq. (2.41), and this leads to

$$\int_\Omega \rho_{ij}\,(E^{-1}_{ijkl}\, \dot{\rho}_{kl} + \dot{\varepsilon}^P_{ij})\, d\Omega = 0 . \tag{2.42}$$

Equations (2.41) and (2.42) are of great importance for the next sections.

3 Shake-down of systems consisting of perfectly plastic material

3.1 General remarks

In this section we consider systems consisting of linear-elastic, perfectly plastic material. The system is subjected to a variable load $\mathbf{P}(t)$, where symbolically $\mathbf{P}(t)$ represents body forces $b_i(t)$, surface tractions $p_i(t)$ and the temperature $T(t)$ at time t, respectively. $\mathbf{P}(t)$ can vary arbitrary with time t, however, it remains within a given load domain $\mathcal{L}$.

The behavior of a system subjected to variable loads can be characterized by one of the following possibilities (cf. [30]):

1. The system remains elastic for all possible loads $\mathbf{P}(t) \in \mathcal{L}$.

2. For a certain load $\mathbf{P}(t_0) \in \mathcal{L}$ the ultimate load of the system is attained and a collapse mechanism develops. Keeping $\mathbf{P}(t_0)$ constant the displacements can grow infinitely.

3. The ultimate load of the system is not attained for any $\mathbf{P}(t) \in \mathcal{L}$, however, at some parts of the system, the remaining displacements and strains, respectively, accumulate during a change of loading. After a certain time the system becomes unserviceable due to large deviation from the initial geometry.

4. At some points (or parts) of the system the plastic strain increments change their sign during a change of loading. Although the remaining displacements are bounded, plastification does not cease. In this case local material failure due to alternating plasticity is present.

5. After initial yielding plastification subsides. Eventually the system behaves elastic.

The behavior of the system according to the first point is not dangerous, since no plastic energy dissipates. However, the capability of the system to support further loads is not fully exploited.

The failure of types 2–4 are characterized by the fact, that plastification does not cease and that the related quantities such as plastic strains ε_{ij}^P and residual stresses ρ_{ij} do not become stationary. Thus, there exist points $\mathbf{x} \in \Omega$ for which

$$\lim_{t \to +\infty} \dot{\varepsilon}_{ij}^P(\mathbf{x}, t) \neq 0 \tag{3.1}$$

and

$$\lim_{t \to +\infty} \dot{\rho}_{ij}(\mathbf{x}, t) \neq 0 \tag{3.2}$$

hold. Additionally there exist points $\mathbf{x} \in \Omega$ for which the plastic energy dissipation per unit volume

$$w_p(\mathbf{x}, t) = \int_0^t \sigma_{ij}(\mathbf{x}, s)\, \dot{\varepsilon}_{ij}^P(\mathbf{x}, s)\, ds \tag{3.3}$$

increases indefinitely with time t.

If the case of point 5 occurs, the system *shakes down* for the given load domain $\mathcal{L}$. It follows that

$$\lim_{t \to +\infty} \dot{\varepsilon}_{ij}^P(\mathbf{x}, t) \to 0 \qquad \forall\ \mathbf{x}\ \in \Omega \tag{3.4}$$

and

$$\lim_{t \to +\infty} \dot{\rho}_{ij}(\mathbf{x}, t) \to 0 \qquad \forall\ \mathbf{x}\ \in \Omega\,, \tag{3.5}$$

respectively.

The foregoing eqs. (3.4) and (3.5) show, that plastification decreases asymptotically and that the system changes from a plastic state into an elastic one. However, they are still too weak to avoid local material failure before shake-down has occurred, since they do not enable us to estimate the plastic energy dissipation per unit volume defined by eq. (3.3). In general, a material point can dissipate only a limited amount of energy. Thus, denoting the maximum possible energy dissipation per unit volume by $c(\mathbf{x})$, in addition to the eqs. (3.4) and (3.5), it would be necessary to require

$$w_p(\mathbf{x}) = \lim_{t \to \infty} w_p(\mathbf{x}, t) = \int_0^{+\infty} \sigma_{ij}(\mathbf{x}, t)\, \dot{\varepsilon}_{ij}^P(\mathbf{x}, t)\, dt \le c(\mathbf{x}) \quad \forall\ \mathbf{x} \in \Omega\,, \tag{3.6}$$

in order to avoid local material failure before the elastic state occurs. However, this makes the formulation of shake-down theorems essentially difficult (cf. [30]). For this reason, we shall restrict ourselves to the weaker conditions (3.4) and (3.5), i.e. we require, that

independent of the loading history the system has to change asymptotically from a plastic state into an elastic one.

By shake-down analysis we will mean that

for a given system and a given load domain $\mathcal{L}$ we want to know how much the domain $\mathcal{L}$ has to be decreased or can be increased, respectively, such that the system will shake down. By introducing a load factor β the problem can be formulated as follows: What is the value of β, such that the system will shake down for the increased ($\beta > 1$) or decreased ($\beta < 1$) load domain $\mathcal{L}_\beta$, respectively?

3.2 The static shake-down theorem of Melan

The necessary shake-down condition for a perfectly plastic material is that there exists at least one residual stress field $\bar{\rho}_{ij}(\mathbf{x})$, such that

$$\Phi\left[\sigma_{ij}^{E}(\mathbf{x},t)+\bar{\rho}_{ij}(\mathbf{x})\right]\leq\sigma_0^2 \qquad \forall\ \mathbf{x}\in\Omega \tag{3.7}$$

is satisfied for all possible loads $\mathbf{P}(t)$ within the given load domain $\mathcal{L}$. This can easily be derived from the requirements that after shake-down the plastic strains and the residual stresses remain stationary (time-independent) and that for all further loading within the load domain $\mathcal{L}$ the entire stresses $\sigma_{ij}(x,t)$, which consist of the load-dependent elastic stresses $\sigma_{ij}^{E}(\mathbf{x},t)$ and the stationary residual stresses, must be purely elastic.

The following static shake-down theorem due to MELAN states, that the necessary shake-down condition (3.7) in the sharper form (3.8) (see below) is also sufficient [38].

Theorem 3.1 (Melan,1938) *If there exist a time–independent residual stress field $\bar{\rho}_{ij}(\mathbf{x})$ and a factor $m>1$, such that*

$$\Phi[m\,(\sigma_{ij}^{E}(\mathbf{x},t)+\bar{\rho}_{ij}(\mathbf{x}))]\leq\sigma_0^2 \tag{3.8}$$

is satisfied for all loads $\mathbf{P}(t)$ in $\mathcal{L}$ and for all $\mathbf{x}$ in Ω, then the structure will shake down under the given load domain $\mathcal{L}$.

Proof: In order to prove Theorem 3.1, we consider the following non-negative functional

$$J(t)=\frac{1}{2}\int\limits_{\Omega}E_{ijkl}^{-1}\left[\rho_{ij}(\mathbf{x},t)-\bar{\rho}_{ij}(\mathbf{x})\right]\left[\rho_{kl}(\mathbf{x},t)-\bar{\rho}_{kl}(\mathbf{x})\right]d\Omega\geq 0\,. \tag{3.9}$$

By differentiation of $J(t)$ with respect to time t one obtains

$$\dot{J}(t)=\int\limits_{\Omega}E_{ijkl}^{-1}\left[\rho_{ij}(\mathbf{x},t)-\bar{\rho}_{ij}(\mathbf{x})\right]\dot{\rho}_{kl}(\mathbf{x},t)\,d\Omega\,. \tag{3.10}$$

Since the difference between two residual stress fields is also a residual stress field, eq. (2.41) yields

$$\int\limits_{\Omega}\left[\rho_{ij}(\mathbf{x},t)-\bar{\rho}_{ij}(\mathbf{x})\right]\left[E_{ijkl}^{-1}\,\dot{\rho}_{kl}(\mathbf{x},t)+\dot{\varepsilon}_{ij}^{P}(\mathbf{x},t)\right]d\Omega=0\,. \tag{3.11}$$

By substitution of eq. (3.11) into eq. (3.10) one gets

$$\begin{aligned}\dot{J}(t)&=-\int\limits_{\Omega}\left[\rho_{ij}(\mathbf{x},t)-\bar{\rho}_{ij}(\mathbf{x})\right]\dot{\varepsilon}_{ij}^{P}(\mathbf{x},t)\,d\Omega\\&=-\int\limits_{\Omega}\left[\sigma_{ij}^{E}(\mathbf{x},t)+\rho_{ij}(\mathbf{x},t)-\sigma_{ij}^{E}(\mathbf{x},t)-\bar{\rho}_{ij}(\mathbf{x})\right]\dot{\varepsilon}_{ij}^{P}(\mathbf{x},t)\,d\Omega\end{aligned}$$

$$
\begin{aligned}
&= -\int_{\Omega} [\,\sigma_{ij}(\mathbf{x},t) - (\sigma^E_{ij}(\mathbf{x},t) + \bar{\rho}_{ij}(\mathbf{x}))\,]\,\dot{\varepsilon}^P_{ij}(\mathbf{x},t)\,d\Omega \\
&\leq\ 0.
\end{aligned}
\tag{3.12}
$$

The inequality $(3.12)_4$ follows from inequality (2.19) and from the fact, that $\sigma^E_{ij}(\mathbf{x},t) + \bar{\rho}_{ij}(\mathbf{x})$ is a safe state of stresses according to shake-down condition (3.8), i.e. $\sigma^E_{ij}(\mathbf{x},t) + \bar{\rho}_{ij}(\mathbf{x}) \in \mathcal{E}$.

Equation $(3.12)_4$ shows, that the functional $J(t)$ is strictly decreasing. On the other hand $J(t)$ is bounded below. Thus, $J(t)$ converges to a stationary value. This means, that the residual stresses $\rho_{ij}(\mathbf{x},t)$ converge to stationary values as well. This implies that

$$
\lim_{t\to\infty} \dot{\rho}_{ij}(\mathbf{x},t) \to 0 \qquad \forall\ \mathbf{x} \in \Omega \tag{3.13}
$$

and

$$
\lim_{t\to\infty} \dot{\varepsilon}^P_{ij}(\mathbf{x},t) \to 0 \qquad \forall\ \mathbf{x} \in \Omega\,. \tag{3.14}
$$

(q.e.d.)

For the case of shake-down the following estimate for the system energy dissipation W_p can be given. By use of inequality (2.20) one obtains

$$
[\,\sigma_{ij}(\mathbf{x},t)\ -\ m\,(\sigma^E_{ij}(\mathbf{x},t) + \bar{\rho}_{ij}(\mathbf{x}))\,]\,\dot{\varepsilon}^P_{ij}(\mathbf{x},t) \geq 0\,. \tag{3.15}
$$

Reformulation of this inequality yields

$$
\begin{aligned}
(m-1)\,\sigma_{ij}(\mathbf{x},t)\,\dot{\varepsilon}^P_{ij}(\mathbf{x},t) &\leq m\,[\,\sigma_{ij}(\mathbf{x},t) - \sigma^E_{ij}(\mathbf{x}) - \bar{\rho}_{ij}(\mathbf{x})\,]\,\dot{\varepsilon}^P_{ij}(\mathbf{x},t) \\
&= m\,[\,\rho_{ij}(\mathbf{x},t) - \bar{\rho}_{ij}(\mathbf{x})\,]\,\dot{\varepsilon}^P_{ij}(\mathbf{x},t).
\end{aligned}
\tag{3.16}
$$

Finally, integration of eq. (3.16) over Ω and the time t yields the result

$$
\begin{aligned}
W_p &= \int_{\Omega}\int_0^{\infty} \sigma_{ij}(\mathbf{x},t)\,\dot{\varepsilon}^P_{ij}(\mathbf{x},t)\,dt\,d\Omega \\
&\leq \int_{\Omega}\int_0^{\infty} \frac{m}{m-1}\,[\,\rho_{ij}(\mathbf{x},t) - \bar{\rho}_{ij}(\mathbf{x})\,]\,\dot{\varepsilon}^P_{ij}\,dt\,d\Omega \\
&= -\frac{m}{m-1}\int_0^{\infty} \dot{J}(t)\,dt = \frac{m}{m-1}\,[J(0) - J(\infty)] \\
&\leq \frac{m}{m-1}\,J(0) = \frac{1}{2}\frac{m}{m-1}\int_{\Omega} E_{ijkl}\,\bar{\rho}_{ij}(\mathbf{x})\,\bar{\rho}_{kl}(\mathbf{x})\,d\Omega.
\end{aligned}
\tag{3.17}
$$

So far nothing was assumed about the load domain $\mathcal{L}$. It should be noted, that degeneration of $\mathcal{L}$ to a point implies monotone loading. Thus the plastic ultimate load theorem is also included in Theorem 3.1.

The static shake-down Theorem 3.1 by MELAN is formulated in terms of stresses and has bound character. A kinematic shake-down theorem, which was firstly given by KOITER [26, 27], is the counterpart to MELAN's theorem. Contrary to the static theorem the kinematic theorem is formulated in terms of kinematic quantities. From the mathematical point of view both theorems are dual to each other. Since in this paper the numerical treatment of shake-down problems is based on the static theorem, the formulation of the kinematic theorem shall not be given here.

4 Shake-down of kinematic hardening materials

4.1 Linear, unlimited kinematic hardening

4.1.1 The hardening model by Melan

In [39] MELAN proposed a model to describe material hardening, which is known as the *kinematic hardening model.* Later a similar model was proposed by PRAGER in [49].

It is assumed that the total stresses σ_{ij} can be decomposed according to

$$\sigma_{ij} = \nu_{ij} + \alpha_{ij} . \tag{4.1}$$

The part ν_{ij} is responsible for yielding. Similar as for perfectly plastic material ν_{ij} has to satisfy the yield condition

$$\Phi(\nu_{ij}) = \Phi(\sigma_{ij} - \alpha_{ij}) \leq \sigma_0^2 . \tag{4.2}$$

The rates of the quantities introduced above are described as follows:

1. If

$$\Phi(\sigma_{ij} - \alpha_{ij}) = \Phi(\nu_{ij}) < \sigma_0^2 , \tag{4.3}$$

and if σ_{ij} is changed by an infinitesimal value $\dot{\sigma}_{ij}\, dt$, then the material point remains elastic. In this case we have

$$\dot{\alpha}_{ij} = 0 \quad \text{and} \quad \dot{\varepsilon}_{ij}^{P} = 0 \tag{4.4}$$

and the rates for ν_{ij} and σ_{ij} are equal, i.e.

$$\dot{\nu}_{ij} = \dot{\sigma}_{ij} . \tag{4.5}$$

The same holds if

$$\Phi(\sigma_{ij} - \alpha_{ij}) = \Phi(\nu_{ij}) = \sigma_0^2 , \tag{4.6}$$

where the rate of σ_{ij} is directed inward or tangential to the yield surface defined by $\Phi(\cdot) = \sigma_0^2$, i.e.

$$\begin{aligned} \Phi(\nu_{ij} + \dot{\nu}_{ij}\, dt) &= \Phi(\sigma_{ij} + \dot{\sigma}_{ij}\, dt - \alpha_{ij}) - \Phi(\sigma_{ij} - \alpha_{ij}) \\ &= \frac{\partial \Phi}{\partial \sigma_{ij}} \dot{\sigma}_{ij} dt \leq 0 . \end{aligned} \tag{4.7}$$

Here we have unloading or neutral change of stress state, respectively.

2. However, if

$$\Phi(\sigma_{ij} - \alpha_{ij}) = \Phi(\nu_{ij}) = \sigma_0^2 \tag{4.8}$$

and

$$\frac{\partial \Phi}{\partial \sigma_{ij}} \dot{\sigma}_{ij} > 0, \tag{4.9}$$

then yielding occurs. This leads to a change of ε_{ij}^P and α_{ij}. During the yielding we require that

$$\Phi(\nu_{ij} + \dot{\nu}_{ij}\, dt) = \Phi(\sigma_{ij} + \dot{\sigma}_{ij}\, dt - \alpha_{ij} - \dot{\alpha}_{ij}\, dt) = \sigma_0^2 \tag{4.10}$$

holds, i.e. the stresses σ_{ij} have to remain on the yield surface. From eq. (4.10) the consistency condition

$$\frac{\partial \Phi}{\partial \sigma_{ij}} \dot{\sigma}_{ij} = \frac{\partial \Phi}{\partial \alpha_{ij}} \dot{\alpha}_{ij} \tag{4.11}$$

for $\dot{\sigma}_{ij}$ and $\dot{\alpha}_{ij}$ can be derived.

For $\dot{\varepsilon}_{ij}^P$ the associated flow rule

$$\dot{\varepsilon}_{ij}^P = \dot{\lambda} \frac{\partial \Phi}{\partial \sigma_{ij}} \tag{4.12}$$

is assumed as for elastic, perfectly plastic material. For $\dot{\alpha}_{ij}$ MELAN proposed the evolution law

$$\dot{\alpha}_{ij} = H' \, \dot{\varepsilon}_{ij}^P, \tag{4.13}$$

where H' denotes the constant hardening modulus of the one-dimensional $\sigma - \varepsilon$ curve. By use of the associated flow rule (4.12), eq. (4.13) can be written as

$$\dot{\alpha}_{ij} = \dot{\lambda}\, H' \frac{\partial \Phi}{\partial \sigma_{ij}} \,. \tag{4.14}$$

The quantities α_{ij} describe a shift of the initial yield surface in the stress spaces due to plastic deformations (see Figure 4.1). They are called *backstresses*.

For the one-dimensional stress state the hardening model of MELAN can describe the BAUSCHINGER-effect if unloading occurs. The geometrical interpretation is, that during the yielding the initial yield surface experiences a translation and moves from the origin to the point α_{ij} in the stress space without changing its form and size. Stresses, which are inside the shifted yield surface cause no further yielding, i.e. they are elastic.

Since the translation of the yield surface is not bounded, the model describes a linear, unlimited kinematic hardening. The linearity is mainly due to the fact, that the hardening modulus H' is constant. For the proposed model MELAN has formulated a static shake-down theorem in the paper [39], which shall be presented next.

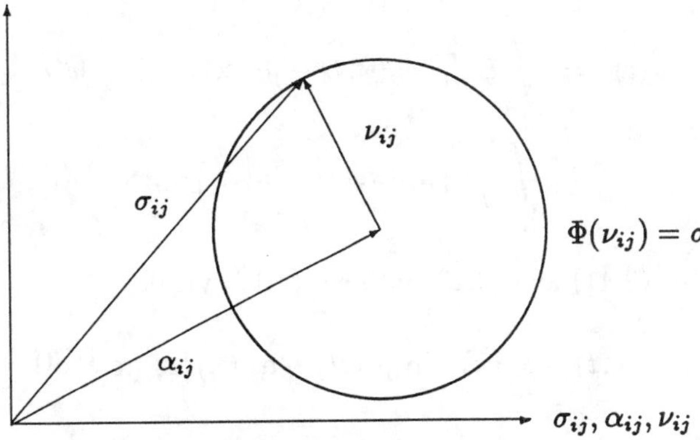

Figure 4.1: Shift of the initial yield surface in the stress space

4.1.2 A static shake-down theorem for linear, unlimited kinematic hardening material

The formulation given below is partly different to that one presented in the original paper by MELAN ([39]).

Theorem 4.1 (Melan,1938) *If there exist a time-independent residual stress field $\bar{\rho}_{ij}(\mathbf{x})$, a time-independent backstress field $\bar{\alpha}_{ij}(\mathbf{x})$ and a factor $m > 1$, such that the condition*

$$\Phi\left[\, m\left(\sigma_{ij}^E(\mathbf{x},t) + \bar{\rho}_{ij}(\mathbf{x}) - \bar{\alpha}_{ij}(\mathbf{x})\right)\right] \leq \sigma_0^2 \tag{4.15}$$

holds for all $\mathbf{P}(t)$ in $\mathcal{L}$ and for all $\mathbf{x}$ in Ω, then the system will shake down.

Let us notice, that the residual stresses $\bar{\rho}_{ij}(\mathbf{x})$ have to satisfy the homogeneous equilibrium condition (2.31) and boundary condition (2.32), while the backstresses $\bar{\alpha}_{ij}(\mathbf{x})$ are fully unconstrained.

Proof: In order to prove Theorem 4.1 we consider the following non-negative functional

$$\begin{aligned} J(t) &= \frac{1}{2}\int_\Omega E^{-1}_{ijkl}\left[\rho_{ij}(\mathbf{x},t) - \bar{\rho}_{ij}(\mathbf{x})\right]\left[\rho_{kl}(\mathbf{x},t) - \bar{\rho}_{kl}(\mathbf{x})\right] d\Omega \; + \\ & \quad \frac{1}{2}\int_\Omega \frac{1}{H'}\left[\alpha_{ij}(\mathbf{x},t) - \bar{\alpha}_{ij}(\mathbf{x})\right]\left[\alpha_{ij}(\mathbf{x},t) - \bar{\alpha}_{ij}(\mathbf{x})\right] d\Omega \\ &\geq 0\,, \end{aligned} \tag{4.16}$$

where $\rho_{ij}(\mathbf{x},t)$ and $\alpha_{ij}(\mathbf{x},t)$ are the actual residual stresses and backstresses at time t, respectively.

By differentiation of $J(t)$ with respect to time t one obtains

$$\dot{J}(t) = \int_\Omega E^{-1}_{ijkl}\,[\,\rho_{ij}(\mathbf{x},t) - \bar{\rho}_{ij}(\mathbf{x})\,]\,\dot{\rho}_{kl}(\mathbf{x},t)d\Omega + \int_\Omega \frac{1}{H'}\,[\,\alpha_{ij}(\mathbf{x},t) - \bar{\alpha}_{ij}(\mathbf{x})\,]\,\dot{\alpha}_{ij}(\mathbf{x},t)\,d\Omega\,. \tag{4.17}$$

Substitution of eqs. (2.41) and (4.13) into eq. (4.17) yields

$$\begin{aligned}\dot{J}(t) &= -\int_\Omega [\,\rho_{ij}(\mathbf{x},t) - \bar{\rho}_{ij}(\mathbf{x}) - \alpha_{ij}(\mathbf{x},t) + \bar{\alpha}_{ij}(\mathbf{x})\,]\,\dot{\varepsilon}^P_{ij}(\mathbf{x},t)\,d\Omega \\ &= -\int_\Omega [\,\sigma^E_{ij}(\mathbf{x},t) + \rho_{ij}(\mathbf{x},t) - \alpha_{ij}(\mathbf{x},t) - \sigma^E_{ij}(\mathbf{x},t) - \bar{\rho}_{ij}(\mathbf{x}) + \bar{\alpha}_{ij}(\mathbf{x})\,]\,\dot{\varepsilon}^P_{ij}\,d\Omega \\ &= -\int_\Omega [\,\sigma_{ij}(\mathbf{x},t) - \alpha_{ij}(\mathbf{x}) - (\sigma^E_{ij}(\mathbf{x},t) + \bar{\rho}_{ij}(\mathbf{x}) - \bar{\alpha}_{ij}(\mathbf{x}))\,]\,\dot{\varepsilon}^P_{ij}(\mathbf{x},t)\,d\Omega \\ &\le 0\,.\end{aligned} \tag{4.18}$$

The inequality $(4.18)_4$ follows from the fact, that $\sigma^E_{ij}(\mathbf{x},t)+\bar{\rho}_{ij}(\mathbf{x})-\bar{\alpha}_{ij}(\mathbf{x})$ defines a safe state of stresses with respect to the shifted yield surface according to the shake-down condition (4.15).

Since the functional $J(t)$ is bounded below, it follows from eq. $(4.18)_4$, that $J(t)$ converges to a stationary value for $t \to +\infty$. Therefore, the quantities $\rho_{ij}(\mathbf{x},t)$ and $\alpha_{ij}(\mathbf{x},t)$ also become stationary (q.e.d.).

The total plastic energy dissipation can be estimated as follows

$$\begin{aligned}W_p &= \int_\Omega \int_0^\infty \sigma_{ij}\,\dot{\varepsilon}^P_{ij}\,d\Omega\,dt \\ &\le \frac{1}{2}\frac{m}{m-1}\int_\Omega [\,E^{-1}_{ijkl}\,\bar{\rho}_{ij}(\mathbf{x})\,\bar{\rho}_{kl}(\mathbf{x}) + \frac{1}{H'}\,\bar{\alpha}_{ij}(\mathbf{x})\,\bar{\alpha}_{ij}(\mathbf{x})\,]\,d\Omega\,.\end{aligned} \tag{4.19}$$

An extension of Theorem 4.1 to materials with combined non-linear isotropic hardening can be found in a paper of MANDEL [33].

4.2 Necessary shake-down conditions for a limited kinematic hardening material

The model by MELAN showed that strain-hardening with BAUSCHINGER-effect can be described by introducing a shift of the initial yield surface in the stress space $\mathcal{S}$. For the MELAN model the shift of the initial yield surface is unconstrained and, therefore, the hardening is unlimited.

Apart from eq. (4.13) there are further rules which describe the shift of the yield surface such as ZIEGLER's rule [64] or MRÓZ's rule [40]. The MRÓZ's rule is able to describe a piecewise linear kinematic hardening. The assumption, that during the movement the yield surface remains its form and size, is common to these rules.

The backstresses will be constrained, if the value for the maximum hardening σ_Y is finite. In this case it is restricted by the requirement that the shifted initial yield surface remains within the hardening surface defined by $\Phi(\sigma_{ij}) = \sigma_Y^2$. Since the elastic domain $\mathcal{E}_{\alpha_{ij}}$ contained within the shifted initial yield surface is determined by

$$\mathcal{E}_{\alpha_{ij}} \equiv \{ \sigma_{ij} \mid \Phi(\sigma_{ij} - \alpha_{ij}) < \sigma_0^2 \}, \qquad (4.20)$$

mathematically the above requirement can be expressed by the condition

$$\Phi(\sigma_{ij}) < \sigma_Y^2 \qquad \forall\ \sigma_{ij} \in \mathcal{E}_{\alpha_{ij}}\,. \qquad (4.21)$$

The following two lemmata state that the condition

$$\Phi(\alpha_{ij}) \leq (\sigma_Y - \sigma_0)^2 \qquad (4.22)$$

is necessary and sufficient to ensure, that

i. the shifted elastic domain $\mathcal{E}_{\alpha_{ij}}$ is contained within the hardening surface,

ii. to each possible stress tensor σ_{ij} satisfying

$$\Phi(\sigma_{ij}) \leq \sigma_Y^2\,, \qquad (4.23)$$

there exist corresponding backstresses α_{ij} with property (4.22), such that $\sigma_{ij} \in \mathcal{E}_{\alpha_{ij}}$ holds.

Lemma 4.1 *Stresses σ_{ij}, which are contained in $\mathcal{E}_{\alpha_{ij}}$, satisfy condition* (4.21) *if the corresponding backstresses α_{ij} fulfill condition* (4.22).

Proof: With respect to the inequalities (2.30) and (4.22) we obtain

$$\begin{aligned} \sqrt{\Phi(\sigma_{ij})} &= \sqrt{\Phi(\sigma_{ij} - \alpha_{ij} + \alpha_{ij})} \\ &\leq \sqrt{\Phi(\sigma_{ij} - \alpha_{ij})} + \sqrt{\Phi(\alpha_{ij})} \\ &< \sigma_0 + (\sigma_Y - \sigma_0) \\ &= \sigma_Y \qquad \forall\ \sigma_{ij} \in \mathcal{E}_{\alpha_{ij}}\,. \end{aligned} \qquad (4.24)$$

(q.e.d.)

Lemma 4.2 *To all possible stresses σ_{ij}, which are contained in the hardening surface, i.e. which satisfy eq.* (4.23), *it is possible to find corresponding backstresses α_{ij} satisfying eq.* (4.22), *such that σ_{ij} are also contained in $\mathcal{E}_{\alpha_{ij}}$.*

Proof: To each possible stress tensor σ_{ij} with $\Phi(\sigma_{ij}) \leq \sigma_Y^2$ we construct backstresses α_{ij} of the form

$$\alpha_{ij} = \frac{\sigma_Y - \sigma_0}{\sigma_Y} \sigma_{ij} \,. \tag{4.25}$$

On one hand we have

$$\Phi(\alpha_{ij}) = \Phi\left[\frac{\sigma_Y - \sigma_0}{\sigma_Y}\sigma_{ij}\right] = \left[\frac{\sigma_Y - \sigma_0}{\sigma_Y}\right]^2 \Phi(\sigma_{ij}) \leq (\sigma_Y - \sigma_0)^2 \tag{4.26}$$

and on the other hand

$$\Phi(\sigma_{ij} - \alpha_{ij}) = \Phi\left[\frac{\sigma_0}{\sigma_Y}\sigma_{ij}\right] = \left[\frac{\sigma_0}{\sigma_Y}\right]^2 \Phi(\sigma_{ij}) \leq \sigma_0^2 \,. \tag{4.27}$$

(q.e.d.).

The results given by Lemmata 4.1 and 4.2 are also illustrated in Figure 4.2.

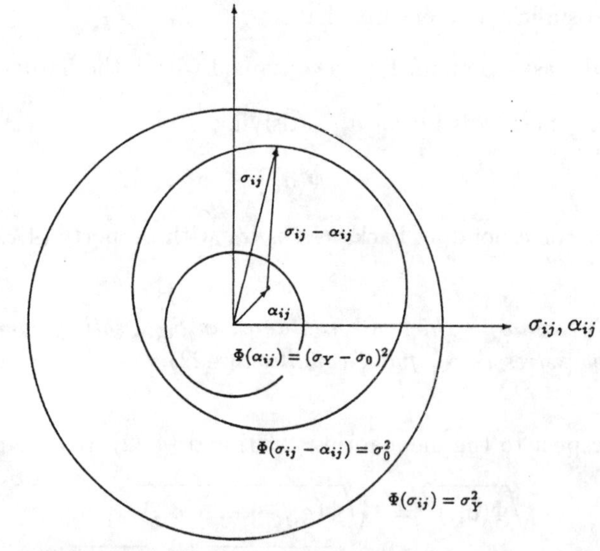

Figure 4.2: Limited kinematic hardening

After shake-down of a system consisting of limited kinematic hardening material the entire stresses, the sum of the load-dependent elastic stresses $\sigma_{ij}^E(\mathbf{x}, t)$ and the stationary residual stresses, must be fully contained in a shifted initial yield surface. Due to limitation of hardening, additionally, the stationary backstresses have to satisfy the condition (4.22).

For these reasons the following necessary shake-down conditions can easily be derived for a limited kinematic hardening material:

There must exist at least a time-independent backstress field $\bar{\alpha}_{ij}(\mathbf{x})$ satisfying

$$\Phi\left(\bar{\alpha}_{ij}(\mathbf{x})\right) \leq (\sigma_Y - \sigma_0)^2 \qquad (4.28)$$

and a time-independent residual stress field $\bar{\rho}_{ij}(\mathbf{x})$, such that

$$\Phi\left[\sigma^E_{ij}(\mathbf{x},t) + \bar{\rho}_{ij}(\mathbf{x}) - \bar{\alpha}_{ij}(\mathbf{x})\right] \leq \sigma_0^2 \qquad (4.29)$$

holds for all loads $\mathbf{P}(t)$ in $\mathcal{L}$ and for all $\mathbf{x}$ in Ω.

4.3 A 3-dimensional overlay model

In order to describe material hardening MASING [35] developed the following overlay-model: A rod (macrorod) with unit area consists of several rods (microrods), which behave elastic, perfectly plastic. All microrods have the same E-modulus, however they have different yield stresses. They are connected in parallel, i.e they have the same kinematics. Therefore the behavior of the macrorod is uniquely determined by the behavior of the microrods. In this way, if the number of microrods is finite, the overlay-model describes piecewise linear kinematic hardening. For description of non-linear kinematic hardening the number of microrods has to go to infinity.

In [45] NEAL formulated a static shake-down theorem for one-dimensional stress states by use of the overlay-model of MASING for description of non-linear kinematic hardening. NEAL used this indirect formulation, since a direct proof that for non-linear limited kinematic hardening material the necessary shake-down conditions are also sufficient ones, is not a trivial task.

In what follows we present a 3-dimensional overlay (microelement)-model (cf. STEIN et al. [56, 57]), which reduces to the model of MASING if the stress state is uniaxial. A similar model (so-called "fraction model") was introduced by BESSELING [5], however, there different theoretical hypotheses have been made.

It is assumed, that a given structure is composed of material points denoted by a vector variable $\mathbf{x} \in \Omega$, and each material point $\mathbf{x}$ is assumed to be composed of a dense spectrum of microelements numbered with a scalar variable $\xi \in [0,1]$.

Stresses and strains are defined separately for the material points (macroelement) as well as for the microelements according to Table 4.1.

Table 4.1: Macro- und microquantities

	macroquantities	microquantities
stresses	$\sigma_{ij}(\mathbf{x})$	$\psi_{ij}(\mathbf{x},\xi)$
strains	$\varepsilon_{ij}(\mathbf{x})$	$\eta_{ij}(\mathbf{x},\xi)$

The macroscopic stress field $\sigma_{ij}(\mathbf{x})$ must satisfy the static equilibrium condition (2.7) and the static boundary condition (2.8). The macroscopic strain field ε_{ij} is associated with the displacements u_i by the linear kinematic relation (2.4).

The basic relations between micro- and macroquantities are given by

$$\sigma_{ij}(\mathbf{x}) = \int_0^1 \psi_{ij}(\mathbf{x},\xi)d\xi \tag{4.30}$$

and

$$\varepsilon_{ij}(\mathbf{x}) = \eta_{ij}(\mathbf{x},\xi) \qquad \forall\ \xi \in [0,1]\,, \tag{4.31}$$

i.e., the microstrains in each microelement are equal to the macrostrains of the macroscopic material point, whereas the macrostresses are the resultants of the microstresses.

Equations (4.30) and (4.31) indicate that

$$\int_0^1 \psi_{ij}(\mathbf{x},\xi)\,\eta_{ij}(\mathbf{x},\xi)\,d\xi = \sigma_{ij}(\mathbf{x})\,\varepsilon_{ij}(\mathbf{x}) \tag{4.32}$$

holds and therefore, by substitution of eq. (4.32) into (2.9) one obtains

$$\int_\Omega\int_0^1 \psi_{ij}(\mathbf{x},\xi)\eta_{ij}(\mathbf{x},\xi)d\xi d\Omega = \int_\Omega b_i(\mathbf{x})\,u_i(\mathbf{x})\,d\Omega + \int_{\partial\Omega_\sigma+\partial\Omega_u} p_i(\mathbf{x})\,u_i(\mathbf{x})\,da\,. \tag{4.33}$$

The relation between microstrains and microstresses is assumed to be elastic, perfectly plastic. Let the yield stresses of the microelements be denoted by $k(\xi)$, then the following relations can be derived (for simplicity of notation in the following discussion the argument $\mathbf{x}$ of the fields is partly omitted):

$$\eta_{ij}(\xi) = \eta^E_{ij}(\xi) + \eta^P_{ij}(\xi) \tag{4.34}$$
$$\eta^E_{ij}(\xi) = E^{-1}_{ijkl}\psi_{kl}(\xi) \tag{4.35}$$
$$\Phi(\psi_{ij}) \leq k^2(\xi) \tag{4.36}$$
$$\dot{\eta}^P_{ij}(\xi) = \dot{\lambda}(\xi)\cdot\partial\Phi(\psi_{ij})/\partial\psi_{ij}\,, \quad \dot{\lambda}(\xi) \geq 0 \tag{4.37}$$
$$\dot{\lambda}(\xi)\,[\Phi(\psi_{ij}(\xi)) - k^2(\xi)] = 0\,. \tag{4.38}$$

Furthermore, it will be assumed that $k(\mathbf{x},\xi)$ is a monotone growing function of ξ (see Figure 4.3a). Thus we have

$$k(\mathbf{x},0) = k_0(\mathbf{x}) \leq k(\mathbf{x},1) = k_1(\mathbf{x})\,, \tag{4.39}$$

where $k_0(\mathbf{x}) = \sigma_0$ is the initial yield stress of the macroelement.

The function $k(\mathbf{x}, \xi)$ is uniquely determined for a given macroscopic $\sigma - \varepsilon$ function and vice versa. In the elastic-plastic range σ and ε are related with the function $k(\xi)$ by

$$\sigma(\xi) = \int_0^{\xi} k(\bar{\xi})\, d\bar{\xi} + (1 - \xi)\, k(\xi) \tag{4.40}$$

and

$$\varepsilon(\xi) = \frac{1}{E}\, k(\xi)\,. \tag{4.41}$$

Differentiation of eq. (4.40) with respect to ξ gives

$$\frac{d\sigma}{d\xi} = (1 - \xi)\, \frac{dk}{d\xi} \qquad \text{with} \qquad \frac{dk}{d\xi} \geq 0\,. \tag{4.42}$$

Eq. (4.42) shows that, similar to $k(\xi)$, σ is a monotone growing function of the variable ξ. The maximum value of σ (denoted by σ_Y) is obtained by setting $\xi = 1$ in (4.40), i.e.

$$\sigma_Y = \sigma(\xi = 1) = K = \int_0^1 k(\xi)\, d\xi\,. \tag{4.43}$$

Therefore $K (= \sigma_Y)$ can be identified as the ultimate stress of the macroelement. A possible $k(\xi)$-function and the resulting $\sigma - \varepsilon$ curve are represented in Figure 4.3a and Figure 4.3b, respectively.

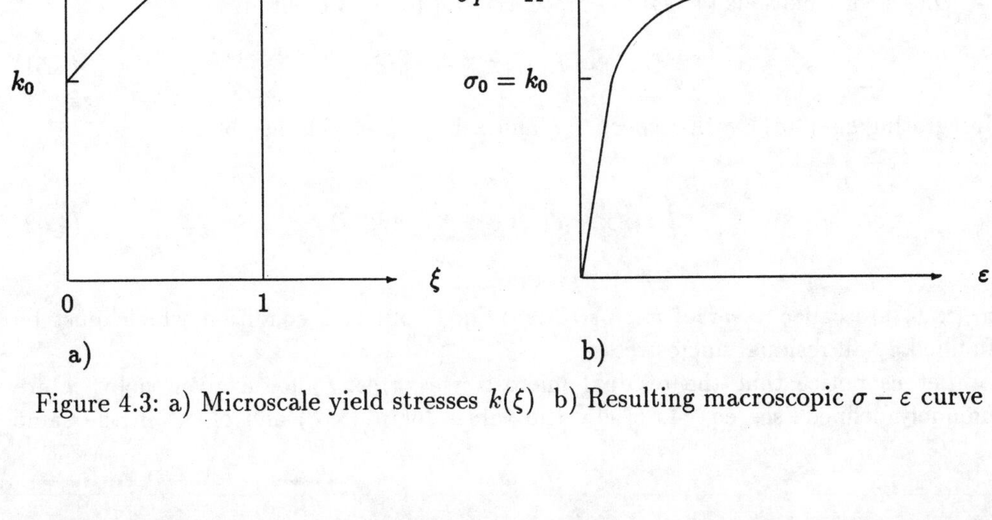

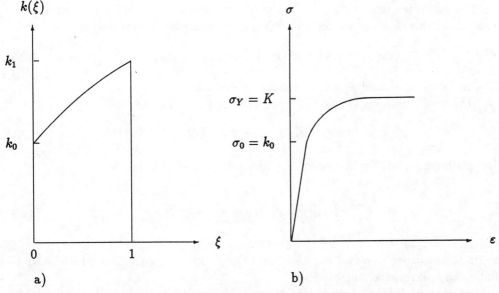

Figure 4.3: a) Microscale yield stresses $k(\xi)$ b) Resulting macroscopic $\sigma - \varepsilon$ curve

The following relations can be derived from equations (4.30), (4.31), (4.34) and (4.35)

$$\psi_{ij}(\xi) = E_{ijkl}\,[\,\varepsilon_{kl} - \eta^P_{kl}(\xi)\,] \tag{4.44}$$

$$\sigma_{ij} = E_{ijkl}\,[\,\varepsilon_{kl} - \int_0^1 \eta^P_{kl}(\xi)\,d\xi\,] \tag{4.45}$$

$$\varepsilon_{ij} = E^{-1}_{ijkl}\,\sigma_{kl} + \int_0^1 \eta^P_{ij}(\xi)\,d\xi \tag{4.46}$$

and therefore, with respect to the relation $\varepsilon^E_{ij} = E^{-1}_{ijkl}\sigma_{kl}$, the plastic macrostrains ε^P_{ij} are given by

$$\varepsilon^P_{ij} = \int_0^1 \eta^P_{ij}(\xi)\,d\xi\,. \tag{4.47}$$

In the absence of plastic microstrains within the whole body, we have the following relations between micro- and macroquantities:

$$\psi_{ij}(\mathbf{x},\xi) = \sigma^E_{ij}(\mathbf{x}) \tag{4.48}$$

$$\eta_{ij}(\mathbf{x},\xi) = \varepsilon^E_{ij}(\mathbf{x})\,. \tag{4.49}$$

Here σ^E_{ij}, ε^E_{ij} are the elastic solutions for prescribed b_i, p_i and u°_i.

The microstresses ψ_{ij} differ from the macrostresses σ_{ij} if plastic microstrains are present. In this case any microstress field $\psi_{ij}(\mathbf{x},\xi)$ can be decomposed as

$$\psi_{ij}(\mathbf{x},\xi) = \sigma_{ij}(\mathbf{x}) + \pi_{ij}(\mathbf{x},\xi) = \sigma^E_{ij}(\mathbf{x}) + \rho_{ij}(\mathbf{x}) + \pi_{ij}(\mathbf{x},\xi)\,. \tag{4.50}$$

The meaning of σ^E_{ij} and ρ_{ij} was explained in Section 2.3.

On use of equations (4.44), (4.45), (4.47) and (4.50) we obtain

$$\pi_{ij}(\xi) = E_{ijkl}\,[\,\varepsilon^P_{kl} - \eta^P_{kl}(\xi)\,]\,. \tag{4.51}$$

Integrating eq. (4.51) with respect to ξ and using eq. (4.47) one obtains

$$\int_0^1 \pi_{ij}(\mathbf{x},\xi)d\xi = 0 \qquad \forall\ \mathbf{x} \in \Omega\,. \tag{4.52}$$

$\pi_{ij}(\mathbf{x},\xi)$ are called *residual microstresses.* Eq. (4.52) is a condition which must be fulfilled by all residual microstresses.

Let us notice that the residual microstresses $\pi_{ij}(\mathbf{x},\xi)$ for a given point $\mathbf{x}$ are uniquely defined (see eq. (4.51)) by the values for $\eta^P_{ij}(\mathbf{x},\xi)$ and $\varepsilon^P_{ij}(\mathbf{x})$ in the same point whereas the residual macrostress field $\rho_{ij}(\mathbf{x})$ is uniquely defined by the whole plastic macrostrain field $\varepsilon_{ij}^P(\mathbf{x})$ (cf. KOITER [27]).

Taking account of the decomposition (4.50) for the microstresses, the micro- and macrostrains η_{ij} and ε_{ij} can be given the following form:

$$\eta_{ij}(\mathbf{x},\xi) = \varepsilon_{ij}(\mathbf{x}) = E_{ijkl}^{-1}\sigma_{kl}^E(\mathbf{x}) + E_{ijkl}^{-1}\rho_{kl}(\mathbf{x}) + E_{ijkl}^{-1}\pi_{kl}(\mathbf{x},\xi) + \eta_{ij}^P(\mathbf{x},\xi)\,. \qquad (4.53)$$

It is easy to show that the relations

$$\frac{1}{2}\left(u_{i,j}^E + u_{j,i}^E\right) = E_{ijkl}^{-1}\,\sigma_{kl}^E \qquad (4.54)$$

and

$$\begin{aligned} \frac{1}{2}\left(u_{i,j}^R + u_{j,i}^R\right) &= E_{ijkl}^{-1}\,\rho_{kl} + E_{ijkl}^{-1}\,\pi_{kl} + \eta_{ij}^P \\ &= E_{ijkl}^{-1}\,\rho_{kl} + \int_0^1 \eta_{ij}^P(\xi)\,d\xi \\ &= E_{ijkl}^{-1}\,\rho_{kl} + \varepsilon_{ij}^P \end{aligned} \qquad (4.55)$$

hold, where $u_i^E(\mathbf{x})$ and $u_i^R(\mathbf{x})$ are the elastic and residual part of the displacements (see Section 2.3). Note that the strain fields defined by eqs. (4.54) and (4.55) are kinematically compatible.

The model described above is an extension of the one-dimensional overlay model proposed by MASING [35]. It is also an extension of the model developed by BESSELING [5]. For complex load paths the elastic domain of the macroelement may not only be shifted but also shrink since the macroscale elastic domain is defined as the intersection of all elastic domains of the microelements represented by inequality (4.36). Nevertheless, this fact does not have any influence on the shake-down loads as the following static shake-down theorem shows.

4.4 A static shake-down theorem for the overlay model

For the overlay model presented above the following static shake-down theorem can be shown (cf. STEIN et al. [57, 58]):

Theorem 4.2 *If there exist a time-independent field $\bar{\alpha}_{ij}(\mathbf{x})$ satisfying*

$$\Phi\left[m\,\bar{\alpha}_{ij}(\mathbf{x})\right] \le \left[K(\mathbf{x}) - k_0(\mathbf{x})\right]^2, \qquad (4.56)$$

a time-independent residual macrostress field $\bar{\rho}_{ij}(\mathbf{x})$ and a factor $m > 1$ such that

$$\Phi\,\left[m\left[\sigma_{ij}^E(\mathbf{x},t) + \bar{\rho}_{ij}(\mathbf{x}) - \bar{\alpha}_{ij}(\mathbf{x})\right]\right] \le k_0^2(\mathbf{x}) \qquad (4.57)$$

holds for all possible loads $\mathbf{P}(t)$ contained in the given load domain $\mathcal{L}$ and for all points $\mathbf{x}$ in Ω, then the system will shake down.

It should be emphasized that eqs. (4.56) and (4.57) are a sharper form of the necessary shake-down conditions (4.28) and (4.29) for a general non-linear kinematic hardening material.

Proof: The proof of Theorem 4.2 is given in two steps.

(1) We show that for a given field $\bar{\alpha}_{ij}(\mathbf{x})$ satisfying eqs. (4.56) and (4.57) we can find a residual microstress field $\bar{\pi}_{ij}(\mathbf{x},\xi)$ which, together with $\bar{\rho}_{ij}(\mathbf{x})$, fulfills the condition

$$\Phi\,\left[m\,[\sigma_{ij}^E(\mathbf{x},t)+\bar{\rho}_{ij}(\mathbf{x})+\bar{\pi}_{ij}(\mathbf{x},\xi)\,]\,\right]\leq k^2(\mathbf{x},\xi)\qquad\forall\xi\in[0,1]\,. \tag{4.58}$$

For this purpose we restrict ourselves to a special class of residual microstresses

$$\bar{\pi}_{ij}(\mathbf{x},\xi)=\mu\,[\frac{k(\mathbf{x},\xi)}{K(\mathbf{x})}-1]\,\bar{\alpha}_{ij}(\mathbf{x})\,, \tag{4.59}$$

where

$$\mu=\frac{K(\mathbf{x})}{K(\mathbf{x})\;-\;k_0(\mathbf{x})}\,. \tag{4.60}$$

By setting $\xi\,=\,0$ in eq. (4.59) we obtain the following relation between $\bar{\alpha}_{ij}(\mathbf{x})$ and $\bar{\pi}_{ij}(\mathbf{x},0)$

$$\bar{\pi}_{ij}(\mathbf{x},0)=-\bar{\alpha}_{ij}(\mathbf{x})\,. \tag{4.61}$$

For the constructed residual microstresses $\bar{\pi}_{ij}(\mathbf{x},\xi)$ we get, on use of eqs. (2.30), (4.56) and (4.57)

$$\begin{aligned}
&\sqrt{\Phi\,\left[m[\sigma_{ij}^E(\mathbf{x},t)+\bar{\rho}_{ij}(\mathbf{x})+\bar{\pi}_{ij}(\mathbf{x},\xi)]\right]}\\
=\;&\sqrt{\Phi\,\left[m[\sigma_{ij}^E(\mathbf{x},t)+\bar{\rho}_{ij}(\mathbf{x})-\bar{\alpha}_{ij}(\mathbf{x})]+m[\bar{\pi}_{ij}(\mathbf{x},\xi)+\bar{\alpha}_{ij}(\mathbf{x})]\right]}\\
\leq\;&\sqrt{\Phi\,\left[m[\sigma_{ij}^E(\mathbf{x},t)+\bar{\rho}_{ij}(\mathbf{x})-\bar{\alpha}_{ij}(\mathbf{x})]\right]}\;+\;\sqrt{\Phi\,\left[m[\bar{\pi}_{ij}(\mathbf{x},\xi)+\bar{\alpha}_{ij}(\mathbf{x})]\right]}\\
\leq\;&k_0(\mathbf{x})+\sqrt{\Phi\left[m\,\frac{k(\mathbf{x},\xi)-k_0(\mathbf{x})}{K(\mathbf{x})-k_0(\mathbf{x})}\,\bar{\alpha}_{ij}(\mathbf{x})\right]}\\
=\;&k_0(\mathbf{x})+\frac{k(\mathbf{x},\xi)-k_0(\mathbf{x})}{K(\mathbf{x})-k_0(\mathbf{x})}\sqrt{\Phi\,[m\,\bar{\alpha}_{ij}(\mathbf{x})]}\\
\leq\;&k_0(\mathbf{x})+k(\mathbf{x},\xi)-k_0(\mathbf{x})\;=\;k(\mathbf{x},\xi)\qquad\qquad\forall\xi\in[0,1]\,,
\end{aligned} \tag{4.62}$$

which proves inequality (4.58).

The construction of the residual microstresses in eq. (4.59) has a nice geometrical interpretation: To each shifted initial yield surface characterized by the backstresses $\bar{\alpha}_{ij}$ and contained within the maximal hardening surface – defined by $\Phi(\sigma_{ij})\;=\;K^2$ – we can find a corresponding residual microstress distribution $\bar{\pi}_{ij}(\xi)$ such that the elastic domain of the macroelement, defined as the intersection of elastic domains of all microelements, coincides with the shifted initial elastic domain.

(2) Now, to prove the static shake-down Theorem 4.2, let us consider the following non-negative functional

$$J(t) = \frac{1}{2}\int_{\Omega}\int_{0}^{1} E^{-1}_{ijkl}[\rho_{ij}(\mathbf{x},t) - \bar{\rho}_{ij}(\mathbf{x}) + \pi_{ij}(\mathbf{x},\xi,t) - \bar{\pi}_{ij}(\mathbf{x},\xi)] \times$$
$$[\rho_{kl}(\mathbf{x},t) - \bar{\rho}_{kl}(\mathbf{x}) + \pi_{kl}(\mathbf{x},\xi,t) - \bar{\pi}_{kl}(\mathbf{x},\xi)]\, d\xi\, d\Omega \geq 0\,, \tag{4.63}$$

where $\rho_{ij}(\mathbf{x},t)$ and $\pi_{ij}(\mathbf{x},\xi,t)$ denote the actual residual macro- and microstress fields, respectively.

The time derivative of the functional $J(t)$ is

$$\dot{J}(t) = \int_{\Omega}\int_{0}^{1} E^{-1}_{ijkl}(\rho_{ij} - \bar{\rho}_{ij} + \pi_{ij} - \bar{\pi}_{ij})\,(\dot{\rho}_{kl} + \dot{\pi}_{kl})\, d\xi\, d\Omega\,. \tag{4.64}$$

In view of the definition of residual stresses and with respect to eqs. (4.33), (4.52) and (4.55) the following equality holds

$$\int_{\Omega}\int_{0}^{1} [\rho_{ij} - \bar{\rho}_{ij} + \pi_{ij} - \bar{\pi}_{ij}][E^{-1}_{ijkl}(\dot{\rho}_{kl} + \dot{\pi}_{kl}) + \dot{\eta}^{P}_{ij}]\, d\xi\, d\Omega = 0\,. \tag{4.65}$$

This implies that

$$\begin{aligned} \dot{J}(t) &= -\int_{\Omega}\int_{0}^{1} (\rho_{ij} - \bar{\rho}_{ij} + \pi_{ij} - \bar{\pi}_{ij})\,\dot{\eta}^{P}_{ij}\, d\xi\, d\Omega \\ &= -\int_{\Omega}\int_{0}^{1} [(\sigma^{E}_{ij} + \rho_{ij} + \pi_{ij}) - (\sigma^{E}_{ij} + \bar{\rho}_{ij} + \bar{\pi}_{ij})]\,\dot{\eta}^{P}_{ij}\, d\xi\, d\Omega \\ &= -\int_{\Omega}\int_{0}^{1} [\psi_{ij} - (\sigma^{E}_{ij} + \bar{\rho}_{ij} + \bar{\pi}_{ij})]\,\dot{\eta}^{P}_{ij}\, d\xi\, d\Omega \\ &\leq 0\,. \end{aligned} \tag{4.66}$$

The inequality $(4.66)_4$ follows from the convexity of the yield function $\Phi(\cdot)$ and from the associated flow rule (4.37) in view of the inequality (4.58). Therefore we have

$$J(0) \geq J(t) \qquad \forall\, t > 0 \tag{4.67}$$

and

$$\dot{J} \to 0, \;\; J \to \text{const. for} \;\; t \to \infty\,. \tag{4.68}$$

The relation (4.68) implies that the actual residual stresses $\rho_{ij}(\mathbf{x},t)$ and $\pi_{ij}(\mathbf{x},\xi,t)$ will be stationary. As a consequence the system will shake down (q.e.d.).

An upper bound to the total plastic dissipation energy can also be established. The inequality (4.58) implies that

$$[\psi_{ij} - m(\sigma_{ij}^E + \bar{\rho}_{ij} + \bar{\pi}_{ij})]\,\dot{\eta}_{ij}^P \geq 0 \tag{4.69}$$

holds, which after some reformulations gives

$$\begin{aligned}(m-1)\,\psi_{ij}\dot{\eta}_{ij}^P &\leq m[\sigma_{ij}^E + \rho_{ij} + \pi_{ij} - \sigma_{ij}^E - \bar{\rho}_{ij} - \bar{\pi}_{ij}]\,\dot{\eta}_{ij}^P \\ &= m\,(\rho_{ij} - \bar{\rho}_{ij} + \pi_{ij} - \bar{\pi}_{ij})\,\dot{\eta}_{ij}^P\,. \end{aligned} \tag{4.70}$$

By integrating eq. (4.70) over all micro- and macroelements we obtain

$$(m-1)\int_\Omega\int_0^1 \psi_{ij}\,\dot{\eta}_{ij}^P\,d\xi\,d\Omega \leq m\,(-\dot{J})\,. \tag{4.71}$$

Integrating inequality (4.71) with respect to time t we get the final result

$$\begin{aligned} W_p &= \int_0^t\int_\Omega\int_0^1 \psi_{ij}\,\dot{\eta}_{ij}^P\,d\xi\,d\Omega\,dt \leq \frac{m}{m-1}\,[J(0) - J(t)] \\ &\leq \frac{m}{m-1}J(0) = \frac{m}{2(m-1)}\int_\Omega\int_0^1 E_{ijkl}^{-1}\,(\bar{\rho}_{ij}\bar{\rho}_{kl} + \bar{\pi}_{ij}\bar{\pi}_{kl})\,d\xi\,d\Omega \\ &= \frac{m}{2(m-1)}\int_\Omega E_{ijkl}^{-1}\,[\bar{\rho}_{ij}\bar{\rho}_{kl} \\ &+ \bar{\alpha}_{ij}\bar{\alpha}_{kl}(K-k_0)^{-2}(\int_0^1 k^2(\xi)d\xi - K^2)]\,d\Omega\,. \end{aligned} \tag{4.72}$$

In order to ensure that W_p is limited the pathological case $\lim\limits_{\xi\to 1} k(\xi) \to +\infty$ must be excluded. Otherwise we would have a macroscopic $\sigma - \varepsilon$ curve for which the maximum hardening σ_Y can only be reached with an infinite strain (cf. eq. (4.41)).

Concerning the relation (4.61) between $\bar{\alpha}_{ij}(\mathbf{x})$ and $\bar{\pi}_{ij}(\mathbf{x},0)$ we can reformulate the above theorem in the following way:

Corollar 4.1 *If there exist a time-independent residual microstress field $\bar{\pi}_{ij}(\mathbf{x},\xi)$ satisfying*

$$\Phi\,[\,m\,\bar{\pi}_{ij}(\mathbf{x},0)] \leq [\,K(\mathbf{x}) - k_0(\mathbf{x})\,]^2, \tag{4.73}$$

a time-independent residual macrostress field $\bar{\rho}_{ij}(\mathbf{x})$ and a factor $m > 1$ such that for all possible loads $\mathbf{P}(t)$ within the load domain $\mathcal{L}$ and for all $\mathbf{x} \in \Omega$, the condition

$$\Phi\ \left[m[\sigma_{ij}^E(\mathbf{x},t) + \bar{\rho}_{ij}(\mathbf{x}) + \bar{\pi}_{ij}(\mathbf{x},0)]\right]\ \leq k_0^2(\mathbf{x}) \qquad \forall\ \mathbf{x} \in \Omega \tag{4.74}$$

is fulfilled, then the system consisting of the proposed material model (overlay-model) will shake down.

The static shake-down Theorem 4.1 for an unlimited, linear kinematic hardening material is included in Theorem 4.2. This is shown by setting $K(\mathbf{x}) \to +\infty$ in eq. (4.56). In this case the constraint (4.56) imposed on the backstresses $\bar{\alpha}_{ij}(\mathbf{x})$ is dropped. A similar theorem for one-dimensional stress states was given by NEAL [45]. Therefore, the theorem presented here can also be considered as a generalization of NEAL's theorem to three-dimensional stress states.

An important conclusion can be drawn from the formulation of shake-down conditions in this section. Namely, the shake-down load for a structure consisting of the proposed material model does not depend on the function $k(\xi)$, and therefore it does not depend on the $\sigma-\varepsilon$ curve but, solely, on the magnitudes of σ_0 and σ_Y. However, the plastic energy dissipated before a state of shake-down has been attained, may depend on the shape of the $\sigma - \varepsilon$ curve (cf. eq. (4.72)). The consequence of this conclusion will be shown in connection with the numerical approach.

5 Finite-element-discretization

5.1 General considerations

The maximal enlargement of the load domain $\mathcal{L}$ (characterized by the load factor β) for systems consisting of linear-elastic, perfectly plastic material can be determined by solving the following optimization problem (based on the static shake-down theorem):

$$\beta \to \max \tag{5.1}$$

subject to

$$\int_\Omega \delta\varepsilon_{mn}(\mathbf{x})\, \rho_{mn}(\mathbf{x}) d\Omega = 0 \tag{5.2}$$

$$\int_\Omega \delta\varepsilon_{mn}(\mathbf{x})\sigma_{mn}^E(\mathbf{x},t) d\Omega = \int_\Omega \delta u_m(\mathbf{x})\, b_m(\mathbf{x},t)\, d\Omega + \int_{\partial\Omega_\sigma} \delta u_m(\mathbf{x})\, p_m(\mathbf{x},t)\, da\,, \tag{5.3}$$

$$\Phi[\beta\sigma_{mn}^E(\mathbf{x},t) + \rho_{mn}(\mathbf{x})] \leq \sigma_0^2(\mathbf{x}) \qquad \forall\ \mathbf{x} \in \Omega\,. \tag{5.4}$$

Here

$$\delta\varepsilon_{mn} = \frac{1}{2}(\delta u_{m,n} + \delta u_{n,m}) \quad \text{with} \quad \delta u_m = 0 \quad \text{on} \quad \partial\Omega_u \tag{5.5}$$

represents an arbitrary virtual kinematically admissible strain field.

The above static formulation (5.1)–(5.4) has bounding character, namely, the load factor β provides a lower bound to the exact load factor, if we are able to find a certain residual stress field $\rho_{mn}(\mathbf{x})$ such that the MELAN condition (5.4) is satisfied for all $\mathbf{x} \in \Omega$. It should be emphasized that the bounding character is valid only if the following three conditions are satisfied simultaneously:

1. Exactness of the elastic stresses $\sigma^E_{mn}(\mathbf{x}, t)$.

2. Fulfillment of the static equilibrium condition (2.31) and the static boundary condition (2.32) for the residual stress distribution $\rho_{mn}(\mathbf{x})$ for all $\mathbf{x}$ in Ω and on the surface $\partial\Omega_\sigma$.

3. Fulfillment of the yield condition (5.4) for all $\mathbf{x}$ in Ω.

Some authors, e.g. BELYTSCHKO [3], WEICHERT and GROSS-WEEGE [61], when analyzing plane stress problems, used the AIRY stress function to construct the required elastic and residual stress fields and hoped, in this way, to obtain a load factor β, which is a lower bound to the exact shake-down load factor. However, this approach is not always successful due to the following difficulties:

1. In general the elastic stresses $\sigma^E_{mn}(\mathbf{x}, t)$ determined by use of the finite-element-method are not exact.

2. If the chosen stress functions are complex, the MELAN condition (5.4) can hardly be satisfied for all $\mathbf{x} \in \Omega$.

Thus, apart from some exceptions, load factors determined numerically possess no bounding character. This holds also for hardening material.

In the sequel numerical shake-down investigations are based on the static shake-down theorems. This avoids the non-differentiability of the objective function, which arises by use of the kinematic formulation (cf. [62]).

5.2 Discretization of systems with perfectly plastic material

As mentioned above, in general no numerical method preserves the bounding character of the determined load factor. In this context it is irrelevant whether a displacement method or a stress method is used. However, concerning the quality of the approximation or convergence, respectively, both methods can differ considerably.

Since the static shake-down theorems are formulated in terms of stresses, it may be meaningful to use stress-methods [3] [61]. For these methods, special stress functions have to be chosen, in order to satisfy a priori the equilibrium conditions for the stresses. However, in doing so the method is restricted, since e.g. for plate and shell problems it may be difficult to find appropriate stress functions. Thus, in order to make the numerical approach as general as possible we prefer a displacement method.

We proceed as follows:

In order to solve the dual problem

$$\begin{aligned} W(\boldsymbol{\mu}, \boldsymbol{\lambda}) &\rightarrow \max \\ \boldsymbol{\lambda} &\geq 0, \end{aligned} \tag{6.25}$$

a projection method according to the iteration process

$$\begin{bmatrix} \boldsymbol{\mu}^{j+1} \\ \boldsymbol{\lambda}^{j+1} \end{bmatrix} = \mathcal{P}\left\{ \begin{bmatrix} \boldsymbol{\mu}^{j} \\ \boldsymbol{\lambda}^{j} \end{bmatrix} + \alpha \mathbf{H}^j \nabla W^j \right\} \tag{6.26}$$

is used, where the projection operator $\mathcal{P}$ is defined as

$$\mathcal{P}\begin{bmatrix} \boldsymbol{\mu} \\ \boldsymbol{\lambda} \end{bmatrix} = \begin{bmatrix} \boldsymbol{\mu} \\ \max(0, \lambda_1) \\ \vdots \\ \max(0, \lambda_{m_2}) \end{bmatrix} . \tag{6.27}$$

Note that the gradient of $W(\boldsymbol{\mu}, \boldsymbol{\lambda})$ is given by

$$\nabla W(\boldsymbol{\mu}, \boldsymbol{\lambda}) = \begin{bmatrix} \mathbf{C}\Delta\boldsymbol{\rho}(\boldsymbol{\mu}, \boldsymbol{\lambda}) \\ \boldsymbol{\Phi}(\Delta\boldsymbol{\beta}(\boldsymbol{\lambda}), \Delta\boldsymbol{\rho}(\boldsymbol{\mu}, \boldsymbol{\lambda})) \end{bmatrix} , \tag{6.28}$$

where $\Delta\boldsymbol{\beta}(\boldsymbol{\lambda})$ and $\Delta\boldsymbol{\rho}(\boldsymbol{\mu}, \boldsymbol{\lambda})$ solve the minimization problem (6.22).

If the iteration matrix $\mathbf{H}^j$ in eq. (6.26) is the unity matrix, then the method is known as the UZAWA-Algorithm. However, in this case convergence would be very poor and instead we use an algorithm which is due to BERTSEKAS [4]. For this purpose let us introduce the working-set

$$\mathcal{I}_w(\boldsymbol{\mu}, \boldsymbol{\lambda}) = \left\{ j \mid \lambda_j = 0 \, , \frac{\partial W}{\partial \lambda_j} < 0 \right\} . \tag{6.29}$$

Given a matrix $\mathbf{H}^*$ we say that a symmetrical $m \times m$ matrix $\mathbf{H}$ with elements H_{ij} is diagonal with respect to a subset $\mathcal{I}_w$ if

$$\mathbf{H} = \begin{cases} 0 & \text{, if } i \neq j \text{ and } (i \in \mathcal{I}_w \text{ or } j \in \mathcal{I}_w) \\ H^*_{ij} & \text{, otherwise.} \end{cases} \tag{6.30}$$

In [4] it is shown that, provided the line-search parameter α^j is chosen properly, the algorithm serves as an ascent-method. Concerning the choice of $\mathbf{H}^*$ in eq. (6.30) a good candidate would be the inverse of

$$\nabla^2 W = \begin{bmatrix} \mathbf{0} \; \mathbf{C} \\ \nabla\boldsymbol{\Phi}^T \end{bmatrix} [\mathbf{A}^{-1}] \begin{bmatrix} \mathbf{0}^T & \nabla\boldsymbol{\Phi} \\ \mathbf{C}^T & \end{bmatrix} . \tag{6.31}$$

The structure of this matrix (and using the comments on Figure 6.2) can be seen in Figure 6.3. Note that the marked area corresponds to the stiffness matrix of the linear

and

$$\boldsymbol{\rho}^T = [\boldsymbol{\rho}_1^T, \dots, \boldsymbol{\rho}_i^T, \dots, \boldsymbol{\rho}_{NG}^T], \tag{5.11}$$

respectively.

Actually, the MELAN condition (5.4) has to be satisfied at all points of the discretized structure. However, since the load factor loses its bounding character due to the FE-discretization, we shall only require that this condition is fulfilled at all Gaussian points, i.e.

$$\Phi[\beta\boldsymbol{\sigma}_i^E(t) + \boldsymbol{\rho}_i] \le \sigma_{i0}^2 \qquad \forall\ i \in \mathcal{I} = [1, 2, \dots, NG]. \tag{5.12}$$

By σ_{i0} we denote the initial yield stress at the i-th Gaussian point. Note that this can be different from one Gaussian point to another. In this way it is easy to take into account inhomogeneous materials. For systems with a homogeneous material the index i in σ_{i0} may be suppressed.

Now we are able to present the discretized static formulation for determination of the maximal shake-down load factor β for problems with homogeneous linear-elastic, perfectly plastic material as follows:

$$\beta \rightarrow \max \tag{5.13}$$

$$\sum_{i=1}^{NG} \mathrm{C}_i \cdot \boldsymbol{\rho}_i = \mathrm{C} \cdot \boldsymbol{\rho} = 0 \tag{5.14}$$

$$\Phi[\beta\boldsymbol{\sigma}_i^E(t) + \boldsymbol{\rho}_i] \le \sigma_0^2 \qquad \forall\ i \in \mathcal{I}. \tag{5.15}$$

Equations (5.13)–(5.15) represent a non-linear mathematical optimization problem with constraints. In eq. (5.13) β is the objective function to be maximized and equations (5.14)–(5.15) represent linear equality constraints and non-linear inequality constraints, respectively. The load factor β and the residual stress vector $\boldsymbol{\rho}$ are the unknowns of problem (5.13)–(5.15).

5.3 Discretization of systems consisting of linear, unlimited kinematic hardening material

For discretization of systems consisting of unlimited kinematic hardening material we start from the static shake-down Theorem 4.1. Note that the backstresses are unconstrained quantities.

The discretization of the equilibrium conditions for the residual stresses is done in the same way as for systems consisting of linear-elastic, perfectly plastic material. By introducing a backstress vector $\boldsymbol{\alpha}_i$ at each Gaussian point, the optimization problem for shake-down investigation of systems consisting of linear, unlimited kinematic hardening material is given as follows:

$$\beta \rightarrow \max \tag{5.16}$$

6.1.3.4 A special SQP-algorithm

In order to avoid the difficulties as described in the previous section, next, we will present a special SQP-algorithm, which takes into account the special structure of problem P.

The Lagrangian of P is given in eq. (6.7). Calculating the first and second derivatives ∇l and $\nabla^2 l = \mathbf{L}$, it can be observed, that a simple structure arises. This is demonstrated in Figure (6.2), where for the case $M = 1$, $NSK = 3$, $NFR = 8$, $NEL = 3$ and $NG = 2 \times 3 = 6$ the structures of both derivatives are shown. Note that if a Gaussian point is not active, the Hessian $\mathbf{L}^k$ is singular. In Figure (6.2) this case corresponds to the second Gaussian point in the second element.

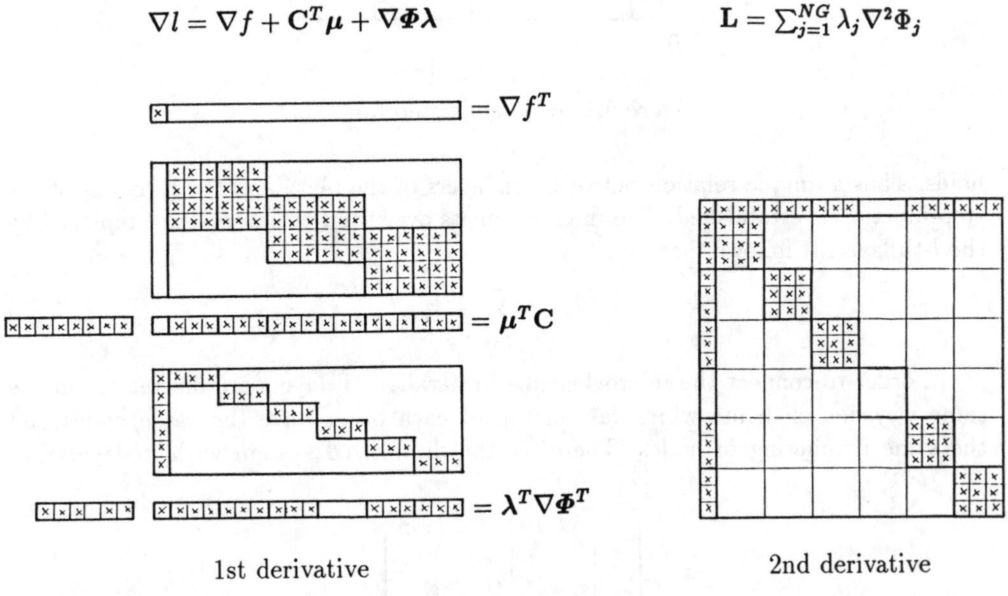

Figure 6.2: First and second derivative of the LAGRANGE-function

Formulation of a quadratic, separable subproblem

Instead of using $\mathbf{L}^k$ for construction of a subprolem, we use a matrix $\mathbf{A}^k$ with diagonal elements L^k_{ss}, $s = 0, \ldots, n$. Those elements, for which $L^k_{ss} = 0$, are replaced by small values $\varepsilon\sqrt{\sum_{s\in\mathcal{M}} L^2_{ss}}$, $\mathcal{M} = \{s = 0, ..., n | L_{ss} \neq 0\}$ (e.g. $\varepsilon = 10^{-4}$) to guarantee the positive definiteness of $\mathbf{A}^k$. Thus the LAGRANGE-function of Q^k may be written as

$$l^k = +\boldsymbol{\mu}^T\mathbf{C}\boldsymbol{\rho}^k + \boldsymbol{\lambda}^T\boldsymbol{\Phi}^k + \sum_{s=0}^{n} l^k_s, \tag{6.17}$$

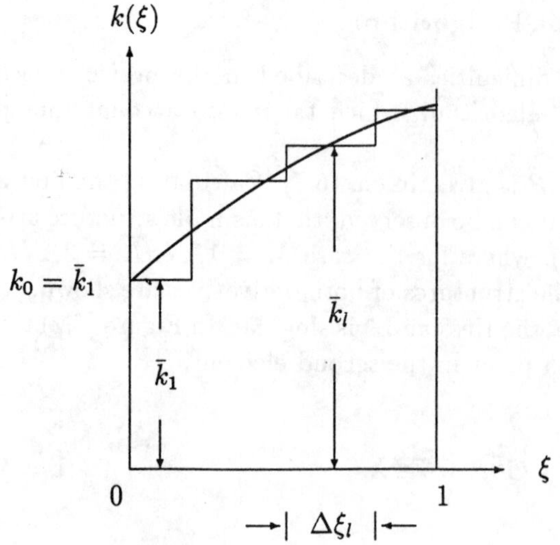

Figure 5.1: A possible step-function

holds. Thus a unique relation between the layers of the plate and the intervals of the step-function is established. The microelements over the l-th interval are replaced by the l-th layer. It follows that

$$K = \sum_{l=1}^{m} \frac{t_l}{t} \bar{k}_l \,. \tag{5.25}$$

In order to connect the microelements in parallel, all layers are discretized in the same way, i.e. elements which lay on top of each other have the same nodes and the same numbering of nodes. Therefore the discretized system with m layers has

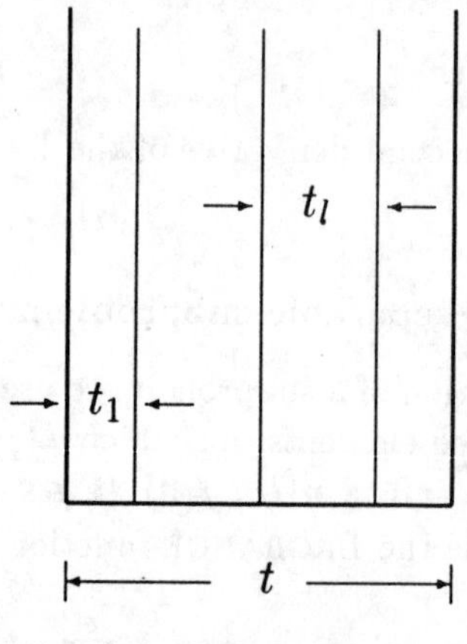

Figure 5.2: Dividing of the thickness of the plate

$(m-1)$ times more elements than the system consisting of linear-elastic, perfectly plastic material with the same discretization.

The investigations in Section 4.3 showed that the stresses $\psi_{mn}(\mathbf{x},\xi)$ in a microelement can be decomposed as

$$\psi_{mn}(\mathbf{x},\xi) = \sigma^E_{mn}(\mathbf{x}) + \rho_{mn}(\mathbf{x}) + \pi_{mn}(\mathbf{x},\xi)\,, \tag{5.26}$$

where $\rho_{mn}(\mathbf{x})$ and $\pi_{mn}(\mathbf{x},\xi)$ represent residual macro- and microstresses, respectively. If $k(\xi)$ is a step-function, then the residual macro- and microstresses have constant values in an interval $\Delta\xi_l$.

The total residual stresses in a microelement are the sum of the residual micro- and macrostresses. In many cases the total residual stresses at the microelements are known, and we want to determine the corresponding residual micro- and macrostresses. This shall be demonstrated for a discretized system with m layers:

For presentation some new notations are introduced. Let NG be the total number of Gaussian points of the discretized macroscopic system. The index i denoting Gaussian points $\mathbf{x}_i$ is going from 1 to NG. To distinguish the different microelements at the i-th Gaussian point the pair of indices (i,l) is introduced, where the index l refers to the microelements of the l-th layer (or equivalently of the l-th interval).

The vector of the total residual stresses of the l–th layer at the i–th Gaussian point shall be denoted by $\boldsymbol{\rho}_{i,l}$ and the related residual micro- and macrostresses are described by $\boldsymbol{\pi}_{i,l}$ and $\boldsymbol{\rho}_i$, respectively. According to eq. (4.52) the integration of the residual microstresses over the interval $[0,1]$ vanishes, i.e.

$$\sum_{l=1}^{m} \boldsymbol{\pi}_{i,l}\,\Delta\xi_l = \sum_{l=1}^{m} \boldsymbol{\pi}_{i,l}\,\frac{t_l}{t} = \mathbf{0}\,. \tag{5.27}$$

By use of

$$\boldsymbol{\rho}_{i,l} = \boldsymbol{\rho}_i + \boldsymbol{\pi}_{i,l} \tag{5.28}$$

and

$$\sum_{l=1}^{m} \boldsymbol{\rho}_{i,l}\,\Delta\xi_l = \sum_{l=1}^{m} \boldsymbol{\rho}_{i,l}\,\frac{t_l}{t} = \sum_{l=1}^{m} (\boldsymbol{\rho}_i + \boldsymbol{\pi}_{i,l})\,\frac{t_l}{t} = \boldsymbol{\rho}_i + \sum_{l=1}^{m} \frac{t_l}{t}\,\boldsymbol{\pi}_{i,l} = \boldsymbol{\rho}_i \tag{5.29}$$

one obtains

$$\boldsymbol{\rho}_i = \sum_{l=1}^{m} \boldsymbol{\rho}_{i,l}\,\frac{t_l}{t}\,. \tag{5.30}$$

Finally, from eqs. (5.28) and (5.30) we derive

$$\boldsymbol{\pi}_{i,l} = \boldsymbol{\rho}_{i,l} - \boldsymbol{\rho}_i = \boldsymbol{\rho}_{i,l} - \sum_{s=1}^{m} \boldsymbol{\rho}_{i,s}\,\frac{t_s}{t}\,. \tag{5.31}$$

Using the above relations the discretized static formulation for shake-down analysis of systems made of materials with limited kinematic hardening can be written as

$$\beta \rightarrow \max \tag{5.32}$$

$$\sum_{i=1}^{NG} \mathrm{C}_i \cdot \sum_{s=1}^{m} \frac{t_s}{t} \boldsymbol{\rho}_{i,s} = \sum_{i=1}^{NG} \mathrm{C}_i \cdot \boldsymbol{\rho}_i = \mathbf{C} \cdot \boldsymbol{\rho} = 0 \tag{5.33}$$

$$\Phi(\beta\, \boldsymbol{\sigma}_i^E(t) + \boldsymbol{\rho}_{i,1}) \leq \bar{k}_1^2 = k_0^2 \qquad \forall\ i \in \mathcal{I} \tag{5.34}$$

$$\Phi(\boldsymbol{\pi}_{i,1}) \leq (K - \bar{k}_1)^2 = (K - k_0)^2 \qquad \forall\ i \in \mathcal{I}. \tag{5.35}$$

In eq. (5.33) the matrix **C** is the same as for a system consisting of a perfectly plastic material with the same discretization. The constraints (5.34) and (5.35) represent the shake-down conditions (4.74) and (4.73), respectively.

6 Numerical treatment of discretized shake-down problems

6.1 Treatment of systems consisting of perfectly plastic material

6.1.1 Elimination of the variable t

Up to now, for formulation of shake-down problems no restriction has been made to the load domain $\mathcal{L}$. Thus $\mathcal{L}$ can be of arbitrary form. However, in most practical cases the loads may vary independently of each other between some given bounds. If the number of independent loads is n, then the corresponding load domain is an n-dimensional polyhedron. In Figure 6.1 a possible load domain is shown for $n = 2$.

For subsequent considerations we shall restrict ourselves to load domains with a form of a convex polyhedron. The vertices of the load domain are numbered with an

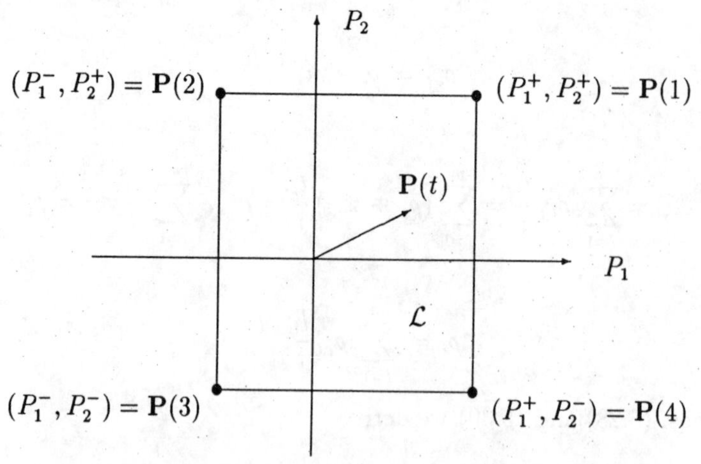

Figure 6.1: A possible load domain ($n = 2$)

index j. For a polyhedron with M vertices we have $1 \leq j \leq M$. By $\mathcal{J} = [1, 2, \ldots, M]$ we mean the set which contains all indices of the vertices. The load, which corresponds to the j-th vertex of $\mathcal{L}$, is denoted symbolically by $\mathbf{P}(j)$ $(j \in \mathcal{J})$. The elastic stress vector at the i-th Gaussian point, which corresponds to the load $\mathbf{P}(j)$, shall be denoted by $\boldsymbol{\sigma}_i^E(j)$.

In view of the convexity of the yield function $\Phi(\cdot)$ and due to the above assumption about the load domain $\mathcal{L}$ it can easily be shown that

$$\Phi[\beta \boldsymbol{\sigma}_i^E(t) + \boldsymbol{\rho}_i] \leq \sigma_0^2 \tag{6.1}$$

is fulfilled at any time t, if

$$\Phi[\beta \boldsymbol{\sigma}_i^E(j) + \boldsymbol{\rho}_i] \leq \sigma_0^2 \tag{6.2}$$

holds for all $j \in \mathcal{J}$. Thus, one can replace problem (5.13)–(5.15) by the following one:

$$\beta \rightarrow \max \tag{6.3}$$

subject to

$$\sum_{i=1}^{NG} \mathbf{C}_i \cdot \boldsymbol{\rho}_i = \mathbf{C} \cdot \boldsymbol{\rho} = \mathbf{0} \tag{6.4}$$

$$\Phi\,[\beta \boldsymbol{\sigma}_i^E(j) + \boldsymbol{\rho}_i] \leq \sigma_0^2 \qquad \forall\ (i,j) \in \mathcal{I} \times \mathcal{J}\,. \tag{6.5}$$

In the above form the time t is eliminated.

The number of unknowns of problem (6.3)–(6.5) is $NG \times NSK + 1$ and the number of constraints is $M \times NG + NFR$, where NG is the number of Gaussian points of the FEM-structure, NSK is the dimension of the stress vector at each Gaussian point and NFR denotes the degrees of freedom of displacements of the discretized structure. Thus, a discretized plane stress problem with 100 elements, 4 Gaussian points per element, 120 degrees of freedom and 4 vertices of the load domain $\mathcal{L}$ has $(4 \times 100) \times 3 + 1 = 1201$ unknowns and the total number of constraints is $100 \times 4 \times 4 + 120 = 1702$. This simple example demonstrates, that small or medium sized FE-problems can lead to a high dimension for the shake-down problem. Therefore, classical methods as the Sequential Quadratic Programming (SQP) method by use of the BFGS-matrix (see eg. [31] [53]) cannot be used as a black box because of high storage requirements. Instead, it is crucial to use optimization techniques which take account of the special structure for the matrices of problem (6.3)–(6.5). In what follows we give a brief overview of Sequential Quadratic Programming methods and propose two algorithms, a special SQP-algorithm and a reduced basis technique to solve the discretized shake-down problem (6.3)–(6.5).

6.1.2 First order optimal condition

In order to facilitate the notation, we replace problem (6.3)–(6.5) by the following set of equations

$$P: \qquad -\beta \ \rightarrow \ \min$$

$$
\begin{aligned}
\mathbf{C}\boldsymbol{\rho} &= 0 \\
\boldsymbol{\Phi}(\beta,\boldsymbol{\rho}) &\leq 0,
\end{aligned}
\tag{6.6}
$$

where $\boldsymbol{\Phi}(\beta,\boldsymbol{\rho}) \in \Re^{m_i}$, $m_i = M \times NG$ assembles all inequality constraints $\Phi(\beta\sigma_j^e(j) + \boldsymbol{\rho}_i) - \sigma_0^2$, $i = 1,\ldots,NG, j = 1,\ldots,M$. Thus a Lagrangian function to P can be defined as

$$
l(\beta,\boldsymbol{\rho},\boldsymbol{\mu},\boldsymbol{\lambda}) = -\beta + \boldsymbol{\mu}^T\mathbf{C}\boldsymbol{\rho} + \boldsymbol{\lambda}^T\boldsymbol{\Phi}(\beta,\boldsymbol{\rho}), \tag{6.7}
$$

where $\boldsymbol{\mu} = [\mu_1,\ldots,\mu_{m_e}]^T, m_e = NFR$ and $\boldsymbol{\lambda} = [\lambda_1,\ldots,\lambda_{m_i}]^T$ are LAGRANGE-multipliers for the equality and inequality constraints, respectively. The KUHN-TUCKER conditions, which are necessary for an optimal solution, are

$$
\begin{aligned}
\nabla l &= 0 && (6.8)\\
\mathbf{C}\boldsymbol{\rho} &= 0 && (6.9)\\
\boldsymbol{\Phi}(\beta,\boldsymbol{\rho}) &\leq 0 && (6.10)\\
\boldsymbol{\lambda} &\geq 0 && (6.11)\\
\boldsymbol{\lambda}^T\boldsymbol{\Phi}(\beta,\boldsymbol{\rho}) &= 0, && (6.12)
\end{aligned}
$$

where $\nabla(.)$ means $\nabla(.) = \left[\frac{\partial(.)}{\partial\beta}, \frac{\partial(.)}{\partial\rho_1}, \ldots \frac{\partial(.)}{\partial\rho_n}\right]$, $n = NG \times NSK$. For the subsequent analysis we will also use $m = m_i + m_e$.

6.1.3 Sequential quadratic programming

6.1.3.1 The basic idea

Analogously to problems without constraints problem P is solved iteratively by correction of a current iteration point $\mathbf{z}^{k^T} = [\beta^k, \boldsymbol{\rho}^{k^T}, \boldsymbol{\mu}^{k^T}, \boldsymbol{\lambda}^{k^T}]$. The correction $\Delta\mathbf{z}^k = [\Delta\beta^k, \Delta\boldsymbol{\rho}^{k^T}, \Delta\boldsymbol{\mu}^{k^T}, \Delta\boldsymbol{\lambda}^{k^T}]$ is calculated by solving the linearized system of KUHN-TUCKER equations (6.8)–(6.12) in the point $\mathbf{z}^k$. It can be shown that this is equivalent to solve a quadratic optimization subproblem

$$
\begin{aligned}
Q^k: \qquad & -\Delta\beta^k + \tfrac{1}{2}[\Delta\beta^k\,\Delta\boldsymbol{\rho}^{k^T}]\mathbf{L}(\mathbf{z}^k)\begin{bmatrix}\Delta\beta^k \\ \Delta\boldsymbol{\rho}^k\end{bmatrix} \longrightarrow \min \\
& \mathbf{C}\boldsymbol{\rho}^k + \mathbf{C}\Delta\boldsymbol{\rho}^k = 0 \\
& \boldsymbol{\Phi}(\beta^k,\boldsymbol{\rho}^k) + \nabla\boldsymbol{\Phi}^{k^T}\begin{bmatrix}\Delta\beta^k \\ \Delta\boldsymbol{\rho}^k\end{bmatrix} \leq 0,
\end{aligned}
\tag{6.13}
$$

where $\mathbf{L}(\mathbf{z}^k)$ is the Hessian of the LAGRANGE-function $l(\mathbf{z}^k)$ at the current iteration point. The new iteration point $\mathbf{z}^{k+1}$ is given by

$$
\mathbf{z}^{k+1} = \mathbf{z}^k - \alpha^k\Delta\mathbf{z}^k\,. \tag{6.14}
$$

The evaluation of the step-size α^k is described in the next section.

6.1.3.2 Line search and merit-functions

In order to determine α^k a line-search with a merit-function must be used. This is necessary to control global convergence of the algorithm. A merit-function is defined for the sole purpose of guiding and measuring the progress of an algorithm [31]. It has the basic feature that its minimum is attained at the solution to the objective function. Furthermore, under suitable conditions, it will serve as descent function for an algorithm, decreasing in value at each step. There exist a great variety of merit functions for optimization algorithms (see [31],[23],[53]), however in the context of SQP-methods the one of augmented Lagrangian type seems to perform best, both theoretically and numerically [53]:

$$\begin{aligned} P(\mathbf{z}) &= -\beta + \boldsymbol{\mu}^T \mathbf{C}\boldsymbol{\rho} + \frac{1}{2} r \boldsymbol{\rho}^T \mathbf{C}^T \mathbf{C}\boldsymbol{\rho} \\ &+ \sum_{j=1}^{m_i} \begin{bmatrix} \lambda_j \Phi_j(\beta, \boldsymbol{\rho}) + \frac{1}{2} r \Phi_j^2(\beta, \boldsymbol{\rho}) & , \text{if} \Phi_j > -\dfrac{\lambda_j}{r} \\ -\dfrac{1}{2}\dfrac{\lambda_j^2}{r} & , \text{otherwise.} \end{bmatrix} \end{aligned} \tag{6.15}$$

An update technique for the penalty parameter r can be found in [53]. For the line-search a simple backtracking strategy of ARMIJO-type might be used as follows:

Evaluate the smallest non-negative integer j_k among all j with

$$P(\mathbf{z}^k - \mu^j \mathbf{z}^k) \leq P(\mathbf{z}^k) - \tau \mu^j \nabla P(\mathbf{z}^k)^T \mathbf{z}^k \,, \tag{6.16}$$

where $0 < \tau < 1$ and $0 < \mu < 1$. An alternative strategy, which uses quadratic and cubic interpolation techniques can be found in [13].

6.1.3.3 The Han-Powell Method

For construction of the subproblem Q^k, eq. (6.13), the Hessian $\mathbf{L}^k$ might be used. However, as we will show in the next section, $\mathbf{L}^k$ is not necessarily positive-definite. Yet, for the descent property to hold for the chosen merit function eq. (6.15) and sequential quadratic programming, a positive-definite iteration matrix must be used. A remedy to overcome this conflict is to replace $\mathbf{L}^k$ by a BFGS-matrix (see [31] for more details).

As shown in the introduction to this section, the number of unknowns for the discretized optimization problem can be very large. In this case two difficulties arise if $\mathbf{L}^k$ is replaced by an BFGS-update formula (see [31] for more details). Firstly, the solution of the subproblem Q^k, which is usually attained by an active set strategy, may become very time consuming. Secondly, the storage requirements for the BFGS-update formula may become too large.

$$\sum_{i=1}^{NG} \mathbf{C}_i \cdot \boldsymbol{\rho}_i = \mathbf{C} \cdot \boldsymbol{\rho} = \mathbf{0} \tag{5.17}$$

$$\Phi\,[\,\beta\boldsymbol{\sigma}_i^E(t) + \boldsymbol{\rho}_i - \boldsymbol{\alpha}_i\,] \le \sigma_0^2 \qquad \forall\; i \in \mathcal{I}\,. \tag{5.18}$$

5.4 Incorporation of microelements

In view of the results obtained in Section 4.4 there are two possibilities for calculation of shake-down problems with limited, kinematic hardening material.

In the first case macroscopic backstresses α_{ij} are used. Different to an unlimited kinematic hardening material an additional condition occurs, which has to be satisfied by the backstresses α_{ij}. Thus, the discretized formulation for determination of the shake-down load factor is given by

$$\beta \rightarrow \max \tag{5.19}$$

$$\sum_{i=1}^{NG} \mathbf{C}_i \cdot \boldsymbol{\rho}_i = \mathbf{C} \cdot \boldsymbol{\rho} = \mathbf{0} \tag{5.20}$$

$$\Phi\,[\,\beta\boldsymbol{\sigma}_i^E(t) + \boldsymbol{\rho}_i - \boldsymbol{\alpha}_i\,] \le \sigma_0^2 \qquad \forall\; i \in \mathcal{I} \tag{5.21}$$

$$\Phi(\boldsymbol{\alpha}_i) \le (\sigma_Y - \sigma_0)^2 \qquad \forall\; i \in \mathcal{I}\,. \tag{5.22}$$

In the second case we exploit some properties of the microelement-model developed in Section 4.3.

A conclusion of Corollar 4.1 is, that for the developed material model the shake-down load is determined only by the initial yield stress σ_0 and the maximum hardening σ_Y $(= K)$. It is not dependent on the particular σ–ε curve, and thus, it does not depend on the particular $k(\xi)$ function. For this reason the function $k(\xi)$ can be replaced by a step-function. Its choice is arbitrary, however, the minimum value must be equal to k_0 $(= \sigma_0)$ and the area below the step-function must be equal to K $(= \sigma_Y)$, i.e.

$$K = \sum_{l=1}^{m} \Delta\,\xi_l\,\bar{k}_l\,. \tag{5.23}$$

The meaning of the notations is explained in Figure 5.1.

It should be mentioned, that for numerical investigations it is sufficient to choose the number of intervals m for the step-function equal to 2, since the results do not depend on this.

If plane stress problems are treated, the microelements can be incorporated in the following way:

The thickness t of the plate is divided into several $(m \ge 2)$ layers (see Figure 5.2), each with thickness $t_l, l = 1, 2, \ldots, m$. Each layer behaves linear-elastic, perfectly plastic and has a corresponding yield stress $\bar{k}_l$ (one value of the step-function). The thicknesses of the layers have to be chosen such that

$$\frac{t_l}{t} = \Delta\,\xi_l \qquad \forall\; l \in [1, 2, \ldots, m] \tag{5.24}$$

where

$$l_o^k = -\Delta\beta + \frac{1}{2}\Delta\beta^2 a_{oo} + \Delta\beta \sum_{j=1}^{m_i} \lambda_j \frac{\partial \Phi_j^k}{\partial \beta} \tag{6.18}$$

$$l_s^k = \frac{1}{2}\Delta\rho_s^2 a_{ss}^k + \Delta\rho_s \left(\sum_{j=1}^{m_e} \mu_j C_{js} + \sum_{j=1}^{m_i} \lambda_j^k \frac{\partial \Phi_j^k}{\partial \rho_s} \right), \quad s = 1, \ldots, n. \tag{6.19}$$

The eqs. (6.17) – (6.19) demonstrate, that Q^k is separable.

Solution of the subproblem by a dual method

In the following the index k shall be dropped. The properties of the subproblem being convex and separable make it very attractive to solve it by a dual method [31]. Dual methods do not attack the original constrained problem directly but, instead, attack an alternative problem, the dual problem, whose unknowns are the Lagrangian multipliers of the primal problem.

Such alternative problems can be constructed by means of a saddle function $l(\Delta\beta, \Delta\boldsymbol{\rho}, \boldsymbol{\mu}, \boldsymbol{\lambda})$, which characterizes the optimal solution $l(\Delta\beta^*, \Delta\boldsymbol{\rho}^*, \boldsymbol{\mu}^*, \boldsymbol{\lambda}^*)$ according to

$$l(\Delta\beta, \Delta\boldsymbol{\rho}, \boldsymbol{\mu}^*, \boldsymbol{\lambda}^*) \le l(\Delta\beta^*, \Delta\boldsymbol{\rho}^*, \boldsymbol{\mu}^*, \boldsymbol{\lambda}^*) \le l(\Delta\beta^*, \Delta\boldsymbol{\rho}^*, \boldsymbol{\mu}, \boldsymbol{\lambda})$$
$$\forall\, (\Delta\beta, \Delta\boldsymbol{\rho}) \in \Re^{n+1}, \boldsymbol{\mu} \in \Re^{m_e}, \boldsymbol{\lambda} \in \Re_+^{m_i}. \tag{6.20}$$

The dual function which has to be maximized for each subproblem Q^k then will be given by

$$W(\boldsymbol{\mu}, \boldsymbol{\lambda}) = \min_{\Delta\beta, \Delta\boldsymbol{\rho}} l(\Delta\beta, \Delta\boldsymbol{\rho}, \boldsymbol{\mu}, \boldsymbol{\lambda}). \tag{6.21}$$

Note that since l is separable the solution of the minimization problem

$$l(\Delta\beta(\boldsymbol{\lambda}), \Delta\boldsymbol{\rho}(\boldsymbol{\lambda}, \boldsymbol{\mu})) = \min_{\Delta\beta, \Delta\boldsymbol{\rho}} l(\Delta\beta, \Delta\boldsymbol{\rho}, \boldsymbol{\mu}, \boldsymbol{\lambda}) \tag{6.22}$$

can be obtained explicitly from

$$\Delta\beta(\boldsymbol{\lambda}) = \frac{1 - \sum_{j=1}^{m_i} \lambda_j \frac{\partial \Phi_j}{\partial \beta}}{a_{oo}} \tag{6.23}$$

$$\Delta\rho_s(\boldsymbol{\mu}, \boldsymbol{\lambda}) = \frac{-\boldsymbol{\mu}^T \mathbf{C}_s + \sum_{j=1}^{m_i} \lambda_j^k \frac{\partial \Phi_j}{\partial \rho_s}}{a_{ss}}, \quad s = 1, \ldots, n, \tag{6.24}$$

where now $\mathbf{C}_s$ is the s-th row of the matrix $\mathbf{C}$. In the context of a FE-method, the term $\boldsymbol{\mu}^T\mathbf{C}$ can be calculated separately for each element (analogously to strains which are obtained from the displacements via $\mathbf{B}$-matrices).

The load dependent elastic stress vectors $\boldsymbol{\sigma}_i^E(t) = \boldsymbol{\sigma}^E(\mathbf{x}_i, t)$ at Gaussian points are calculated by means of a displacement method, where the index i refers to the i-th Gaussian point. The corresponding coordinate vector shall be denoted by $\mathbf{x}_i$.

The virtual displacements $\delta u_m(\mathbf{x})$ for an element are approximated by

$$\delta u_m(\mathbf{x}) = \sum_{k=1}^{NK} \delta u_m^k \, N_k(\mathbf{x}) \,, \tag{5.6}$$

where $N_k(\mathbf{x})$ and δu_m^k denote the k-th shape function and the virtual displacement at the k-th node of the element in direction of x_m, respectively. NK denotes the number of nodes of the considered element. The virtual strains $\delta\varepsilon_{mn}(\mathbf{x})$ are derived by substitution of eq. (5.6) into eq. (5.5), i.e.

$$\delta\boldsymbol{\varepsilon}(\mathbf{x}) = \mathbf{B}(\mathbf{x}) \, \delta\mathbf{u} \,. \tag{5.7}$$

Here $\mathbf{B}(\mathbf{x})$ is a matrix dependent on coordinates, and $\delta\mathbf{u}$ assembles all components of virtual displacements δu_m^k.

The integral (5.2) is divided into integrals over the elements. For a specific element with number Ie, the corresponding integral

$$\int_{\Omega_{Ie}} \delta\varepsilon_{mn}(\mathbf{x}) \, \rho_{mn}(\mathbf{x}) \, d\Omega = \int_{\Omega_{Ie}} \delta\boldsymbol{\varepsilon}^T(\mathbf{x}) \cdot \boldsymbol{\rho}(\mathbf{x}) \, d\Omega \approx \delta\mathbf{u}^T \sum_i \mathbf{B}^T(\mathbf{x}_i) \cdot \boldsymbol{\rho}_i \, \omega_i \tag{5.8}$$

is calculated numerically by using the GAUSS-LEGENDRE integration technique, where Ω_{Ie} denotes the set of all points of the element, $\boldsymbol{\rho}_i$ denotes the unknown residual stress vector at the i-th Gaussian point $\mathbf{x}_i$, and ω_i denotes the weighting factor for the i-th Gaussian point. In eq. (5.8) the summation index i runs over all Gaussian points of the Ie-th element. To obtain eq. $(5.8)_2$, equation (5.7) was used.

By summation of the contributions of all elements and by variation of the virtual node-displacements with respect to geometrical boundary conditions, finally one gets the linear system of equations

$$\sum_{i=1}^{NG} \mathbf{C}_i \cdot \boldsymbol{\rho}_i = \mathbf{C} \cdot \boldsymbol{\rho} = \mathbf{0} \,. \tag{5.9}$$

Eq. (5.9) represents the discretized equilibrium conditions for the residual stresses, where NG is the total number of Gaussian points of the discretized structure.

In eq. (5.9) $\mathbf{C}$ is a constant, system dependent matrix uniquely defined by the chosen discretization and the geometrical boundary conditions. $\boldsymbol{\rho}$ is the global residual stress vector of the discretized structure. The relations between $\mathbf{C}, \mathbf{C}_i, \boldsymbol{\rho}$ and $\boldsymbol{\rho}_i$ are given by

$$\mathbf{C} = [\mathbf{C}_1, \ldots, \mathbf{C}_i, \ldots, \mathbf{C}_{NG}] \tag{5.10}$$

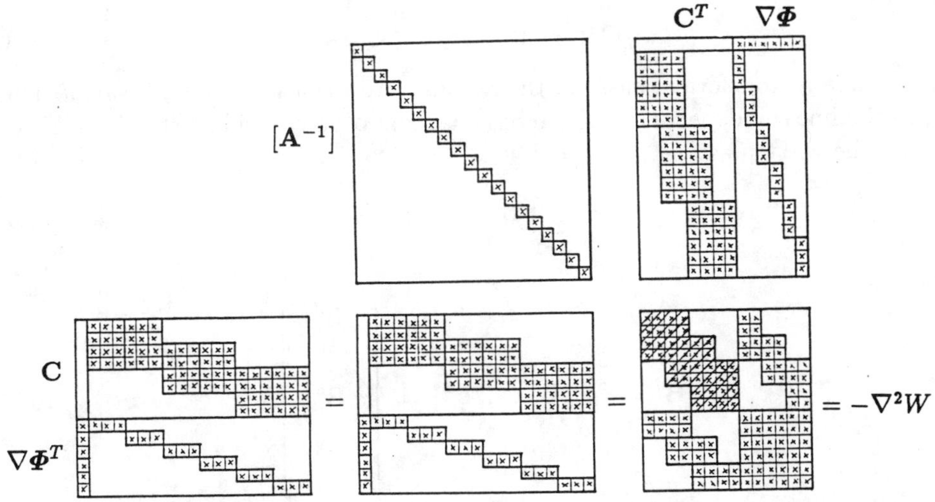

Figure 6.3: Newton–matrix of the dual-function

FE-calculation. This fact must be exploited because of storage requirements. However, since for large sized problems the storage of the whole matrix $\nabla^2 W$ still would be too expensive, a simple alternative is to construct $\mathbf{H}^*$ from the above mentioned fictitious stiffness matrix and the remaining diagonal elements of $\nabla^2 W$. Let us denote this matrix by $\mathbf{H}^{**}$.

To make the algorithm even more effective for large sized problems, one can use a Quasi-Newton method, where the matrix $\mathbf{H}^{**}$ is used as a precondition matrix and where the BFGS-update formula corresponds to the set

$$\mathcal{I}_2(\boldsymbol{\mu}, \boldsymbol{\lambda}) = \{j | j = 1, \ldots, m \quad \backslash \mathcal{I}_w(\boldsymbol{\mu}, \boldsymbol{\lambda})\} . \tag{6.32}$$

Note that for $\mathcal{I}_2^j(\boldsymbol{\mu}, \boldsymbol{\lambda}) \neq \mathcal{I}_2^{j-1}(\boldsymbol{\mu}, \boldsymbol{\lambda})$ a restart has to be carried out. The implementation of BFGS-algorithms for unconstrained minimization problems in the context of an FE-algorithm is described in [36]. The papers [42], [43],[46] give further hints, if computer storage is restricted.

6.1.4 Reduction techniques

6.1.4.1 Basic idea

In this section we describe how reduction techniques can be used by elimination of the equality constraints of problem (6.3)–(6.5).

All possible residual stresses form a vector space which is called the residual stress space. It is uniquely determined by the matrix $\mathbf{C} \in \Re^{m_e \times n}$. The row vectors of $\mathbf{C}$ are linearly independent and thus there exist $d = n - m_e$ vectors $\mathbf{b}^l \in \Re^n$ $(l = 1, \ldots, d)$

such that

$$\mathbf{C}\,\mathbf{b}^l = \mathbf{0} \qquad \forall\ 1 \le l \le d \tag{6.33}$$

holds. These vectors form a basis for the residual stress space $\mathcal{B}^d$ of dimension d. Thus, due to the linear independence of the basis vectors $\mathbf{b}^l$, any residual stress $\boldsymbol{\rho} \in \mathcal{B}^d$ can uniquely be written as

$$\boldsymbol{\rho} = \sum_{l=1}^{d} \mathbf{b}^l x_l = \mathbf{B}^d \cdot \mathbf{X}_d\,, \tag{6.34}$$

where

$$\mathbf{B}^d = [\,\mathbf{b}^1, \mathbf{b}^2, \ldots, \mathbf{b}^d\,] = \begin{bmatrix} \mathbf{B}_1^d \\ \vdots \\ \mathbf{B}_i^d \\ \vdots \\ \mathbf{B}_{NG}^d \end{bmatrix} \tag{6.35}$$

and

$$\mathbf{X}_d^T = [\,x_1, x_2, \ldots, x_d\,]\,. \tag{6.36}$$

Using the basis $\mathbf{B}^d$ problem (6.3)–(6.5) can be replaced by the reduced problem

$$\begin{aligned} \mathcal{P}_{\text{red.}}: \quad & \beta \rightarrow \max \\ & \Phi\,[\,\beta \boldsymbol{\sigma}_i^E(j) + \mathbf{B}_i^d \cdot \mathbf{X}_d\,] \le \sigma_0^2 \qquad \forall\ (i,j) \in \mathcal{I} \times \mathcal{J}\,. \end{aligned} \tag{6.37}$$

In the above form the equality constraints are eliminated.

Problem $\mathcal{P}_{\text{red.}}$ can be solved e.g. by the classical SQP-algorithm described in Section 6.1.3. However, for large numbers of d the effectiveness of this strategy is still very poor.

6.1.4.2 The solution algorithm using a reduced basis technique

The following algorithm is an iterative method which solves problem (6.3)–(6.5) iteratively in a sequence of reduced residual stress spaces of very low dimensions. In this way large finite element systems of different structures have been approached successfully.

Sequential iteration in reduced subspaces

For the sake of simplicity let us restrict ourselves to the case $M = 1$ (limit analysis). The iteration index, indicating each subproblem $\mathcal{P}^k$ in a reduced residual stress space, is denoted by k ($k = 1, 2, \ldots$).

At the beginning of the k-th subproblem $\mathcal{P}^k$ we have a known state characterized by a load factor $\beta^{(k-1)}$ and a residual stress distribution $\boldsymbol{\rho}^{(k-1)}$ with

$$\Phi\,[\,\beta^{(k-1)} \boldsymbol{\sigma}_i^E + \boldsymbol{\rho}_i^{(k-1)}\,] \le \sigma_0^2 \qquad \forall\ i \in \mathcal{I}\,. \tag{6.38}$$

Here the index j is omitted due to the restriction $M = 1$. Eq. (6.38) indicates that $[\beta^{(k-1)}, \boldsymbol{\rho}^{(k-1)}]$ is a feasible point of the optimization problem (6.3)–(6.5). Therefore $\beta^{(k-1)}$ is a lower bound to the shake-down factor of the discretized system (but not necessarily a lower bound to the original problem).

For $k = 1$ we set $\beta^{(0)} = \beta_e$ and $\boldsymbol{\rho}^{(0)} = \mathbf{0}$, where β_e is the elastic limit. It is important to note that during the iteration process there are always points which are on the yield surfaces. For further considerations only these points are of interest since the remaining ones can support a small load increment without violating the yield condition.

Now, we select a few ($r \ll d$) basis vectors $\mathbf{b}^{p,k}$ ($p = 1, 2, \ldots, r$) from the residual stress space $\mathcal{B}^d$. They form a subspace $\mathcal{B}^{r,k}$ (or reduced space) of $\mathcal{B}^d$. Here r denotes the dimension of the subspace, i.e. the number of the chosen basis vectors.

The improved state (k) is determined by solving the following reduced optimization problem:

$$\mathcal{P}^k: \qquad \beta^{(k)} \to \max \tag{6.39}$$

$$\Phi\,[\,\beta^{(k)}\,\boldsymbol{\sigma}_i^E + \boldsymbol{\rho}_i^{(k-1)} + \mathbf{B}_i^{r,k} \cdot \mathbf{X}_r^k\,] \le \sigma_0^2 \qquad \forall\; i \in \mathcal{I}\,. \tag{6.40}$$

In (6.40) we have

$$\mathbf{B}^{r,k} = [\,\mathbf{b}^{1,k}, \ldots, \mathbf{b}^{p,k}, \ldots, \mathbf{b}^{r,k}\,] = \begin{bmatrix} \mathbf{B}_1^{r,k} \\ \vdots \\ \mathbf{B}_i^{r,k} \\ \vdots \\ \mathbf{B}_{NG}^{r,k} \end{bmatrix}, \tag{6.41}$$

and $\mathbf{X}_r^k$ is an r-dimensional vector with components x_p^k, $p = 1, 2, \ldots, r$. $\mathbf{X}_r^k$ and $\beta^{(k)}$ are the primary unknowns of the actual subproblem $\mathcal{P}^k$ (6.39)–(6.40).

Let us notice that $\boldsymbol{\rho}^{(k-1)}$ is a residual stress which can also be taken as a basis vector for the reduced space $\mathcal{B}^{r,k}$. In this way we have to solve the following slightly modified problem

$$\mathcal{P}^k: \qquad \beta^{(k)} \to \max \tag{6.42}$$

$$\Phi\,[\,\beta^{(k)}\,\boldsymbol{\sigma}_i^E + \mathbf{B}_i^{r,k} \cdot \mathbf{X}_r^k\,] \le \sigma_0^2 \qquad \forall\; i \in \mathcal{I}\,. \tag{6.43}$$

After solving problem (6.39)–(6.40) or (6.42)–(6.43), respectively, we obtain the k-th state by the updates

$$\boldsymbol{\rho}^{(k)} = \boldsymbol{\rho}^{(k-1)} + \mathbf{B}^{r,k} \cdot \mathbf{X}_r^k \tag{6.44}$$

or

$$\boldsymbol{\rho}^{(k)} = \mathbf{B}^{r,k} \cdot \mathbf{X}_r^k\,, \tag{6.45}$$

respectively.

The increment $\Delta\beta^{(k)}$ for the load factor is easily determined by

$$\Delta\beta^{(k)} = \beta^{(k)} - \beta^{(k-1)}. \tag{6.46}$$

Selecting new reduced basis vectors the process is repeated until the convergence criterion

$$|\,\Delta\beta^{(k)}\,| \leq \text{tolerance} \tag{6.47}$$

is fulfilled.

By using eq. (6.47) as a criterion for convergence it must be stressed that this criterion is not a sufficient one for convergence of the original problem (6.3)–(6.5), since the value of $\Delta\beta^{(k)}$ at each iteration is strongly dependent on the choice of the reduced space $\mathcal{B}^{r,k}$. It may happen that $\Delta\beta^{(k)}$ is very small at the beginning of the iteration while the true load factor is much higher than $\beta^{(k)}$. The check of the KUHN-TUCKER conditions (6.8)–(6.12) of the original problem (6.3)–(6.5) is the only way to determine the quality of the approximation $\beta^{(k)}$ and $\boldsymbol{\rho}^{(k)}$. But usually this is very expensive for large systems.

Construction of basis vectors

Now, we come to the question of how to generate a reduced subspace $\mathcal{B}^{r,k}$. We present a possibility for the generation of basis vectors and thus the reduced spaces.

The entire stress vector $\boldsymbol{\sigma}_i^{(k-1)}$ at the i-th Gaussian point at the beginning of the k-th subproblem $\mathcal{P}^k$ is given by

$$\boldsymbol{\sigma}_i^{(k-1)} = \hat{\boldsymbol{\sigma}}_i^0 = \beta^{(k-1)}\boldsymbol{\sigma}_i^E + \boldsymbol{\rho}_i^{(k-1)}. \tag{6.48}$$

For the case $r = 1$, i.e. if only one basis vector

$$\mathbf{b}^{1,k} = \begin{bmatrix} \mathbf{b}_1^{1,k} \\ \vdots \\ \mathbf{b}_i^{1,k} \\ \vdots \\ \mathbf{b}_{NG}^{1,k} \end{bmatrix} \tag{6.49}$$

is used for the actual subproblem, it can be shown (cf. [63]) that either

$$\frac{\partial\Phi}{\partial\boldsymbol{\sigma}^T}\Big|_{\boldsymbol{\sigma}_i^{(k-1)}} \cdot \mathbf{b}_i^{1,k} < 0 \tag{6.50}$$

or

$$\frac{\partial\Phi}{\partial\boldsymbol{\sigma}^T}\Big|_{\boldsymbol{\sigma}_i^{(k-1)}} \cdot \mathbf{b}_i^{1,k} > 0 \tag{6.51}$$

must be satisfied simultaneously for all $i \in \mathcal{A}$ with

$$\mathcal{A} := [\,i \mid \Phi(\sigma_i^{(k-1)}) = \sigma_0^2 \wedge \frac{\partial\Phi}{\partial\boldsymbol{\sigma}^T}\Big|_{\boldsymbol{\sigma}_i^{(k-1)}} \cdot \boldsymbol{\sigma}_i^E > 0\,] \tag{6.52}$$

such that problem (6.39)–(6.40) or (6.42)–(6.43) possesses a solution $(\Delta\beta^{(k)}, x_1^k)$ with $\Delta\beta^{(k)} = \beta^{(k)} - \beta^{(k-1)} > 0$.

To generate basis vectors, let us add a load increment, defined by $\Delta\bar{\beta}^{(k)} > 0$, to the known load factor $\beta^{(k-1)}$ (cf. [54]). The structure will yield further. We perform an equilibrium iteration procedure. The index q denotes the number of the equilibrium iteration.

During the iteration the discretized equilibrium condition for the first step is

$$\sum_{i=1}^{NG} \mathbf{C}_i \cdot \mathbf{D}^p(\hat{\boldsymbol{\sigma}}_i^0)\,\Delta\boldsymbol{\varepsilon}_i^1 \;=\; (\beta^{(k-1)} + \Delta\bar{\beta}^{(k)})\,\mathbf{P} - \sum_{i=1}^{NG} \mathbf{C}_i \cdot \hat{\boldsymbol{\sigma}}_i^0\,, \tag{6.53}$$

and for the q-th iteration step we have

$$\sum_{i=1}^{NG} \mathbf{C}_i \cdot \mathbf{D}^p(\hat{\boldsymbol{\sigma}}_i^{q-1})\Delta\boldsymbol{\varepsilon}_i^q = (\beta^{(k-1)} + \Delta\bar{\beta}^{(k)})\,\mathbf{P} - \sum_{i=1}^{NG} \mathbf{C}_i \cdot \hat{\boldsymbol{\sigma}}_i^{q-1}\,, \tag{6.54}$$

where $\mathbf{P}$ is the external load vector corresponding to $\beta = 1$ and $\mathbf{D}^p(\hat{\boldsymbol{\sigma}}_i^{q-1})$ is a matrix which is dependent on the employed iteration procedure, e.g. an iteration with the initial stiffness matrix or with the consistent tangent matrix (cf. [55], [22]). The right-hand sides of eqs. (6.53) and (6.54) are the residuals during the equilibrium iteration.

The difference between eqs. (6.53) and (6.54) yields

$$\begin{aligned}&\sum_{i=1}^{NG} \mathbf{C}_i \cdot (\hat{\boldsymbol{\sigma}}_i^0 + \mathbf{D}^p(\hat{\boldsymbol{\sigma}}_i^0)\Delta\boldsymbol{\varepsilon}_i^1 - \hat{\boldsymbol{\sigma}}_i^{q-1} - \mathbf{D}^p(\hat{\boldsymbol{\sigma}}_i^{q-1})\Delta\boldsymbol{\varepsilon}_i^q) = \\ &\sum_{i=1}^{NG} \mathbf{C}_i \cdot \mathbf{b}_i^q = \mathbf{C} \cdot \mathbf{b}^q = \mathbf{0}\,.\end{aligned} \tag{6.55}$$

Eq. (6.55) shows that $\mathbf{b}^q$ is a residual stress vector.

In Figure 6.4 the points $1, 2, \ldots$ represent the elastic predictors during the equilibrium iteration in the stress space at the i-th Gaussian point and the points $1', 2', \ldots$ are traces of the iteration in the stress space.

If the equilibrium iteration converges, it is possible to choose a basis vector $\mathbf{b}^q$ with the property

$$\mathbf{n}_i^T \cdot \mathbf{b}_i^q = \frac{\partial\Phi}{\partial\boldsymbol{\sigma}^T}\Big|_{\hat{\boldsymbol{\sigma}}_i^0} \cdot \mathbf{b}_i^q > 0 \qquad \forall\; i \in \mathcal{A}\,. \tag{6.56}$$

This follows from the convexity of the yield surface and the assumption that $\Delta\boldsymbol{\varepsilon}_i^q$ converges to zero (see Figure 6.4). Eq. (6.56) states that an improvement of the load factor in a reduced subspace $\mathcal{B}^{r,k}$ generated in this way is possible. To ensure that the equilibrium iteration converges, the load increments $\Delta\bar{\beta}^{(k)}$ and the matrices $\mathbf{D}^p(\hat{\boldsymbol{\sigma}}_i^{q-1})$ must be chosen in a suitable way. More details can be found in [63].

Note that all differences $\mathbf{b}^q$ during the equilibrium iteration are residual stress vectors. Practically we construct 3–5 basis vectors at each equilibrium iteration procedure. Correspondingly, the reduced optimization problem (6.39)–(6.40) contains only

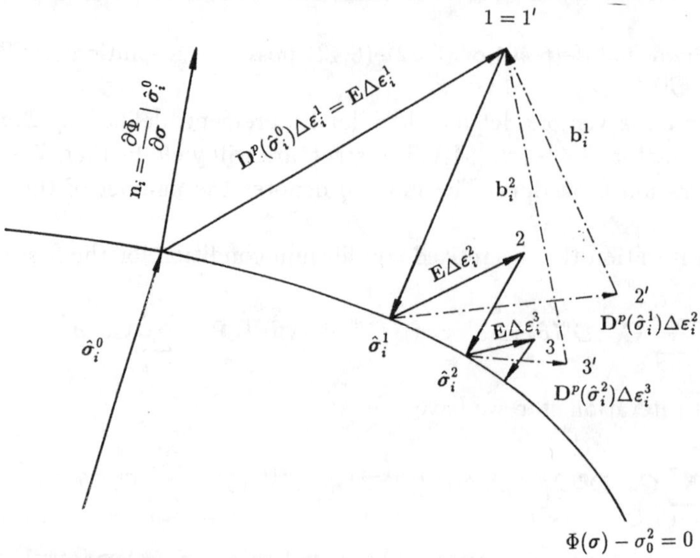

Figure 6.4: Traces of the equilibrium iteration in the stress space

4–6 unknowns (the number of the inequality constraints remains unaltered). Its solution can be obtained very efficiently with an SQP– or a penalty method.

Extension to shake-down problems

The above presented procedure can be extended to the general case $M > 1$. A load vertex j is active if there exists at least one Gaussian point $i \in \mathcal{I}$ with

$$\Phi\,[\,\beta^{(k-1)}\,\boldsymbol{\sigma}_i^E(j) + \boldsymbol{\rho}_i^{(k-1)}\,] = \sigma_0^2\,. \tag{6.57}$$

For every active load vertex, reduced basis vectors have to be constructed. Instead of eq. (6.48) the entire stress vectors at the Gaussian points at the beginning of the equilibrium iteration are defined by

$$\hat{\boldsymbol{\sigma}}_i^0 = \beta^{(k-1)}\boldsymbol{\sigma}_i^E(j) + \boldsymbol{\rho}_i^{(k-1)} = \boldsymbol{\sigma}_i^{(k-1)}, \quad i \in \mathcal{I}\,. \tag{6.58}$$

The load increment for the generation of basis vectors is defined by $\Delta\bar{\beta}^{(k)}\,\mathbf{P}(j)$.

The equilibrium iteration can be performed in the same way as for the case $M = 1$. Analogously, the discretized equilibrium condition at the q-th iteration step is

$$\sum_{i=1}^{NG} \mathbf{C}_i \cdot \mathbf{D}^p(\hat{\boldsymbol{\sigma}}_i^{q-1})\Delta\boldsymbol{\varepsilon}_i^q = (\beta^{(k-1)} + \Delta\bar{\beta}^{(k)})\mathbf{P}(j) - \sum_{i=1}^{NG} \mathbf{C}_i \cdot \hat{\boldsymbol{\sigma}}_i^{q-1}\,. \tag{6.59}$$

Now in a similar way as to eq. (6.55) basis vectors $\mathbf{b}^q(j)$ can be generated. The basis for the reduced space $\mathcal{B}^{r,k}$ of subproblem $\mathcal{P}^k$ consists of all basis vectors generated for all active load vertices.

The (k)-th improved state is determined by solving the reduced non-linear optimization problem

$$\mathcal{P}^k : \qquad \beta^{(k)} \to \max \tag{6.60}$$

$$\Phi\,[\,\beta^{(k)}\,\boldsymbol{\sigma}_i^E(j) + \mathbf{B}_i^{r,k} \cdot \mathbf{X}_r^k\,] \leq \sigma_0^2 \qquad \forall\ (i,j) \in \mathcal{I} \times \mathcal{J}\,. \tag{6.61}$$

6.2 Treatment of systems consisting of unlimited kinematic hardening material

In Section 6.1 systems consisting of linear-elastic, perfectly-plastic material were treated. This section is contributed to problems with unlimited kinematic hardening material. For this class of problems an especially easy and elegant solution procedure can be given.

6.2.1 Solution procedure

With the assumption that the load domain is a convex polyhedron, one has to solve the following non-linear optimization problem

$$\beta \to \max \tag{6.62}$$

$$\sum_{i=1}^{NG} \mathbf{C}_i \cdot \boldsymbol{\rho}_i = \mathbf{C} \cdot \boldsymbol{\rho} = \mathbf{0} \tag{6.63}$$

$$\Phi\,[\,\beta \boldsymbol{\sigma}_i^E(j) + \boldsymbol{\rho}_i - \boldsymbol{\alpha}_i\,] \leq \sigma_0^2 \qquad \forall\ (i,j) \in \mathcal{I} \times \mathcal{J}\,. \tag{6.64}$$

Whereas the number of the equality and inequality constraints of problem (6.62)–(6.64) remains unaltered in comparison to problem (6.3)–(6.5), the number of unknowns has almost been doubled. At first sight its solution seems to be more difficult and more expensive than the solution of problem (6.3)–(6.5).

The following consideration is useful for developing solution procedures for problem (6.62)–(6.64): While the residual stress vectors $\boldsymbol{\rho}_i$ have to fulfill the equilibrium condition (6.63), the backstress vectors $\boldsymbol{\alpha}_i$ are unconstrained quantities (see Section 4.1.2). Recall, that in this case the shift of the initial yield surface is unlimited.

At each Gaussian point $\mathbf{x}_i$ let us introduce a new vector

$$\mathbf{y}_i := \boldsymbol{\rho}_i - \boldsymbol{\alpha}_i\,. \tag{6.65}$$

Since $\boldsymbol{\alpha}_i$ are unconstrained quantities, $\mathbf{y}_i$ are unconstrained as well, and thus the equality constraint (6.63) of problem (6.62)–(6.64) can be dropped. Therefore problem (6.62)–(6.64) can be transferred into the following equivalent form

$$\beta \to \max \tag{6.66}$$

$$\Phi\,[\,\beta \boldsymbol{\sigma}_i^E(j) + \mathbf{y}_i\,] \leq \sigma_0^2 \qquad \forall\ (i,j) \in \mathcal{I} \times \mathcal{J}\,. \tag{6.67}$$

Problem (6.66)–(6.67) has an easy structure. The maximum value of the load factor β_s is determined by

$$\beta_s = \bar{\beta} \quad \text{with} \quad \bar{\beta} = \min_{i \in \mathcal{I}} \beta_i \,, \tag{6.68}$$

where now β_i $(i \in \mathcal{I})$ is defined as the solution of the subproblem for the i–th Gaussian point

$$\beta_i \rightarrow \max \tag{6.69}$$

$$\Phi\left[\beta_i \boldsymbol{\sigma}_i^E(j) + \mathbf{y}_i\right] \leq \sigma_0^2 \qquad \forall\, j \in \mathcal{J} \,. \tag{6.70}$$

Let us notice, that problem (6.69)–(6.70) results from problem (6.66)–(6.67) by dropping all inequality constraints which do not contain the subscript i.

In order to show that eq. (6.68) yields the maximum load factor of problem (6.66)–(6.67), the following lemma of mathematical optimization shall be used:

Lemma 6.1 *Two maximization problems with the same objective function $f(\mathbf{y})$ are considered. The feasible regions of both problems are denoted by $\mathcal{Z}_1$ and $\mathcal{Z}_2$, respectively. Assuming $\mathcal{Z}_1 \subseteq \mathcal{Z}_2$ we have*

$$\max_{\mathbf{y} \in \mathcal{Z}_1} f(\mathbf{y}) \leq \max_{\mathbf{y} \in \mathcal{Z}_2} f(\mathbf{y}) \,. \tag{6.71}$$

It is supposed that $f(\mathbf{y})$ has no local maximum in $\mathcal{Z}_2$.

Note that frequently problem 2 in Lemma 6.1 results from problem 1 by dropping a part of constraints which define the feasible region $\mathcal{Z}_1$ of problem 1.

Now we turn to the proof of the statement (6.68):

On one side, due to Lemma 6.1, the maximum load factor β_s defined by problem (6.66)–(6.67) can not be larger than β_i, and therefore

$$\beta_s \leq \min_{i \in \mathcal{I}} \beta_i = \bar{\beta} \tag{6.72}$$

holds.

On the other side, it can easily be shown that the vector $(\bar{\beta}, \bar{\mathbf{y}}_1, \ldots, \bar{\mathbf{y}}_i, \ldots, \bar{\mathbf{y}}_{NG})$ is a feasible point of problem (6.66)–(6.67), where $\bar{\beta}$ and $\bar{\mathbf{y}}_i$ $(i \in \mathcal{I})$ represent solutions of the problems

$$\bar{\beta}_i \rightarrow \max \tag{6.73}$$

$$\bar{\beta}_i - \bar{\beta} \leq 0 \tag{6.74}$$

$$\Phi\left[\bar{\beta}_i \boldsymbol{\sigma}_i^E(j) + \bar{\mathbf{y}}_i\right] \leq \sigma_0^2 \qquad \forall\, j \in \mathcal{J} \,. \tag{6.75}$$

This means that

$$\beta_s \geq \bar{\beta} \,. \tag{6.76}$$

Finally, from equations (6.72) and (6.76) we obtain $\beta_s = \bar{\beta}$ (q.e.d).

The dimension of problem (6.69)–(6.70) is very low. The number of the unknowns is $NSK + 1$, where NSK denotes the number of stress components at each Gaussian point. Therefore, for 2-d problems the number of unknowns is between 3 and 6. The number of the constraints is identical to the number of vertices of the load domain which in general is also low. Problem (6.69)–(6.70) can efficiently be solved by using an SQP-method [53].

The maximum load factor β_s can also be determined in a slightly modified way. We have

$$\beta_s = \beta_{NG}, \tag{6.77}$$

where β_{NG} is the solution of the last problem of the consecutive subproblems ($i = 2, 3, \ldots, NG$) which are defined as

$$\beta_i \rightarrow \max \tag{6.78}$$

$$\beta_i - \beta_{i-1} \leq 0 \tag{6.79}$$

$$\Phi\,[\,\beta_i\,\boldsymbol{\sigma}_i^E(j) + \mathbf{y}_i\,] \leq \sigma_0^2 \qquad \forall\; j \in \mathcal{J}\,. \tag{6.80}$$

For $i = 1$ the factor β_1 has to be determined by solving the problem

$$\beta_1 \rightarrow \max \tag{6.81}$$

$$\Phi\,[\,\beta_1\,\boldsymbol{\sigma}_1^E(j) + \mathbf{y}_1\,] \leq \sigma_0^2 \qquad \forall\; j \in \mathcal{J}\,. \tag{6.82}$$

In general NG is a large number for 2-d problems. Therefore, solution of the subproblems (6.69)–(6.70) or (6.78)–(6.80) at all Gaussian points can be very expensive. But the following intermediate step can reduce the computation time considerably.

Before solving the problem for the i-th Gaussian point, we determine its elastic limit β_i^e. It is defined by

$$\beta_i^e = \min_{j \in \mathcal{J}} \tilde{\beta}_j\,, \tag{6.83}$$

where $\tilde{\beta}_j$ is calculated from

$$\Phi\,[\,\tilde{\beta}_j\,\boldsymbol{\sigma}_i^E(j)\,] = \sigma_0^2 \qquad \text{with} \qquad \tilde{\beta}_j > 0\,. \tag{6.84}$$

For $\beta_i^e > \beta_{i-1}$ the solution of problem (6.78)–(6.80) for the i-th Gaussian point is unnecessary, since this Gaussian point will not yield under the load factor β_s, if we consider that

$$\beta_s = \beta_{NG} \leq \beta_{i-1} \leq \beta_i^e \tag{6.85}$$

holds.

By setting $\beta_i = \beta_{i-1}$ we turn to the next Gaussian point. The proposed intermediate step can reduce the entire consumption of CPU time by more than 90%.

6.2.2 Geometrical interpretation

In the sequel we give a geometrical interpretation to shake-down behavior of systems consisting of unlimited kinematic hardening material.

We start from the local problem (6.69)–(6.70). The vectors $\boldsymbol{\sigma}_i^E(j)$ represent vertices of the domain $\mathcal{S}_i^E$ in the stress space $\mathcal{S}$, which contains all elastic stress vectors $\boldsymbol{\sigma}_i^E(t)$ at the i-th Gaussian point. If we identify the vector $(-\mathbf{y}_i)$ with the shift of the initial yield surface in $\mathcal{S}$, the solution of problem (6.69)–(6.70) can be given the following geometrical interpretation: Find the maximum possible affine enlargement of $\mathcal{S}_i^E$ and the corresponding shift $(-\mathbf{y}_i)$ of the yield surface on condition that the enlarged elastic domain $\beta_i\mathcal{S}_i^E$ is just contained in the shifted yield surface. Note that the shift of the yield surface is unconstrained. The represented geometrical interpretation is also illustrated in Figure 6.5.

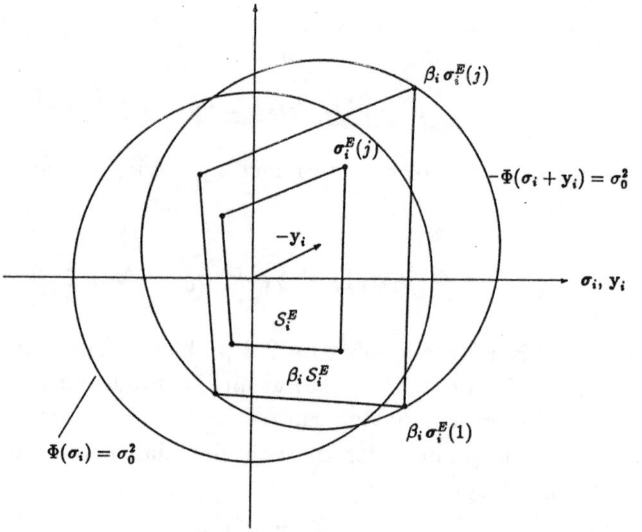

Figure 6.5: Geometrical interpretation of the local problem (6.69)–(6.70)

Equation (6.68) implies that for systems consisting of unlimited kinematic hardening material the shake-down load or the failure, respectively, is determined by that point $\mathbf{x}_{i_p}$, for which the maximum enlargement β_{i_p} of the elastic domain $\mathcal{S}_{i_p}^E$ is the least in comparison to all other points. Thus, $\beta_s = \beta_{i_p}$ holds.

Since the size of the yield surface remains unchanged during the yielding, it is evident, that generally the shake-down load of systems consisting of unlimited kinematic hardening material subjected to variable loading can not be infinite as to be expected for monotone loading. The only exception is that all elastic domains $\mathcal{S}_i^E$ $(\forall\ i \in \mathcal{I})$ are straight lines in the space of principle stresses, which coincide with the principle diagonal of the space of principle stresses. This means that all elastic stresses $\sigma^E_{mn}(\mathbf{x},t)$ are hydrostatic. However, this case is almost meaningless for the practice.

6.2.3 Relationship between perfectly plastic and unlimited kinematic hardening material

According to the investigation in Section 6.2.2 the shake-down load of a system consisting of unlimited kinematic hardening material is determined by that point $\mathbf{x}_{i_p}$, where the maximum possible enlargement of the elastic stress domain $\mathcal{S}^E_{i_p}$ is the least in comparison to other points. Thus, the failure is of local character. This reflects the fact that a system consisting of unlimited kinematic hardening material and subjected to cyclic loading can fail only locally in form of alternating plasticity (cf. [30]). Incremental collapse can not occur since it is connected with a non-trivial, kinematic compatible plastic strain field, which has global character.

The alternating feature of the failure is made clear in Figure 6.6. If the true load factor β is larger than the shake-down load factor β_s, then there is always a portion of the enlarged elastic stress domain $\beta\mathcal{S}^E_{i_p}$, which is outside the shifted yield surface. As a result the material point $\mathbf{x}_{i_p}$ does not cease to yield when the load is changing from one vertex of the load domain to another. During the yielding some components of the plastic strain increments $\Delta\varepsilon^P_{i_p}$ change their sign alternately (see Figure 6.6). Hence the failure is local and is caused by alternating plasticity.

The following observation is useful for demonstrating the relationship between shake-down behavior of a perfectly plastic and a kinematic hardening material: The discretized problem (6.66)–(6.67) for systems consisting of unlimited kinematic hard-

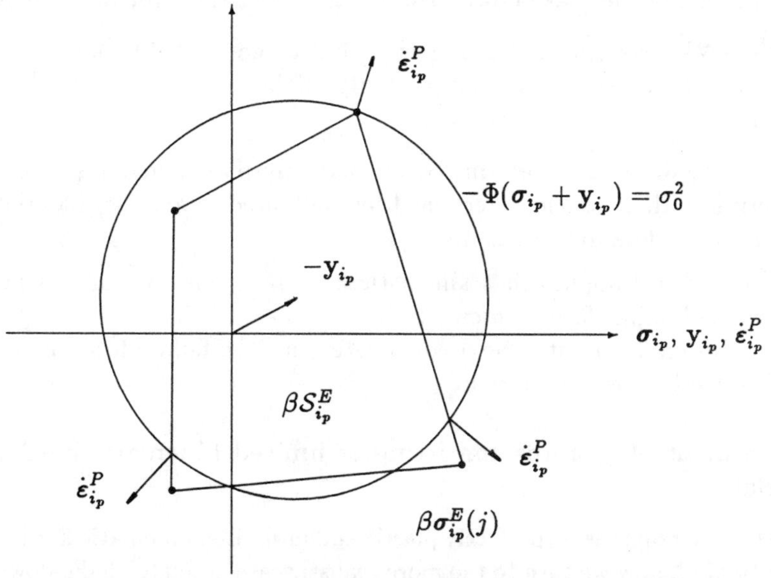

Figure 6.6: Illustration of the alternating feature of the failure

ening material formally also results from problem (6.3)–(6.5) by dropping the equality constraints (6.4) which have to be satisfied by the residual stress vectors $\boldsymbol{\rho}_i$. Thus, according to Lemma 6.1, the shake-down load of a system consisting of perfectly plastic material can not be larger than the shake-down load of the same system, consisting of unlimited kinematic hardening material with the same initial yield stress σ_0.

However, it is possible, that the shake-down loads for perfectly plastic and kinematic hardening material are identical. This is the case only, if alternating plasticity is dominant in both cases. It is easy to show that

$$\boldsymbol{\rho}_{i_p} = \mathbf{y}_{i_p} \tag{6.86}$$

holds. If we interpret $\boldsymbol{\rho}_{i_p}$ as a shift of the yield surface, Figure 6.6 can also be used to show the alternating feature of the failure.

We sum up:

1. A system consisting of unlimited kinematic hardening material and subjected to variable loading can fail only locally in form of alternating plasticity.

2. The kinematic hardening does not increase the shake-down load, if the same system with perfectly plastic material, subjected to the same loading, fails in form of alternating plasticity in such a manner that there exists at least one point $\mathbf{x}_{i_p}$, for which the enlarged elastic stress domain $\beta_s \mathcal{S}^E_{i_p}$ is just contained in the yield surface shifted to $-\boldsymbol{\rho}_{i_p}$. A further shift of the yield surface at this point would cause that a portion of the enlarged elastic domain $\beta_s \mathcal{S}^E_{i_p}$ leaves the yield surface (see Figure 6.6). In the sequel the alternating plasticity with the special character mentioned before will be denoted by APSC. In all other cases an increase of the shake-down load due to kinematic hardening is expected.

3. If the shake-down loads for perfectly plastic and unlimited kinematic material are identical, then alternating plasticity is the dominant failure form in both cases.

4. The shake-down load determined for perfectly plastic material is exact if it is identical with that one determined for unlimited kinematic material provided the latter is determined exactly.

The conclusion 2 implies that kinematic hardening does not always increase the shake-down load unlike for monotone loading.

The conclusion 3 can often be used to determine the failure form of systems consisting of perfectly plastic material.

6.3 Treatment of systems consisting of limited kinematic hardening material

After the two extreme cases (perfectly plastic and unlimited kinematic hardening material) were treated, now we turn to the more realistic case, namely, shake-down behavior of systems consisting of limited kinematic hardening material.

According to Section 5.4 we have to solve the following non-linear optimization problem

$$\beta \rightarrow \max \tag{6.87}$$

$$\sum_{i=1}^{NG} \mathbf{C}_i \cdot \boldsymbol{\rho}_i = \mathbf{C} \cdot \boldsymbol{\rho} = \mathbf{0} \tag{6.88}$$

$$\Phi\left[\beta \boldsymbol{\sigma}_i^E(j) + \boldsymbol{\rho}_i - \boldsymbol{\alpha}_i\right] \leq \sigma_0^2 \qquad \forall\ (i,j) \in \mathcal{I} \times \mathcal{J} \tag{6.89}$$

$$\Phi(\boldsymbol{\alpha}_i) \leq (\sigma_Y - \sigma_0)^2 \qquad \forall\ i \in \mathcal{I}, \tag{6.90}$$

if we work with the macroscopic backstresses $\boldsymbol{\alpha}_i$.

Unlike for unlimited kinematic hardening now there is an additional constraint (6.90) which must be fulfilled by the backstresses $\boldsymbol{\alpha}_i$. The solution procedure developed for unlimited kinematic hardening material can not be employed anymore. This means that problem (6.87)–(6.90) can not be solved locally.

From the physical point of view this corresponds to the fact that failure of systems consisting of limited kinematic hardening material must not be local. Incremental collapse can not be excluded anymore. In fact the solution of problem (6.87)–(6.90) is more expensive than the solution of problem (6.3)–(6.5). For this reason we turn to the microelement-model.

The discretized formulation for the microelement-model was given in Section 5.4. Accordingly the following non-linear optimization problem has to be solved

$$\beta \rightarrow \max \tag{6.91}$$

$$\sum_{i=1}^{NG} \mathbf{C}_i \cdot \sum_{s=1}^{m} \frac{t_s}{t} \boldsymbol{\rho}_{i,s} = \sum_{i=1}^{NG} \mathbf{C}_i \cdot \boldsymbol{\rho}_i = \mathbf{C} \cdot \boldsymbol{\rho} = \mathbf{0} \tag{6.92}$$

$$\Phi\left[\beta \boldsymbol{\sigma}_i^E(j) + \boldsymbol{\rho}_{i,1}\right] \leq k_0^2 = \sigma_0^2 \qquad \forall\ (i,j) \in \mathcal{I} \times \mathcal{J} \tag{6.93}$$

$$\Phi(\boldsymbol{\pi}_{i,1}) \leq (K - k_0)^2 = (\sigma_Y - \sigma_0)^2 \qquad \forall\ i \in \mathcal{I}. \tag{6.94}$$

For the case $m = 2$ problem (6.91)–(6.94) contains as many unknowns as problem (6.87)–(6.90). A direct solution of problem (6.91)–(6.94) implies the same difficulty as for (6.87)–(6.90).

It is known that all microelements behave elastic, perfectly plastic. Therefore we can employ the reduced basis technique developed for perfectly plastic material to solve problem (6.91)–(6.94).

Using the developed reduction technique the entire residual stress vector $\boldsymbol{\rho}_{i,l}$ in the microelements of the l-th interval at the i-th Gaussian point for the subproblem $\mathcal{P}^k$ can be represented as

$$\boldsymbol{\rho}_{i,l}^{(k)} = \mathbf{B}_{i,l}^{r,k} \cdot \mathbf{X}_r \tag{6.95}$$

by means of matrix $\mathbf{B}_{i,l}^{r,k}$ of the basis vectors. With respect to eqs. (5.30) and (5.31) one obtains

$$\boldsymbol{\rho}_i^{(k)} = \sum_{l=1}^{m} \Delta\xi_l\, \boldsymbol{\rho}_{i,l}^{(k)} = \sum_{l=1}^{m} \Delta\xi_l\, \mathbf{B}_{i,l}^{r,k} \cdot \mathbf{X}_r \tag{6.96}$$

and

$$\boldsymbol{\pi}_{i,l}^{(k)} = [\mathbf{B}_{i,l}^{r,k} - \sum_{s=1}^{m} \Delta\xi_s \mathbf{B}_{i,s}^{r,k}] \cdot \mathbf{X}_r^k . \tag{6.97}$$

With the relations (6.95) and (6.97) we get the discretized formulation for the (k)-th iteration

$$\beta^{(k)} \to \max \tag{6.98}$$

$$\Phi[\beta^{(k)} \boldsymbol{\sigma}_i^E(j) + \mathbf{B}_{i,1}^{r,k} \cdot \mathbf{X}_r^k] \leq \bar{k}_1 = k_0^2 \qquad \forall\ (i,j) \in \mathcal{I} \times \mathcal{J} \tag{6.99}$$

$$\Phi(\boldsymbol{\pi}_{i,1}^{(k)}) = \Phi[(\mathbf{B}_{i,1}^{r,k} - \sum_{s=1}^{m} \Delta\xi_s \mathbf{B}_{i,s}^{r,k}) \cdot \mathbf{X}_r^k] \leq (K - \bar{k}_1)^2 = (K - k_0)^2 \qquad \forall\ i \in \mathcal{I}. \tag{6.100}$$

Problem (6.98)–(6.100) contains only $r+1$ unknowns and can efficiently be solved by means of an SQP- or a penalty method .

7 Numerical Examples

In this section numerical examples of different structures are considered. The iteration behavior of the algorithms presented in the foregoing section is shown. The influence of kinematic hardening on the shakedown limits is demonstrated.

7.1 Plane stress problem with a notch

In the first example we consider a plane stress problem which was also investigated by WEICHERT and GROSS-WEEGE in [61]. In order to compare the classical SQP-algorithm to our algorithms for calculation of the ultimate load the finite element discretization was kept very coarse (see Fig. 7.1a). (Recall, that for considerably larger sized problems the classical SQP-algorithm fails to be practical due to high storage requirements.)

The following material constants were used: YOUNG's modulus $E = 21000 kN/cm^2$, POISSON's ratio $\mu = 0.3$ and the yield limit $\sigma_0 = 24 kN/cm^2$. The number of unknowns for the optimization problem is 181, the number of equality constraints is 39 and the number of inequality constraints is 60. The value for the load factor β obtained by the reduced basis technique was $\beta = 18.61$. In Fig. 7.1b the iteration history for the classical SQP-algorithm and the special SQP-algorithm is compared. Obviously, the convergence behavior of the latter is superior.

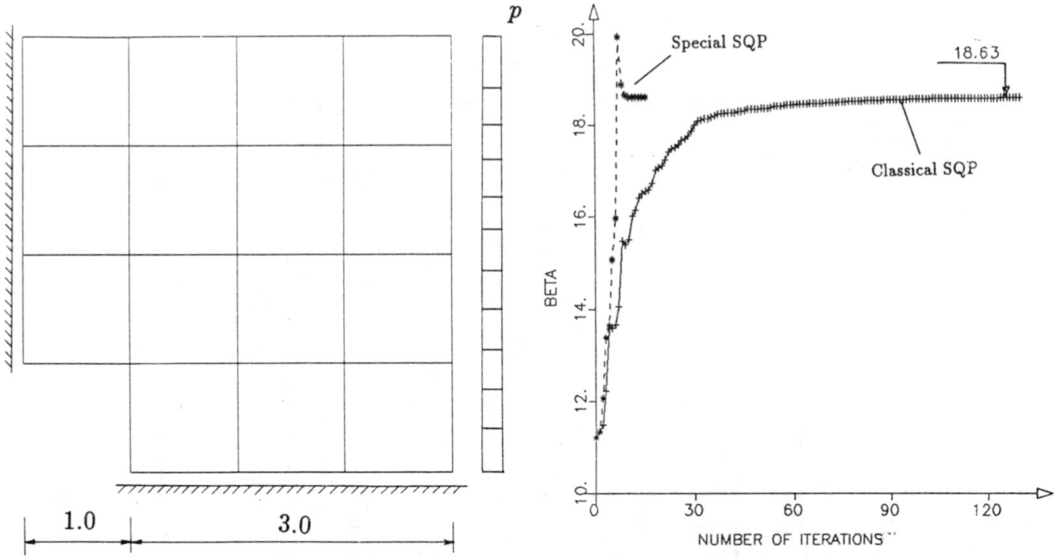

a) FE-Mesh b) Iteration behavior

Figure 7.1: Plane stress problem with a notch

7.2 Square plate with a central circular hole

A square plate with a central circular hole is considered. The length of the plate is L and the ratio between the diameter of the hole and the length of the plate is 0.2. The system is subjected to the biaxial uniform loads p_1 and p_2 (see Figure 7.2). Both can vary independently of each other between zero and certain maximum magnitudes $\bar{p}_1$ and $\bar{p}_2$. The load domains are defined by

$$0 \leq p_1 \leq \beta \gamma_1 \sigma_0 = \bar{p}_1 , \qquad 0 \leq \gamma_1 \leq 1$$
$$0 \leq p_2 \leq \beta \gamma_2 \sigma_0 = \bar{p}_2 , \qquad 0 \leq \gamma_2 \leq 1 ,$$

where σ_0 is the initial yield stress of the employed material and β is the shake-down load factor. Suitable choices of γ_1 and γ_2 enable us to investigate arbitrary load variation domains with different ratios of the load limit $\bar{p}_1$ to $\bar{p}_2$.

The shake-down behavior of this system consisting of elastic, perfectly plastic material firstly was investigated numerically by BELYTSCHKO [3] by use of 26 elements. Later, the same example was also investigated by CORRADI & ZAVELANI [11] by means of linear programming. There, 66 triangular elements were employed.

In the case of an elastic, perfectly plastic matarial our finite element mesh consists of 400 4-nodes isoparametric elements (see Figure 7.3). The special SQP-algorithm (cf. Section 6.1.3) and the reduced basis technique (cf. Section 6.1.4) were used to solve the resulting optimization problem (6.3)–(6.5), which contains 4801 unknowns and

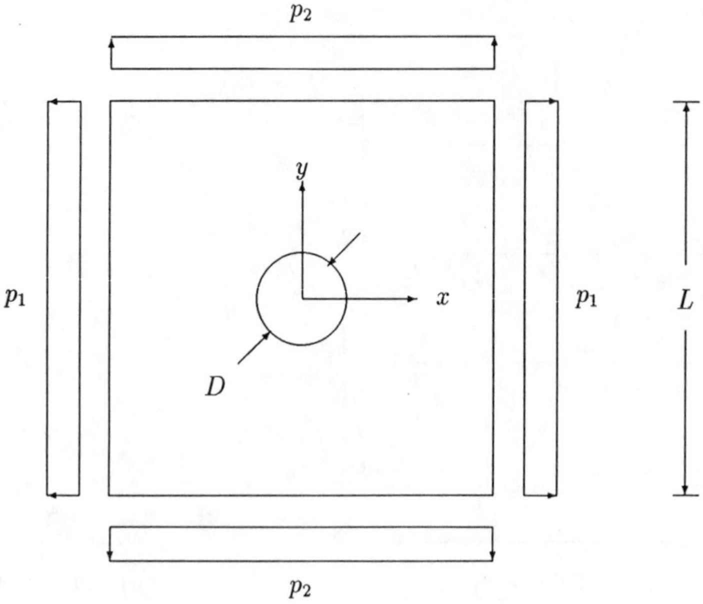

Figure 7.2: Square plate with a central circular hole

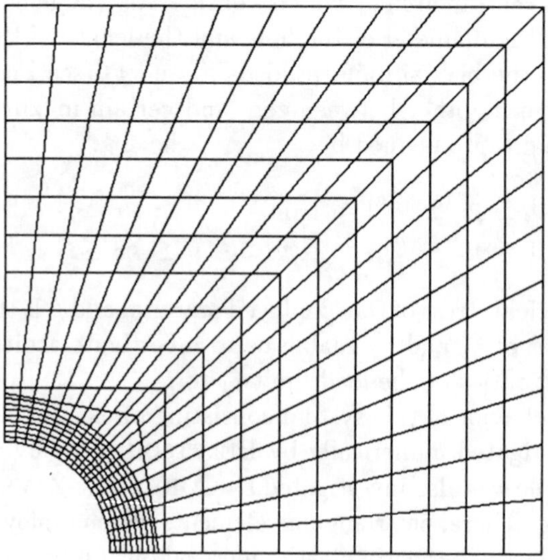

Figure 7.3: FE-mesh for the plate

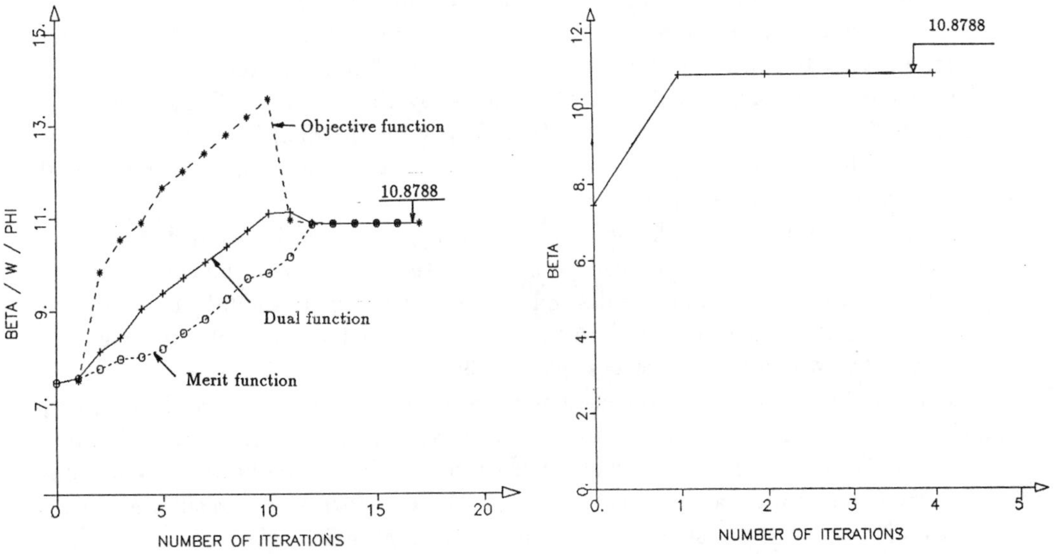

a) SQP b) Reduced basis technique

Figure 7.4: Iteration procedure for the special SQP-algorithm and the reduced basis technique

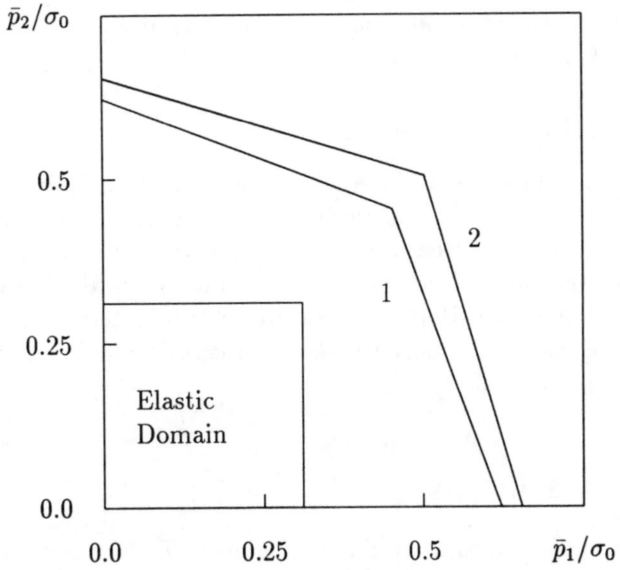

Figure 7.5: Shake-down limits for the square plate with circular hole

7240 linear and non-linear constraints. We have chosen this relatively high number of elements in order to show the efficiency of the presented algorithms.

We consider the calculation for the special load domain $0 \leq p_1 \leq \beta\sigma_0$ and $0 \leq p_2 \leq \beta\sigma_0$ in order to illustrate the iteration behavior of the presented algorithms.

The shake-down load $\bar{p}_1 = \bar{p}_2 = \beta\,\sigma_0$ obtained by the special SQP-algorithm was equal to 10.8788. In Figure 7.4a the iteration history for the dual-function, the merit-function and the objective function is shown. The figure demonstrates that the final solution is reached after less than 20 iterations.

The same value for the shake-down load was also obtained by using the reduced basis technique. The iteration procedure is illustrated in Figure 7.4b. Four iteration steps were necessary to obtain the desired shake-down load. It should be mentioned that the sum of the generated basis vectors in all subproblems is twelve whereas the dimension of the whole residual stress space is 3960.

The shake-down limits of the considered system for different values of γ_1 and γ_2 and, thus, for different load domains are represented by the curve 1 in Figure 7.5. They are about 4.2%–5.2% higher than shake-down limits which were obtained by using a finite element mesh with about 7000 elements. For comparison the results given by CORRADI & ZAVELANI are shown by the curve 2 in Figure 7.5.

In order to investigate the influence of kinematic hardening on the shake-down loads, the same system consisting of an unlimited kinematic hardening material was considered as well. The optimization problem was solved by use of the algorithm developed in Section 6.2.1. The numerical results showed that the kinematic hardening has no influence on the shake-down behavior, i.e. for this system under the given loading the shake-down loads for an elastic, perfectly plastic material are the same as for an unlimited kinematic hardening material. From the discussion in Section 6.2.3 we deduce that APSC is the dominant failure form.

7.3 Thin-walled cylindrical shell

A cylindrical shell is subjected to an internal pressure p and an internal temperature T_i (see Figure 7.6). The external temperature T_e is equal to zero for all time t. The system is assumed to be stress-free if $T_i = T_e$ and $p = 0$. The distribution of the temperature over the thickness of the shell is assumed to be linear what is justified by the small thickness of the shell. Both the pressure and the internal temperature can vary independently of each other between zero and certain maximum magnitudes $\bar{p}$ and $\bar{T}_i$. The load domains are defined by

$$\begin{aligned} 0 \leq p \leq \beta\,\gamma_1\,p_0 &= \bar{p}\,, \qquad 0 \leq \gamma_1 \leq 1 \\ 0 \leq T_i \leq \beta\,\gamma_2\,T_i^0 &= \bar{T}_i\,, \qquad 0 \leq \gamma_2 \leq 1\,. \end{aligned}$$

A similar system subjected to a variable temperature T_i, however with a constant internal pressure p was investigated theoretically by BREE [8] and numerically by GROSS-WEEGE [20].

$R = 1.0\,m$
$d = 0.1\,R$
$\nu = 0.3$
$E = 1.6 \cdot 10^4 \text{ kN/cm}^2$
$\alpha_\vartheta = 2.0 \cdot 10^{-5} \text{ K}^{-1}$
$\sigma_0 = 16.0 \text{ KN/cm}^2$
$p_0 = d\sigma_0 / R$
$T_i^0 = 2(1 - \nu)\sigma_0 / (E\alpha_\vartheta)$

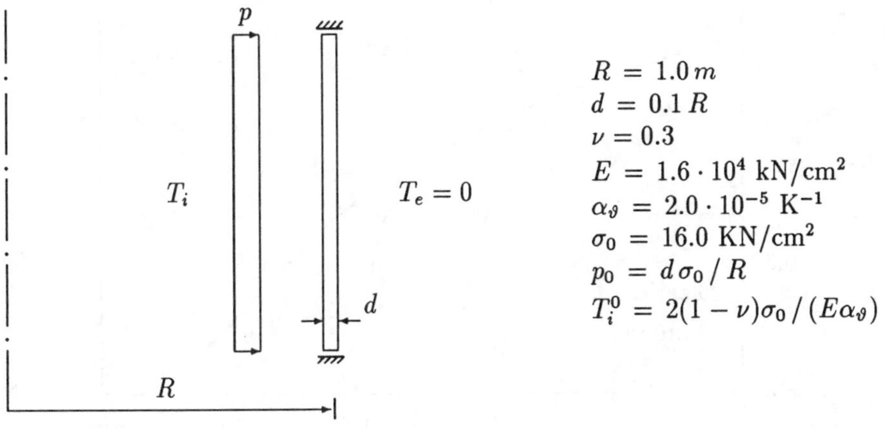

Figure 7.6: Thin-walled cylindrical shell: System and loading

To visualize the influence of kinematic hardening on the shake-down load and temperature limits, the following three cases of the constitutive law have been considered:

1. Elastic, perfectly plastic material with the initial yield stress σ_0.

   The computation has been performed by means of an axi-symmetrical shell element with 2 Gaussian points. To fulfill the yield condition across the shell thickness, 8 layers were introduced. The resulting shake-down limits are given by the curve 1 in Figure 7.7.

2. Material with a limited hardening ($\sigma_Y = 1.35\,\sigma_0$)

   To simulate the non-linear kinematic hardening effect several microelements with different yield stresses were introduced. The weakest $\bar{k}_1$ has the yield stress equal to σ_0 and the subsequent ones have yield stresses $\bar{k}_l > \sigma_0$ ($1 < l \leq m$) in an arbitrary manner, however such that the ultimate stress $\sigma_Y (= K)$ in the macroscale has the value $1.35\,\sigma_0$ according to eq. (5.23). The determined shake-down limits are represented by the curve 2 in Figure 7.7. It has to be stressed that this result depends neither on the number of the microelements employed nor on their yield stresses, provided the magnitude of σ_Y remains the same. This confirms the theoretical consideration given in Section 4.4.

3. Material with an unlimited hardening ($\sigma_Y = +\infty$)

   By assuming the same model as in 2 but with a very high yield stress of the last microelement one can analyze the limit case $\sigma_Y = +\infty$. The result is shown by the curve 3 in Figure 7.7. Again, it depends neither on the number of the employed microelements nor on their yield stresses (provided they are higher than σ_0).

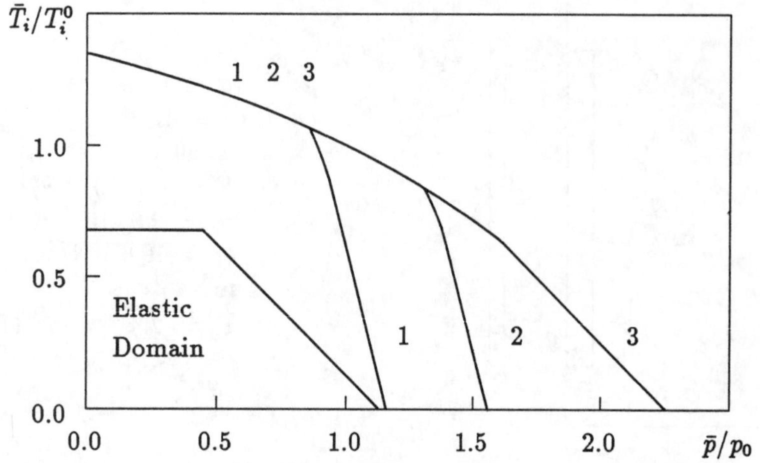

Figure 7.7: Shake-down limits for the cylindrical shell

For comparison, the case of an elastic, plastic material with a linear unlimited kinematic hardening was investigated as well. The results are identical with those of the case 3. This confirms that the form of the $\sigma - \varepsilon$ curve does not influence the shake-down limits. The initial yield stress σ_0 and the ultimate stress σ_Y (= K) define these limits solely.

The curves 1, 2 and 3 in Figure 7.7 show that the kinematic hardening does not always increase shake-down limits. The common part of the curves 1, 2 and 3 represents load domains which lead exclusively to APSC in all three cases, whereas both alternating plasticity and incremental collapse can occur under remaining load domains (see the different parts of the curves 1, 2 and 3 in Figure 7.7).

7.4 Rotating disc

A circular disc of the radius R and thickness d rotates with an angular velocity ω varying between 0 and its maximum value $\bar{\omega}$ (see Figure 7.8). The centrifugal force b_r acting on the system due to the angular velocity ω is given by

$$b_r = \omega^2 \rho r ,$$

where ρ is the density of the mass of the disc.

Moreover, the disc is subjected to a temperature field $T(r)$ with a quadratic distribution along the radius r

$$T(r) = \left[\frac{r}{R}\right]^2 T_R .$$

The boundary temperature T_R can vary between 0 and its maximum value $\bar{T}_R$ independent of the angular velocity ω. The load domains are defined by

$$0 \le \omega^2 \le \beta \gamma_1 \omega_0^2 = \bar{\omega}^2 , \quad 0 \le \gamma_1 \le 1$$

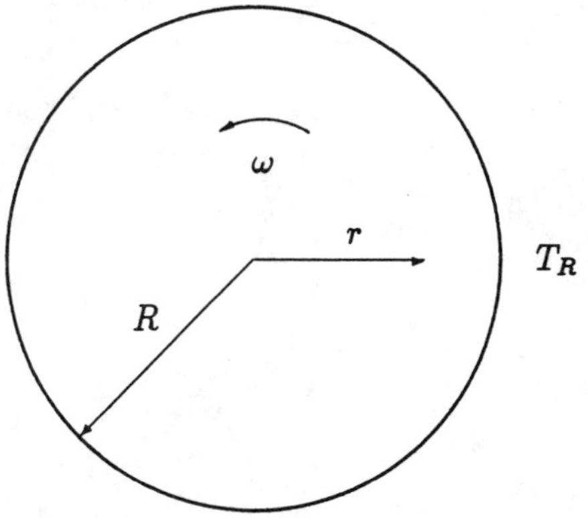

Figure 7.8: Rotating disc

$$0 \le T_R \le \beta \gamma_2 T_0 = \bar{T}_R \,, \qquad 0 \le \gamma_2 \le 1$$

with

$$T_0 = \frac{4\,\sigma_0}{\alpha_\vartheta\, E}\,, \qquad \omega_0^2 = \frac{3\,\sigma_0}{\rho\, R^2}\,.$$

The shake-down behavior of this system consisting of an elastic, perfectly plastic material was investigated first by ROZENBLUM [50] based on the static shake-down theorem. An investigation of the system based on the kinematic shake-down theorem can be found in a book by GOKHFELD & CHERNIAVSKY [19].

The following three different material models have been considered:

1. Elastic, perfectly plastic material with initial yield stress σ_0.
2. Material with a limited hardening, i.e. with the initial yield stress σ_0 and the ultimate stress $\sigma_Y = 1.2\,\sigma_0$.
3. Material with an unlimited hardening ($\sigma_Y = +\infty$).

In case 1 forty axi-symmetrical elements were used. Each element had 3 Gaussian points. The relative error of elastic stresses (in Gaussian points) calculated by means of the finite element method was about 1%.

Calculations for the cases 2 and 3 were performed by introducing different numbers of microelements (layers). As expected, the results were independent of the number of employed layers.

The shake-down loads for the cases 1–3 are represented in Figure 7.9 by the curves 1, 2 and 3. Similar conclusions can be drawn from Figure 7.9 as for Example 7.3.

For comparison an analytical "upper" bound (curve 4′) and an analytical "lower" bound (curve 4) for the elastic, perfectly plastic material (to be found in GOKHFELD & CHERNIAVSKY's book [19]) are also given in the same figure.

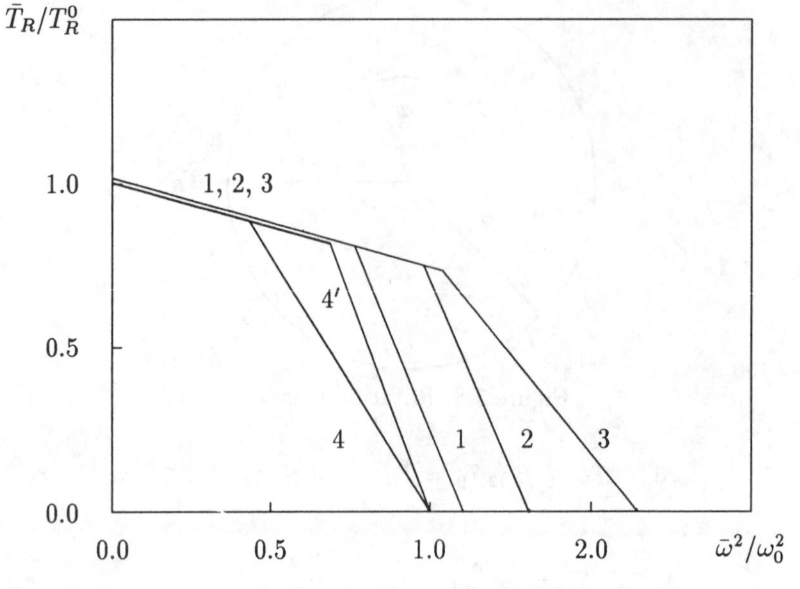

Figure 7.9: Shake-down limits for the rotating disc

The numerical shake-down limits (without hardening) are in good agreement with the analytic solutions (relative error about 1%) in the range where alternating plasticity is dominant (see the common part of the curves 1, 2 and 3). It can be shown that the analytical shake-down temperature $\bar{T}_R$ is exact for $\bar{\omega} = 0$. However, the discrepancies are distinct (about 8%) in that range where both incremental collapse and alternating plasticity can occur (see the different parts of the curves 1, 2 and 3). Since in this range the numerical shake-down limits are essentially higher than the analytical "upper" bound it must be questioned whether the numerical solutions are reliable.

This question can be answered if we take account of the fact that different yield conditions were used for the analytical and numerical solutions. The TRESCA yield condition employed for the analytic solutions has a conservative character in comparison to the v. MISES yield condition. The maximum difference between these two yield conditions is about 15%. Generally, the analytical "upper" bound obtained by employing the TRESCA yield condition loses its bounding character in comparison to the solution obtained by using the v. MISES yield condition.

7.5 Steel girder with a cope

Next, a steel girder with a length of 4 m will be investigated. It consists of an IPB-500 profile, and there is a cope on either side. The girder is simply supported and, in the middle, it is subjected to a concentrated single load P. Both corners of the copes are provided with a round drill hole of radius $r = 8\,mm$ (see Figure 7.10). The material is St 52-3 with an initial yield stress $\sigma_0 = 37.5\,kN/cm^2$ and a maximal hardening $\sigma_Y = 52.0\,kN/cm^2$. The hardening can be regarded as kinematic.

7.5.1 Experimental investigation

At the Institute for Steel Construction of the University of Braunschweig the girder was investigated by experiment [52]. In the first place the behavior of the system subjected to cyclic loading was of interest. Additionally, for comparison, the ultimate load was determined by experiment as well.

Firstly, the girder was subjected to different cyclic load programs. The load program of the first 15 cycles is shown in Table 7.1. Then, the girder was subjected to a load program, where the load varied between 0 and 600 kN with a velocity of 600 kN/min. The number of load cycles which led to a crack (with a length of 1 mm) at a drill hole was 145. The number of load cycles, which led to collapse of the girder, was 372.

Table 7.1: Loading program of the first 15 cycles

Cycles 1 – 5	Cycles 6 – 10	Cycles 11 – 15
$P:\ 0 \rightleftharpoons 540\ kN$	$P:\ 0 \rightleftharpoons 320\ kN$	$P:\ 0 \rightleftharpoons 540\ kN$

After collapse due to cyclic loading, the girder was shortened on either side by 50 cm, and it was recoped as before. Then, for this system the ultimate load was determined as 887 kN. Note that this value can also be regarded as the ultimate load for the former system, since only those cross-sections near the cope are responsible for failure of the system.

7.5.2 Numerical investigation

Due to symmetry of the system and the loading only a quarter of the system was considered for numerical investigation. For the FE-discretization two different types of elements were employed. The web of the girder was discretized by using 8128 isoparametric elements each with 4 nodes (see Figure 7.10). The upper and lower flanges were discretized with 768 and 640 DKT (Discrete Kirchhoff Triangle)–elements [2], respectively. Apart from bending forces the DKT-elements can also be stressed in plane in order to take account of the fact that the flanges are not subjected to pure bending.

For the numerical investigation, both, the ultimate load and the shake-down load were calculated. The solutions were obtained by the reduced basis technique. In order to demonstrate the influence of kinematic hardening on the ultimate load and on the shake-down load, respectively, the calculations were performed both for a perfectly plastic and for a kinematic hardening material. The results are shown in Table 7.2.

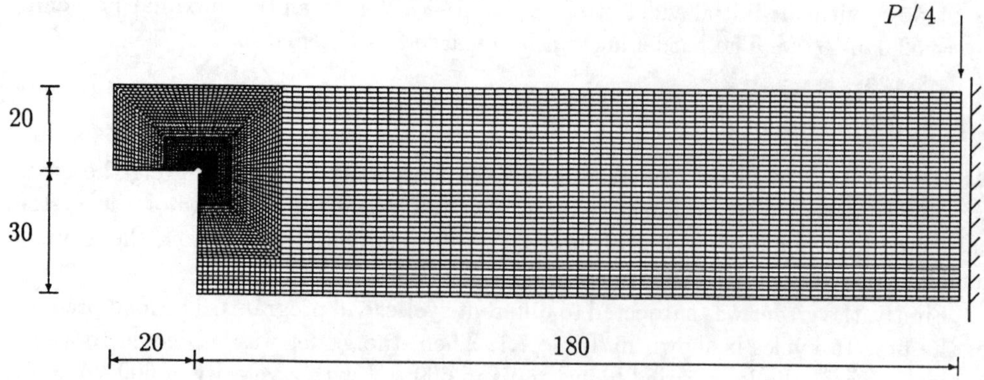

Figure 7.10: Discretization of the web

Note that the value for the ultimate load determined numerically (877.7 kN) was lower by 1% as the value determined by experiment (887 kN).

Table 7.2: Numerical ultimate and shake-down load

Material	Ultimate load [kN]	Shakedown load [kN]
perfectly plastic	633.2	164.2
with kinematic hardening	877.7	164.2

From Table 7.2 it can be seen, that, while the ultimate load increases by a factor of σ_Y/σ_0 due to kinematic hardening, the shake-down load remains unaltered. In this case the girder fails due to alternating plasticity near the drill holes (cf. Section 6.2.3).

It should be pointed out, that, originally, the experiment was not intended to analyze shake-down behavior. During the experiment, the amplitude of the cyclic loads were higher than the theoretical shake-down load. Thus no shake-down behavior could be observed. However, some valuable information can be drawn from the load-strain diagram for the first 5 load cycles shown in Figure 7.11. The strains were measured directly at one of both drill holes:

In Figure 7.11 a region can be observed, where the load P is linearly proportional to the strains, i.e. where the system behaves purely elastic. The amplitude of the region is between 155.2 kN and 179.5 kN. Thus, the experimental shake-down limit is larger than 155.2 kN, but smaller than 179.5 kN. A comparison shows, that the theoretical results are in good agreement with the experiment.

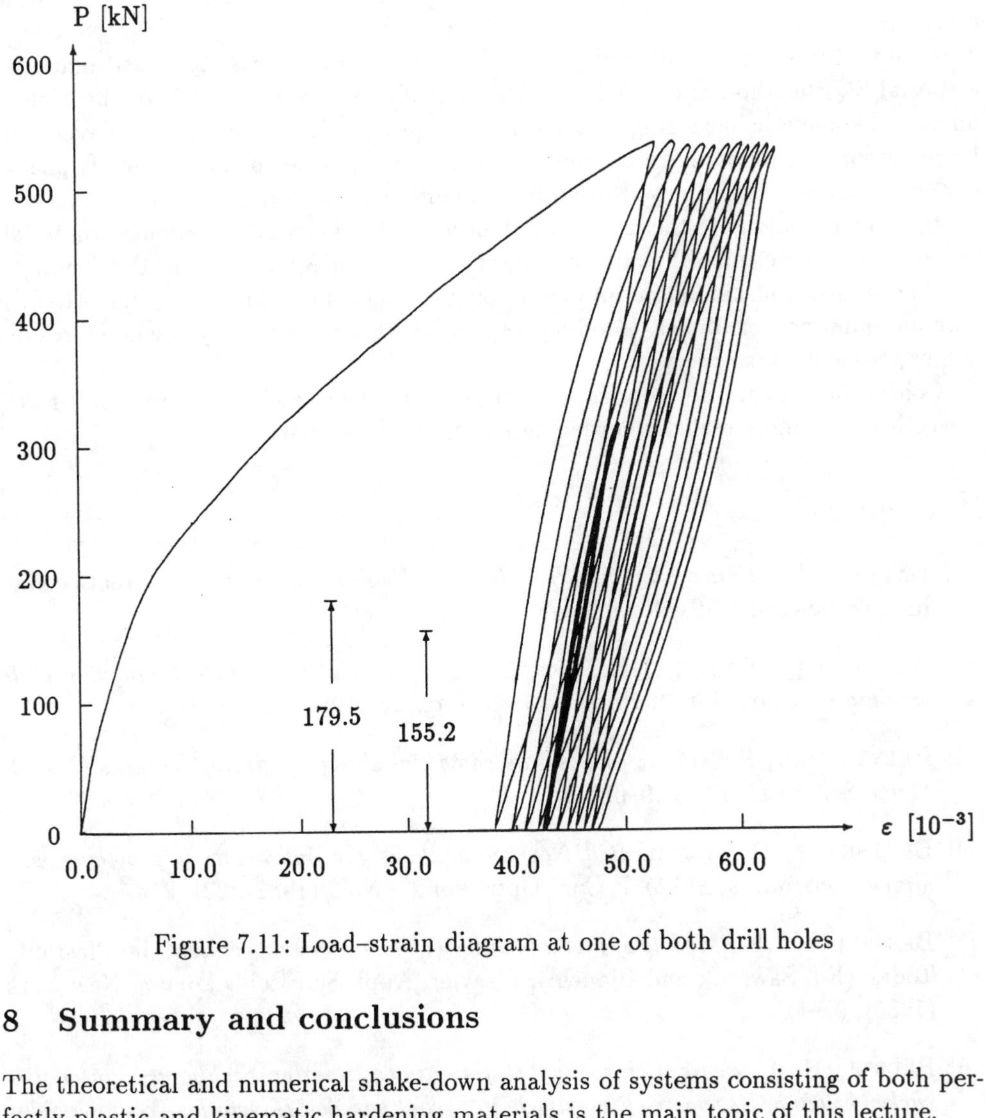

Figure 7.11: Load–strain diagram at one of both drill holes

8 Summary and conclusions

The theoretical and numerical shake-down analysis of systems consisting of both perfectly plastic and kinematic hardening materials is the main topic of this lecture.

After the presentation of the static shake-down theorems by MELAN for perfectly plastic and linear unlimited kinematic hardening materials a three-dimensional overlay-model was proposed, in order to describe non-linear limited kinematic hardening. A static shake-down theorem was formulated for the proposed material model and it was shown that the static shake-down theorem for the overlay-model is a generalization of the MELAN theorems.

The numerical approach based on the static shake-down theorems. In this way non-differentiability of the objective function of the optimization problem is avoided. For discretization a displacement method was chosen, which enables us to treat a wide class of problems.

In order to solve the resulting optimization problem for a perfectly plastic material, a special SQP-method and a reduced basis technique were presented. In the case of unlimited kinematic hardening material a local optimization technique was developed. By means of the proposed overlay-model, the algorithms developed for perfectly plastic materials can also be employed to limited kinematic hardening materials.

In many cases, due to the investigation of unlimited kinematic hardening material, we are able to determine the failure form of systems subjected to a variable loading.

The influence of kinematic hardening on the shake-down limits are illustrated in different numerical examples. Good agreement between theoretically predicted results and experiments was observed.

Concerning further developments attempts should be made to extend the shake-down theory to more realistic hardening materials.

References

[1] BATHE, J.K.: *Finite Element Procedures in Engineering Analysis*, Prentice-Hall Inc., Englewood Cliffs, New Jersey (1982).

[2] BATOZ, J.L., BATHE, J.K., HO, L.W.:*A study of three-node triangular plate bending elements*, Int. J. Numer. Meth. Eng. **15** (1980), 1771–1812.

[3] BELYTSCHKO, T.: *Plane stress shakedown analysis by finite elements*, Int. J. Mech. Sci. **14** (1972), 619–625.

[4] BERTSEKAS, D.P.: *Projected Newton methods for optimization problems with simple constraints*, SIAM J. Con. Opt., Vol 20, No 2 (1982), 221–246.

[5] BESSELING, J.F.: *Models of metal plasticity: theory and experiment*, in: Plasticity Today (Ed. Sawczuk and Bianchi), Elsevier, Appl. Sci. Publ. London-New York (1985), 97–113.

[6] BLEICH, H: *Über die Bemessung statisch unbestimmter Stahlwerke unter Berücksichtigung des elastisch-plastischen Verhaltens des Baustoffes*, Bauingenieur **13** (1932), 261–267.

[7] BORKOWSKI, A., KLEIBER, M.: *On a numerical approach to shakedown analysis of structures*, Comput. Methods Appl. Mech. Eng. **22** (1980), 101–119.

[8] BREE, J.: *Elastic-plastic behavior of thin tubes subjected to internal pressure and intermittent high-heat fluxes with application to fast-nuclear-reactor fuel elements*, J. Strain Analysis **2** (1967), 226–238.

[9] BUCKLEY, A.; LENIR, A.: *QN-like variable storage conjugate gradients.* Mathematical Programming, **27** (1983), 155–175.

[10] CORRADI, L.: *Dynamic shakedown in elastic-plastic bodies*, J. Eng. Mech. Div., Proc. of the American Society of Civil Engineers **106** No EM3 (1980), 481–499.

[11] CORRADI, L., ZAVELANI, I.: *A linear programming approach to shakedown analysis of structures*, Comp. Mech. Appl. Mech. Eng. **3** (1974), 37–53.

[12] ČYRAS, A.: *Optimization Theory in Limit Analysis of a Solid Deformable Body*, Mintis Publ. House, Vilnius (1971).

[13] DENNIS, J.E.; SCHNABEL, R.B.: *Numerical Methods for Unconstrained Optimization and Nonlinear Equations*, Prentice Hall, New Jersey (1983).

[14] DE DONATO, O.: *Second shakedown theorem allowing for cycles of both loads und temperature*, Ist. Lombardo Scienza Lettere (A) **104**, 265–277.

[15] DRUCKER, D.C.: *A more fundamental approach to plasticity stress-strain relations*, Proceedings of the 1st U.S. National Congress of Applied Mechanics, ASME (1951), 487–491.

[16] DRUCKER, D.C.: *A definition of stable inelastic material*, J. App. Mech. **26**, Trans. ASME **81**, Series E (1959), 101–106.

[17] FIACCO, A.V.; MCCORMICK, G.P.: *Nonlinear Programming: Sequential Unconstrained Minimization Techniques*, John Wiley and Sons, New York (1968).

[18] GILL, P.E.; MURRAY, W.; WRIGHT, M.H.: *Practical Optimization*, Academic Press, Inc. London (1981).

[19] GOKHFELD, D.A., CHERNIAVSKY, O.F.: *Limit Analysis of Structures at Thermal Cycling*, Sijthoff & Noordhoff (1980).

[20] GROSS-WEEGE, J.: *Zum Einspielverhalten von Flächentragwerken*, PhD thesis, Inst. für Mech., Ruhr-Universität Bochum (1988).

[21] GRÜNING, M: *Die Tragfähigkeit statisch unbestimmter Tragwerke aus Stahl bei beliebig häufig wiederholter Belastung*, Springer Verlag, Berlin (1926).

[22] GRUTTMANN, F., STEIN, E.: *Tangentiale Steifigkeitsmatrizen bei Anwendung von Projektionsverfahren in der Elastoplastizitätstheorie*, Ing.-Arch. **58** (1987), 15–24.

[23] HAN, S.P.: *A globally convergent method for nonlinear programming*, J. Opt. Th. Appl. **22** (1977), 297–309.

[24] HILL, R.: *Mathematical Theory of Plasticity*, Oxford (1950).

[25] KACHANOV, L.M.: *Foundations of the Theory of Plasticity*, North-Holland Publishing Company, Amsterdam · London (1971).

[26] KOITER, W.T.: *A new general theorem on shake-down of elastic-plastic structures*, Proc. Koninkl. Akad. Wet. **B 59** (1956), 24–34.

[27] KOITER, W.T.: *General theorems for elastic-plastic solids*, in: Progress in Solid Mechanics (Ed. SNEDDON and HILL), North Holland, Amsterdam (1960), 165–221.

[28] KÖNIG, J.A.: *Shakedown of strainhardening structures*, First Conadian Congr. Appl. Mech., Quebec (1967).

[29] KÖNIG, J.A., MAIER, G.: *Shakedown analysis of elastoplastic structures, a review of recent developments*, Nucl. Eng. Design **66** (1981), 81–95.

[30] KÖNIG, J.A.: *Shakedown of Elastic-Plastic Structures*, PWN-Polish Scientific Publishers, Warsaw (1987).

[31] LUENBERGER, D.G.: *Linear and Nonlinear Programming*, 2nd Ed., Addison-Wesley Publishing Company, New York (1984).

[32] MAIER, G.: *A shakedown matrix theory allowing for workhardening and second-order geometric effects*, Proc. Sypm. Foundations of Plasticity, Warsaw (1972).

[33] MANDEL, J.: *Adaptation d'une structure plastique ecrouissable et approximations*, Mech. Res. Comm., **3** (1976), 483.

[34] MARTIN, J.B.: *Plasticity*, MIT Press, Boston (1975).

[35] MASING, G.: *Zur Heyn'schen Theorie der Verfestigung der Metalle durch verborgen elastische Spannungen*, Wissenschaftliche Veröffentlichungen aus dem Siemens-Konzern **3** (1924), 231–239.

[36] MATTHIES, H., STRANG, G.: *The solution of nonlinear finite element equations*, Int. J. Num. Meths. Eng., 14, (1979), 1613–1623.

[37] MELAN, E.: *Theorie statisch unbestimmter Systeme aus ideal plastischem Baustoff*, Sitzber. Akad. Wiss. Wien IIa **145** (1936), 195–218.

[38] MELAN, E.: *Der Spannungszustand eines Mises-Henckyschen Kontinuums bei veränderlicher Belastung*, Sitzber. Akad. Wiss. Wien IIa **147** (1938), 73–78.

[39] MELAN, E.: *Zur Plastizität des räumlichen Kontinuums*, Ing.-Arch. **8** (1938), 116–126.

[40] MRÓZ, Z.: *On the description of anisotropic workhardening*, J. Mech. Phys. Solids **15** (1967), 163–175.

[41] MÜLLER-HOEPPE, N., PAULUN, J., STEIN, E.: *Einspielen ebener Stabtragwerke unter Normalkrafteinfluss*, Bauingenieur **61** (1986), 23–26 .

[42] NAZARETH, L.: *A relationship between the BFGS and conjugate gradient algorithms and its implications for new algorithms*, SIAM J. Numer. Analysis, **16** (1980), 773–782.

[43] NAZARETH, L.; NOCEDAL, J.: *Conjugate direction methods with variable storage*, Mathematical Programming, **23** (1982), 326–340.

[44] NGUYEN DANG, H., MORELLE, P.: *Numerical shakedown analysis of plates and shells of revolution*, Proceedings of 3rd world congress and exhibition on FEM's, Beverly Hills (1981).

[45] NEAL, B.G.: *Plastic collapse and shake-down theorems for structures of strain-hardening material*, J. Aero. Sci. **17** (1950), 297–306.

[46] NOCEDAL, J.: *Updating quasi-Newton matrices with limited storage*, Mathematical Programming, **35** (1980), 773-782.

[47] POWELL, M.J.D.: *A Fast Algorithm for Nonlinear Constrained Optimization Calculations*, In: Watson, G.A.: Numerical Analysis; Proc. of the Biennial Conference held at Dundee (June 1977).

[48] PONTER, A.R.S.: *A general shakedown theorem for elastic-plastic bodies with workhardening*, Proc. 3rd SMIRT Conf. London (1975), paper L 5/2.

[49] PRAGER, W.: *The theory of plasticity: a survey of recent achievements*, Proc. Inst. Mech. Engrs. **169** (1955), 41.

[50] ROZENBLUM, V.I.: *On shakedown theory of elastic-plastic bodies (in Russian)*, Izw. AN SSSR, OTN **6** (1958).

[51] SAWCZUK, A.: *On incremental collapse of shells under cyclic loading*, Second IUTAM Symp. on Theory of Thin Shells, Copenhagen 1967, Springer Verlag, Berlin (1969), 328–340.

[52] SCHEER, J., SCHEIBE, H.J., KUCH, D.: *Untersuchung von Trägerschwächungen unter wiederholter Belastung bis in den plastischen Bereich*, Bericht Nr. 6099, Institut für Stahlbau, TU Braunschweig (1990).

[53] SCHITTKOWSKI, K.: *The nonlinear programming method of Wilson, Han, and Powell with an augmented Larangian type line search function*, Numer. Math. **38** (1981), 83–127.

[54] SHEN, W.P.: *Traglast- und Anpassungsanalyse von Konstruktionen aus elastisch, ideal plastischem Material*, Dissertation, Inst. für Computeranwendung, Uni Stuttgart (1986).

[55] SIMO, J.C., TAYLOR, R.L.: Consistent tangent operators for rate–independent elastoplasticity, *Comput. Meth. Appl. Mech. Eng.* **48** (1985), 101–118.

[56] STEIN, E., ZHANG, G., MAHNKEN, R., KÖNIG, J.A.: *Micromechanical modelling and computation of shakedown with nonlinear kinematic hardening including examples for 2-D problems*, Proc. of CSME Mechanical Engineering Forum, Toronto (1990), 425–430.

[57] STEIN, E., ZHANG, G., KÖNIG, J.A.: *Micromechanical modelling and computation of shakedown with nonlinear kinematic hardening including examples for 2-D problems*, in: Recent Developments in Micromechanics (Ed. AXELRAD, D.R. and MUSCHIK, W.), Springer Verlag, Berlin (1990).

[58] STEIN, E., ZHANG, G., KÖNIG, J.A.: *Shakedown with nonlinear hardening including structural computation using finite element method*, to appear in Int. J. Plasticity.

[59] WASHIZU, K.: *Variational Methods in Elasticity & Plasticity*, 3rd edition, Pergamon Press (1982).

[60] WEICHERT, D.: *On the influence of geometrical nonlinearities on the shakedown of elastic-plastic structures*, Int. J. Plasticity **2** (1986), 135–148.

[61] WEICHERT, D., GROSS-WEEGE, J.: *The numerical assessment of elastic-plastic sheets under variable mechanical and thermal loads using a simplified two-surface yield condition*, Int. J. Mech. Sci. **30** (1988), 757–767.

[62] ZHANG, G., BISCHOFF, D.: *Traglastberechnung mit Hilfe der kinematischen Formulierung*, Z. angew. Math. Mech. **68** (1988), T267–T268.

[63] ZHANG, G.: *Einspielen und dessen numerische Behandlung für ideal plastische bzw. kinematisch verfestigende Materialien*, PhD thesis, Universität Hannover (1991).

[64] ZIEGLER, H.: *A modification of Prager's hardening rule*, Quart. Appl. Math. **17**, 55–65.

[65] ZIENKIEWICZ, O.C.: *The Finite Element Method*, 3rd Ed., McGRAW-HILL Book Company London (1979).

CONTINUUM MECHANICS, NONLINEAR FINITE ELEMENT TECHNIQUES AND COMPUTATIONAL STABILITY

P. Wriggers
T. H. Darmstadt, Darmstadt, Germany

Abstract

This three lectures course will give a modern concept of finite–element– analysis in nonlinear solid mechanics using material (Lagrangian) and spatial (Eulerian) coordinates. Elastic response of solids is treated as an essential example for the geometrically and material nonlinear behavior. Furthermore a brief introduction in stability analysis and the associated numerical algorithms will be given.

A main feature of these lectures is the derivation of consistent linearizations of the weak form of equilibrium within the same order of magnitude, taking also into account the material laws in order to get Newton-type iterative algorithms with quadratic convergence.

The lectures are intended to introduce into effective discretizations and algorithms based on a well founded mechanical and mathematical analysis.

1 Survey of nonlinear continuum mechanics

1.1 Introductory Remarks

In this section a short summary of the continuum mechanics of solids is given. In detail we will discuss the ingredients which must be provided for any theory used for the description of an isothermal mechanical process. These are the kinematical relations, the balance laws and the constitutive equations. We will restrict ourselves to the Lagrangian and spatial description of the motion and the related strain measures. Furthermore we will introduce the weak forms of the balance laws which will be the basis for the finite element formulation derived in section 2. As an example for constitutive laws the isotropic hyperelastic materials of Neo–Hookean type have been chosen to show besides geometrically nonlinear effects also the implications of nonlinear material behavior.

However due to the restricted space only a few equations and discussions concerning continuum mechanics can be presented. Thus, the reader who wishes to get a deeper insight may consult standard textbooks on continuum mechanics and the mathematical background (*Malvern* (1969), *Truesdell and Noll* (1965), *Marsden and Hughes* (1983), *Ogden* (1984), *Ciarlet* (1987)).

1.2 Kinematics

Let $\mathcal{B}$ be the reference configuration of a continuum body with particles $\mathbf{X} \in \mathcal{B}$. A deformation is described by the one–to–one mapping $\boldsymbol{\varphi}$ which defines the current configuration $\boldsymbol{\varphi}(B)$ as shown in Fig. 1.

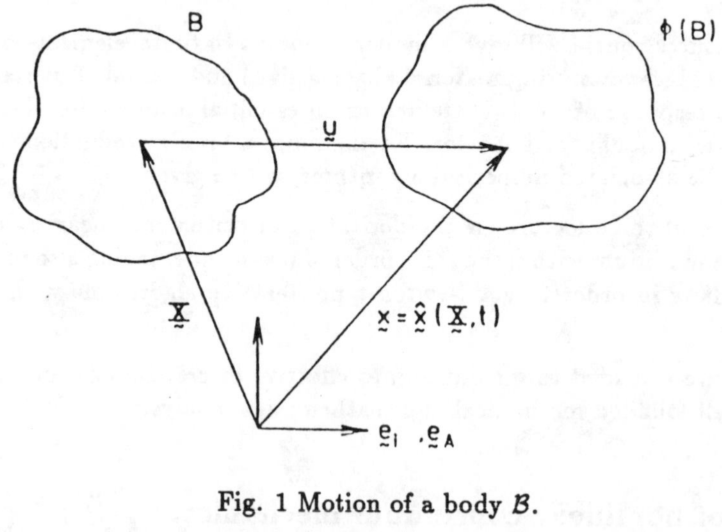

Fig. 1 Motion of a body $\mathcal{B}$.

A particle X in the undeformed configuration is decribed by the position vector $\mathbf{X} = X_A\, \mathbf{e}_A$. At time t the position occupied by the particle X is given by

$$\mathbf{x} = \boldsymbol{\varphi}(\mathbf{X}, t) = x_i\, \mathbf{e}_i. \tag{1.1}$$

The deformation gradient $\mathbf{F}$ is defined as the tensor which associates an infinitesimal vector $\mathbf{dX}$ at $\mathbf{X}$ with a vector $\mathbf{dx}$ at $\mathbf{x}$: $\mathbf{dx} = \mathbf{F}\mathbf{dX}$. Therefore, the components of $\mathbf{F}$ are the partial derivatives $\frac{\partial x_i}{\partial X_A} = x_{i,A}$. With (1.1) we obtain

$$\begin{aligned} \mathbf{F} &= F_{iA}\, \mathbf{e}_i \otimes \mathbf{e}_A = \operatorname{Grad} \mathbf{x}\,, \\ F_{iA} &= x_{i,A}\,. \end{aligned} \tag{1.2}$$

By $J = \det \mathbf{F}$ we define the determinant of the deformation gradient. Throughout our considerations we restrict the deformation such that the local impenetrability condition $J > 0$ is always fulfilled. The equations of continuum mechanics may be formulated with respect to the deformed or the undeformed configuration of the body. Within this lecture we refer all quantities either to the undeformed configuration B (material or Lagrangian description) or to the current configuration $\varphi(B)$. It should be remarked that the chosen description of course does not change the theory. Thus whether one uses the material or spatial description is a matter of tast. However there can be strong implication coming from numerical solution procedures regarding efficiency.

Knowing the deformation gradient $\mathbf{F}$ we can introduce the right and left Cauchy–Green tensors

$$\begin{aligned} \mathbf{C} &= \mathbf{F}^T\,\mathbf{F}\,, \\ \mathbf{b} &= \mathbf{F}\,\mathbf{F}^T\,. \end{aligned} \tag{1.3}$$

These strain measures are classically used to define the Green-Lagrangian strain tensor $\mathbf{E}$

$$\begin{aligned} \mathbf{E} &= E_{AB}\,\mathbf{e}_A \otimes \mathbf{e}_B = \frac{1}{2}\,(\,\mathbf{C} - \mathbf{1}\,), \\ E_{AB} &= \frac{1}{2}\,(\,C_{AB} - \delta_{AB}\,) \end{aligned} \tag{1.4}$$

and its spatial equivalent the Almansi strain tensor

$$\begin{aligned} \mathbf{e} &= e_{ik}\mathbf{e}_i \otimes \mathbf{e}_k = \frac{1}{2}\,(\,\mathbf{1} - \mathbf{b}^{-1}\,), \\ e_{ik} &= \frac{1}{2}\,(\,\delta_{ik} - (b^{-1})_{ik}\,) \end{aligned} \tag{1.5}$$

The equivalence of both tensors can be shown by using the notion of *push forward* operations, see e.g. *Marsden, Hughes* (1983). From (1.4) and (1.5) we deduce easily the following relation

$$\mathbf{E} = \mathbf{F}^T\,\mathbf{e}\,\mathbf{F}\,. \tag{1.6}$$

The Green strain tensor is often used for the theoretical formulation within engineering applications, however, it is not the only possible strain measure. A generalization for strain measures related to the reference configuration may be found e.g. in *Ogden* (1984)

$$\mathbf{E}^{\alpha} = \frac{1}{\alpha}(\mathbf{U}^{\alpha} - \mathbf{1}), \quad \alpha \in \mathbb{R}. \tag{1.7}$$

Its spatial counterpart in $\boldsymbol{\varphi}(B)$ is given by

$$\mathbf{e}^{\alpha} = \frac{1}{\alpha}(\mathbf{V}^{\alpha} - \mathbf{1}), \quad \alpha \in \mathbb{R},. \tag{1.8}$$

For the definition of these strain measures one has to apply the well known polar decomposition of the deformation gradient in a rotation tensor $\mathbf{R}$ and the right and left stretch tensors $\mathbf{U}, \mathbf{V}$ which are needed in (1.7) and (1.8)

$$\mathbf{F} = \mathbf{R}\,\mathbf{U} = \mathbf{V}\,\mathbf{R}. \tag{1.9}$$

Furthermore one has to consider that $\mathbf{U}^{\alpha}$ and $\mathbf{V}^{\alpha}$ are only defined via the spectral decomposition of the tensors $\mathbf{U}^{\alpha} = \sum_{i=1}^{3} \lambda_i^{\alpha} \mathbf{N}_i \otimes \mathbf{N}_i$ or $\mathbf{V}^{\alpha} = \sum_{i=1}^{3} \lambda_i^{\alpha} \mathbf{n}_i \otimes \mathbf{n}_i$. The eigenvalues λ_i which appear in the decomposition are the principal values of $\mathbf{U}$ or $\mathbf{V}$.

As a special case of (1.7) we obtain with $\alpha = 2$ the Green strains. Furthermore, equations (1.8) yields with $\alpha = -2$ the Almansi strains (1.5).

At the end of this section we like to discuss deformations which are restricted due to constraints. This is for example true if incompressibility of a rubberlike material or in plasticity plays a roll in the formulation. The condition associated with this constraint is $det\,\mathbf{F} = J = 1$. This constraint can be incorporated in the theory with the following multiplicative decomposition of the deformation gradient $\mathbf{F}$ which has been introduced by *Flory* (1961)

$$\mathbf{F} = J^{\frac{1}{3}}\,\widehat{\mathbf{F}}, \tag{1.10}$$

Here J is the dilatoric part. $\widehat{\mathbf{F}}$ denotes the volume preserving part which leads to an a priori fulfillment of the constant volume constraint since $\det \widehat{\mathbf{F}} \equiv 1$.

1.2 Balance laws

In this section we will summarize differential equations expressing locally the conservation of mass and momentum. These equations are associated with the fundamental postulates of continuum mechanics. A detailed derivation may be found e.g. in *Malvern* (1969, Ch. 5). Within the material description the conservation of mass is given by $\rho_o = J\,\rho$ where ρ is the density. From this equation we obtain a relation between the volume elements in the deformed and undeformed configuration

$$dv = \frac{\rho_o}{\rho}\,dV = J\,dV. \tag{1.11}$$

The conservation of momentum or the local form of the equation of motion is given with respect to the reference configuration by

$$\begin{aligned} \mathrm{Div}\mathbf{P} + \rho_o \mathbf{b} &= \rho_o \ddot{\mathbf{x}} \,, \\ P_{Ai,A} + \rho_o b_i &= \rho_o \ddot{x}_i \,. \end{aligned} \tag{1.12}$$

ere $\mathbf{P}$ denotes the unsymmetric first Piola-Kirchhoff stress tensor which associates the stress vector $\mathbf{t}$ with a normal vector $\mathbf{n}$ by $\mathbf{t} = \mathbf{P}\,\mathbf{n}$. The latter relation is well known as the Cauchy theorem. $\rho_o \mathbf{b}$ is the body force (e.g. gravity force) and $\rho_o \ddot{\mathbf{x}}$ denotes the inertial force which will be neglected in the following. The balance of moment of momentum leads to the following condition for the 1. Piola-Kirchhoff stress tensor

$$\mathbf{P}\,\mathbf{F}^T = \mathbf{F}\,\mathbf{P}^T \tag{1.13}$$

which shows the nonsymmetry. Obviously it is more convenient to work with symmetric stress tensors. Therefore we may introduce other stress tensors. Since a given stress vector does not change its magnitude when referred to the undeformed or deformed configuration we may write

$$\int_{\partial B} \mathbf{t}\, dA = \int_{\partial B} \mathbf{P}\,\mathbf{n}_o\, dA = \int_{\partial B} \frac{1}{J} \mathbf{P}\,\mathbf{F}^T \mathbf{n}\, da = \int_{\partial B} \boldsymbol{\sigma}\,\mathbf{n}\, da \tag{1.14}$$

which defines the symmetric Cauchy stress tensor $\boldsymbol{\sigma}$ referred to the current configuration

$$\begin{aligned} \boldsymbol{\sigma} &= \frac{1}{J} \mathbf{P}\,\mathbf{F}^T \\ \sigma_{ik} &= \frac{1}{J} F_{iA}\, P_{Ak} \,. \end{aligned} \tag{1.15}$$

The symmetric second Piola-Kirchhoff stress tensor is referred to B and defined by

$$\begin{aligned} \mathbf{S} &= \mathbf{F}^{-1}\,\mathbf{P} \,, \\ S_{AB} &= F_{Ai}\, P_{Bi} \,. \end{aligned} \tag{1.16}$$

$\mathbf{S}$ itself does not represent a physically meaningful stress but plays an important role in constitutive theory because it is the conjugated stress measure to the Green–Lagrangian strain tensor (1.4).

1.3 Constitutive laws

The constitutive theory describes the macroscopic behaviour of a material. By the constitutive equations strains are related to the stresses within a body. Since real materials behave in very complex ways we have to approximate physical observations of the real material's response. This approximations can be done seperately for different material responses (e.g. elastic, plastic or viscoplastic behavior).

In this section we restrict ourselves to pure elastic material behavior. The constitutive equation for the elastic response of the second Piola-Kirchhoff stress tensor may be derived from the hyperelastic potential ψ by

$$\mathbf{S} = \rho_o \frac{\partial \psi(\mathbf{E})}{\partial \mathbf{E}} ; \qquad S_{AB} = \rho_o \frac{\partial \psi(\mathbf{E})}{\partial E_{AB}} . \tag{1.17}$$

In the current configuration we may construct a potential which yields the Cauchy stresses when differentiated with respect to the covariant metric tensor of the current configuration $\mathbf{g} = g_{ik} \mathbf{e}^i \otimes \mathbf{e}^k$

$$\boldsymbol{\sigma} = 2 \rho \frac{\partial \psi}{\partial \mathbf{g}} = \rho \frac{\partial \psi}{\partial \mathbf{e}} . \tag{1.18}$$

The form $(1.18)_1$ is due to *Doyle, Erickson* (1956). Further versions can be found in *Simo, Marsden* (1984).

In case of isotropic materials relation (1.18) can be specialized by using the representation theorem for isotropic tensor functions. This leads to a function which only depends on the invariants of the stretch tensors $\mathbf{U}$ or $\mathbf{V}$: $\psi_V = \psi_V (I_V, II_V, III_V)$. Due to $\mathbf{V}^2 = \mathbf{b}$ we may also write $\psi_b = \psi_b (I_b, II_b, III_b)$. Furthermore we can write the invariants in terms of the principal stretches λ_i, se e.g. *Truesdell, Noll* (1965), which leads to the following functional $\psi(\mathbf{U}) \equiv \psi(\mathbf{V}) = \psi_\lambda (\lambda_1, \lambda_2, \lambda_3)$. Representaion ψ_b yields the following expression for the Cauchy stresses when the chain rule and the Cayley–Hamilton theorem are applied

$$\boldsymbol{\sigma} = \rho \left[\left(\frac{\partial \psi_b}{\partial J} + \frac{2}{J} \frac{\partial \psi_b}{\partial II_b} II_b \right) \mathbf{g}^{-1} + \frac{2}{J} \frac{\partial \psi_b}{\partial I_b} \mathbf{b} - 2 J \frac{\partial \psi_b}{\partial II_b} \right) \mathbf{b}^{-1} \right] . \tag{1.19}$$

Equation (1.19) may also be written in terms of the reference configuration using the second Piola–Kirchhoff stresses and the right Cauchy Green tensor. Based on equation (1.19) one can now describe elastic materials like rubber. A simple material model, the compressible Neo–Hookean law, will be discussed in this context. It depends only on the first and the third invariant of $\mathbf{b}$ with $J = \sqrt{III_b}$

$$\rho \psi(I_b, J) = g(J) + \frac{1}{2} \mu (I_b - 3) . \tag{1.20}$$

For more complex constitutive relations, see e.g. *Ogden* (1984). If the function $g(J)$ fulfills the polyconvexity conditions then the only known global existence result for finite elasticity applies to this model, see e.g. *Ciarlet* (1988). Such a function has been reported in *Ciarlet* (1988) and *Simo, Hughes* (1990)

$$g(J) = \frac{\lambda}{4}\,(J^2 - 1) - \frac{1}{2}\,(\lambda - 2\,\mu)\ln J\,. \tag{1.21}$$

From (1.19) we obtain

$$\boldsymbol{\sigma} = \frac{\lambda}{2\,J}\,(J^2 - 1)\,\mathbf{1} + \frac{\mu}{J}\,(\mathbf{b} - \mathbf{1})\,. \tag{1.22}$$

Equation (1.22) can now be expressed in terms of the reference configuration. To this purpose we recall the relation between the Cauchy and the second Piola–Kirchhoff stresses $\mathbf{S} = J\,\mathbf{F}^{-1}\,\boldsymbol{\sigma}\,\mathbf{F}^{-T}$. After some algebraic manipulation we derive

$$\mathbf{S} = \frac{\lambda}{2}\,(J^2 - 1)\,\mathbf{C}^{-1} + \mu\,(\mathbf{1} - \mathbf{C}^{-1})\,. \tag{1.23}$$

We like to remark that this constitutive relation has nothing to do with the so called St. Venant material which relates Green strains (1.4) linearly to second Piola–Kirchoff stresses.

$$\mathbf{S} = \lambda\,\mathrm{tr}\,\mathbf{E}\,\mathbf{1} + 2\,\mu\,\mathbf{E} \tag{1.24}$$

This material is often used in engineering applications. However, it does not fulfill the conditions for polyconvexity and can only be applied for small elastic deformations.

It should be remarked that both constitutive laws (1.22) as well as (1.24) reduce to the classical Hooke's law of elasticity in case of small strains.

1.4 Variational principles

The variational or weak form of the equations derived in the previous sections are very useful for the following reasons. They are helpful in the study of existence and uniqueness and furthermore basis for many numerical methods e.g. the finite element method. The energy functional for elasticity problems is given for a displacement formulation by

$$\Pi\,(\boldsymbol{\varphi}) = \int_B \rho_o\,\psi(\mathbf{C})\,dV - \int_B \rho\mathbf{b}\cdot\boldsymbol{\varphi}\,dV - \int_{\partial B_\sigma} \bar{\mathbf{t}}\cdot\boldsymbol{\varphi}\,dA \quad \Rightarrow \quad \text{Min}\,. \tag{1.25}$$

where $\rho_o\,\psi\,(\,\mathbf{C}\,)$ denotes a stored energy function like (1.20). $\Pi\,(\,\boldsymbol{\varphi}\,)$ must be a minimum for $\boldsymbol{\varphi}$ to be a solution of the associated boundary value problem. Thus we have to compute the derivative of Π. The derivative of Π at φ in the direction η is given by

$$D\Pi(\varphi)\cdot\eta = \frac{d}{d\alpha}\Pi(\varphi+\alpha\eta)\Big|_{\alpha=0}$$

and often called first variation of Π, directional derivative or Gateaux–derivative in the literature. Applying this formalism to (1.25) we obtain

$$G(\varphi,\eta) = D\Pi\cdot\eta = \int_B \rho_o \frac{\partial\psi}{\partial \mathbf{C}}\cdot(D\mathbf{C}\cdot\eta)\,dV - \int_B \rho_o \mathbf{b}\cdot\eta\,dV - \int_{\partial B_\sigma} \hat{\mathbf{t}}\cdot\eta\,dA = 0\,. \quad (1.26)$$

The directional derivative $G(\varphi,\eta)$ must vanish for Π to be a minimum at φ.
The first term in (1.26) is often referred to as stress divergence term. The last two terms denote the virtual work of the applied loads. Formulation (1.26) is also called principle of virtual work or weak form of the boundary value problem of elastostatics. The principle of virtual work, however, may be derived from equation (1.12) without using a constitutive relation. The principle of virtual work, starting from (1.12), can be applied to problems for which no potential exist like frictional sliding or non-associated plasticity.

The variation of $\mathbf{C}$ yields

$$\delta\,\mathbf{C} = D\mathbf{C}\cdot\eta = \mathbf{F}^T\,\mathrm{Grad}\,\eta + \mathrm{Grad}^T\eta\,\mathbf{F}\,, \quad (1.27)$$

In general, since (1.26) is nonlinear, iterative numerical methods have to be employed to solve this equation. For this purpose — especially for Newton-type methods — the linearization $D\,G$ of (1.26) at $\bar{\varphi}$ may be necessary

$$G(\bar{\varphi},\eta,\Delta\mathbf{u}) = G(\bar{\varphi},\eta) + D\,G(\bar{\varphi},\eta)\cdot\Delta\mathbf{u} = 0\,. \quad (1.28)$$

From (1.28) one can construct a Newton–Raphson procedure for the solution of the nonlinear boundary value problem (1.26), see section 2.4. The linearization of G yields

$$D\,G(\bar{\varphi},\eta)\cdot\Delta\mathbf{u} = \int_B \{\frac{\partial\mathbf{S}}{\partial\mathbf{C}}[D\mathbf{C}\cdot\Delta\mathbf{u}]\cdot\delta\mathbf{C} + \bar{\mathbf{S}}\cdot D(\delta\,\mathbf{C})\cdot\Delta\mathbf{u}\}\,dV\,. \quad (1.29)$$

Here, we have to evaluate the expression $\frac{\partial\mathbf{S}}{\partial\mathbf{C}}$ which coincides with the standard definition of the elasticity tensor

$$\mathbb{C} = 2\frac{\partial\mathbf{S}}{\partial\mathbf{C}}\,. \quad (1.30)$$

Using the material law (1.23) we obtain explicitly

$$\mathbb{C} = \lambda J^2 \mathbf{C}^{-1} \otimes \mathbf{C}^{-1} + 2\mu[1 - \frac{\lambda}{2\mu}(J^2 - 1)]\mathbb{I}. \tag{1.31}$$

$\mathbb{I}$ denotes a unit tensor with respect to $\mathbf{C}^{-1}$. The elasticity tensor (1.31) reduces to classical Hooke's law of the linear theory when only small strains are present. Then we have $J \approx 1$ and $\mathbf{C}^{-1} \approx \mathbf{1}$ and thus $\mathbb{C}_L = \lambda \mathbf{1} \otimes \mathbf{1} + 2\mu\mathbb{I}$. The same linear constitutive tensor is also obtained for the St. Venant material (1.24).

The Gateaux derivative used for the variation of (1.25) applies also to the linearization of (1.26). Thus $\Delta\mathbf{D} := (D\mathbf{C} \cdot \Delta\mathbf{u})$ has the same structure as $\delta\mathbf{C}$ in (1.27), only the arguments $\boldsymbol{\eta}$ and $\Delta\mathbf{u}$ have to be exchanged. Since also

$$D(\delta\mathbf{C}) \cdot \Delta\mathbf{u} = \delta\Delta\mathbf{C} = \text{Grad}^T \Delta\mathbf{u}\, \text{Grad}\boldsymbol{\eta} + \text{Grad}^T \boldsymbol{\eta}\, \text{Grad}\Delta\mathbf{u} \tag{1.32}$$

is symmetric in $\boldsymbol{\eta}$ and $\Delta\mathbf{u}$ we observe that the linearization (1.29) yields an symmetric tangent operator

$$DG(\bar{\boldsymbol{\varphi}}, \boldsymbol{\eta}) \cdot \Delta\mathbf{u} = \int_B \{\, \delta\mathbf{C} \cdot \mathbb{C}[\Delta\mathbf{C}] + \bar{\mathbf{S}} \cdot \delta\Delta\mathbf{C} \,\}\, dV\,. \tag{1.33}$$

In this section all equations have been derived so far with respect to the reference configuration. Since we also want to provied the spatial formulation for future use we have to transform the associated quantities. We denote by $grad(\bullet) = \frac{\partial(\bullet)}{\partial\mathbf{x}}$ the spatial gradient. The following equation relates the material and the spatial gradients $Grad(\bullet) = \frac{\partial(\bullet)}{\partial\mathbf{X}} = \frac{\partial(\bullet)}{\partial\mathbf{x}}\frac{\partial\mathbf{x}}{\partial\mathbf{X}} = grad(\bullet)\,\mathbf{F}$. Thus the variation of $\mathbf{C}$ can be expressed by

$$D\mathbf{C} \cdot \boldsymbol{\eta} = 2\,\mathbf{F}^T \text{grad}\,\boldsymbol{\eta}\,\mathbf{F}\,, \tag{1.34}$$

and we are able to formulate the first integrand in (1.26) with respect to the current configuration

$$\rho_o \frac{\partial\psi}{\partial\mathbf{C}} \cdot (D\mathbf{C} \cdot \boldsymbol{\eta}) = \frac{1}{2}\,\mathbf{S} \cdot \delta\mathbf{C} = \frac{1}{2}\,J\,\mathbf{F}^{-1}\boldsymbol{\sigma}\mathbf{F}^{-T} \cdot 2\,\mathbf{F}^T \text{grad}\,\boldsymbol{\eta}\,\mathbf{F} = J\,\boldsymbol{\sigma} \cdot \text{grad}\,\boldsymbol{\eta}\,. \tag{1.35}$$

This yields with (1.11) and (1.26)

$$g(\mathbf{u}, \boldsymbol{\eta}) = D\Pi \cdot \boldsymbol{\eta} = \int_{\Phi(B)} \boldsymbol{\sigma} \cdot \text{grad}\boldsymbol{\eta}\, dv - \int_{\Phi(B)} \rho\mathbf{b} \cdot \boldsymbol{\eta}\, dv - \int_{\partial\Phi(B_\sigma)} \hat{\mathbf{t}} \cdot \boldsymbol{\eta}\, da = 0 \tag{1.36}$$

The tangent operator (1.33) can be transformed to the current configuration in an analogous way. Here we may use the result, see e.g. *Marsden, Hughes* (1983)

$$\mathcal{C}[\mathbf{a}] = \frac{1}{J}\,\mathbf{F}\,\mathbb{C}[\,\mathbf{F}^T\,\mathbf{a}\,\mathbf{F}\,]\,\mathbf{F}^T \tag{1.37}$$

which relates the elasticity tensors in the current and reference configuration. Applied to (1.31) we obtain

$$\boldsymbol{C} = \lambda J^2 \mathbf{1} \otimes \mathbf{1} + 2\mu [1 - \frac{\lambda}{2\mu}(J^2 - 1)] \mathbb{I} \tag{1.38}$$

where $\mathbf{1}$ and $\mathbb{I}$ denote 2nd– and 4th– order unit tensors with respect to the current configuration. Using this result we may derive the spatial form of tangent operator (1.33) for large elastic deformations

$$D\, g(\bar{\varphi}, \boldsymbol{\eta}) \cdot \Delta \mathbf{u} = \int_{\Phi(B)} \{ \operatorname{grad} \boldsymbol{\eta} \cdot \boldsymbol{C} [\operatorname{sym}(\operatorname{grad} \Delta \mathbf{u})] + \boldsymbol{\sigma} \cdot \operatorname{grad}^T \boldsymbol{\eta} \operatorname{grad} \Delta \mathbf{u} \}\, dv \tag{1.39}$$

where $sym(grad\,\Delta \mathbf{u}) = \frac{1}{2}(grad\,\Delta \mathbf{u} + grad^T \Delta \mathbf{u})$.

This closes the section of continuum mechanics which was ment to give a short introductory overview. As an example hyperelastic materials undergoing large deformations have been studied and the variational formulation was derived. The following section is devoted to the finite element formulation for this class of problems and the associated algorithmic treatment.

2 Discretization, Finite–Element–Formulation

2.1 Introduction

In this section the finite element approximation of the boundary value problem of elastostatics stated in section 1 will be considered. We derive the finite element matrices based on the formulation in the previous section with respect to the current and to the reference configuration. Since this section is only an introduction to nonlinear finite element calculations we will limit the discussion of different finite element models. Thus we restrict ourselves to pure a displacement formulation. However we will have some remarks regarding mixed and other advanced models which are necessary for a successful treatment of plasticity problems.

2.2 Lagrangian Finite–Element–Formulation

Let us introduce a standard finite element discretization B^h for the body $\mathcal{B}$

$$B^h = \bigcup_{e=1}^{n_e} \Omega_e$$

with n_e elements and the volume of an element denoted by Ω_e.

For two–and three–dimensional solid applications isoparametric finite elements are widely used. Their advantage can be seen in the good modeling of complicated geometries. The isoparametric concept means that geometry $\mathbf{X}$, the displacement field $\mathbf{u}$, the incremental displacement field $\Delta\mathbf{u}$ and the variations $\boldsymbol{\eta}$ are interpolated within an element Ω_e by the same functions (see Figure 2)

$$\begin{aligned} \mathbf{X}_e^h &= \sum_{I=1}^{n} N_I \mathbf{X}_I\,, & \boldsymbol{\eta}_e^h &= \sum_{i=I}^{n} N_I \boldsymbol{\eta}_I\,, \\ \mathbf{u}_e^h &= \sum_{I=1}^{n} N_I \mathbf{v}_I\,, & \Delta\mathbf{u}_e^h &= \sum_{I=1}^{n} N_I \Delta\mathbf{v}_I\,, \end{aligned} \tag{2.1}$$

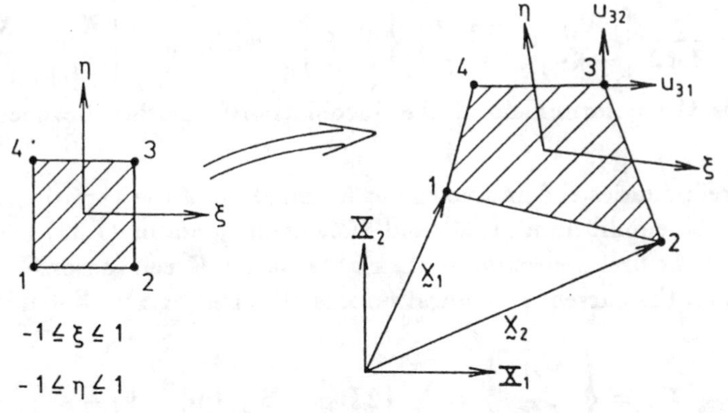

Figure 2 The isoparametric concept for two-dimensional applications.

In (2.1) the functions N_I are called shape functions, n is the number of nodes connected to one element and $(...)_I$ are the nodal values of the interpolated field $(...)^h$. For solid application we have as a requirement for the choice of N_I — the C^o – continuity — which follows from the fact that only first derivatives appear in (1.26). Furthermore N_I should be a complete polynominal in X_1 , X_2 and X_3, see *Zienkiewics, Taylor* (1989) to fulfill the requirements for convergence.

There are several ways for the generation of shape functions. Here we restrict ourselves to the Lagrangian family. The reader may consult standard textbooks on finite elements (e.g. *Zienkiewics, Taylor* (1989), *Hughes* (1987) or *Bathe* (1982)) for different approaches.

A simple example for the choice of shape functions within the Lagrangian family is given for an 8-node brick element by

$$N_I = \frac{1}{2}(1+\xi_i\,\xi)\,\frac{1}{2}(1+\eta_i\,\eta)\,\frac{1}{2}(1+\phi_i\,\phi)\,. \tag{2.2}$$

The last term in (2.2) vanishes in two–dimensional applications. The shape functions are defined in a local coordinate system ξ, η, ϕ therefore a transformation to the global coordinates X_1 , X_2 , X_3 is necessary (see Figure 2). Since the following formulation is the same for two– and three–dimensional problems we will restrict ourselves to the two–dimensional case for the ease of understanding. In the weak form (1.26) the derivatives of the deformation or the variations are used. Within the isoparametric concept we obtain for e.g. the derivative of the approximated displacement field

$$\frac{\partial \mathbf{u}_e^h}{\partial X_A} = \sum_{I=1}^{n} \frac{\partial N_I(\xi,\eta)}{\partial X_A}\,\mathbf{v}_I \tag{2.3}$$

where now the chain rule has to be used for the evaluation of the partial derivative of N_I with respect to X_A (see e.g. *Zienkiewics, Taylor* (1989)

$$\begin{Bmatrix} N_{I,1} \\ N_{I,2} \end{Bmatrix} = \frac{1}{\det \mathbf{J}} \begin{bmatrix} X_{2,\eta} & -X_{2,\xi} \\ -X_{1,\eta} & X_{1,\xi} \end{bmatrix} \begin{Bmatrix} N_{I,\xi} \\ N_{I,\eta} \end{Bmatrix} \quad \text{with} \quad \mathbf{J} = \begin{bmatrix} X_{1,\xi} & X_{2,\xi} \\ X_{1,\eta} & X_{2,\eta} \end{bmatrix} \tag{2.4}$$

where $det\,\mathbf{J}$ is the determinant of the Jacobi transformation between $d\xi$, $d\eta$ and dX_1, dX_2.

With these preliminaries the finite element formulation of the variational equation or weak form of the equilibrium (1.26) and of tangent operator (1.33) can be derived. For this purpose we first approximate the right Cauchy–Green tensor $\mathbf{C}$. Furthermore we will note that the current configuration $\mathbf{x}$ is obtained by $\mathbf{x} = \mathbf{X} + \mathbf{u}$. Then we can write

$$\mathbf{C}_e^h = \begin{Bmatrix} C_{11} \\ C_{22} \\ 2\,C_{12} \end{Bmatrix} = \sum_{I=1}^{n} [\,2\,\mathbf{B}_{0I} + \mathbf{B}_{LI}(\,\mathbf{u}_e^h\,)\,]\,\mathbf{v}_I + \mathbf{e} \tag{2.5}$$

where the so called B–matrices $\mathbf{B}_0$, $\mathbf{B}_L$ and the unit vector $\mathbf{e}$ are defined by

$$\mathbf{B}_{0I} = \begin{bmatrix} N_{I,1} & 0 \\ 0 & N_{I,2} \\ N_{I,2} & N_{I,1} \end{bmatrix} ; \qquad \mathbf{e} = \begin{Bmatrix} 1 \\ 1 \\ 0 \end{Bmatrix}$$
$$\mathbf{B}_{LI} = \begin{bmatrix} u_{1,1}\,N_{I,1} & u_{2,1}\,N_{I,1} \\ u_{1,2}\,N_{I,2} & u_{2,2}\,N_{I,2} \\ u_{1,1}\,N_{I,2} + u_{1,2}\,N_{I,1} & u_{2,2}\,N_{I,1} + u_{2,1}\,N_{I,2} \end{bmatrix} \tag{2.6}$$

Note, that $\mathbf{B}_L$ vanishes for $\mathbf{u} = const.$ The approximation of the virtual right Cauchy–Green strains (1.27) leads with (2.6) to

$$\delta \mathbf{C}_e^h = 2 \sum_{I=1}^{n} [\mathbf{B}_{0I} + \mathbf{B}_{LI}(\mathbf{u}_e^h)] \boldsymbol{\eta}_I \tag{2.7}$$

Equations (2.1) and (2.7) inserted into (1.26) yield the discretized weak form for a finite element Ω_e

$$G_e^h(\mathbf{v}, \boldsymbol{\eta}) = \sum_{I=1}^{n} \boldsymbol{\eta}_I^T \{ \int_{\Omega_e} (\mathbf{B}_{0I} + \mathbf{B}_{LI})^T \mathbf{S}^h \, d\Omega - \int_{\Omega_e} N_I \, \rho \mathbf{b}^h \, d\Omega - \int_{\partial\Omega_e} N_I \, \hat{\mathbf{t}}^h \, d(\partial\Omega) \} \tag{2.8}$$

with $(\mathbf{S}^h)^T = \{ S_{11}, S_{22}, S_{12} \}$.

The assembly process of all element contributions to the global algebraic system of equations for the problem at hand shall be denoted by the operator $\cup$ which incorporates boundary and transition conditions for the displacements into the global structure of the system of equations. With this notation the discretized form of the equilibrium equations is given by

$$\mathbf{G}(\mathbf{v}) = \mathbf{R}(\mathbf{v}) - \mathbf{P} = \mathbf{0} \qquad \forall \mathbf{v} \in \mathbb{R}^N . \tag{2.9}$$

Here N denotes the total number of degrees of freedom. The following definitions have been used in (2.9)

$$\begin{aligned} \mathbf{R}(\mathbf{v}) &= \bigcup_{e=1}^{n_e} \sum_{I=1}^{n} \int_{\Omega_e} (\mathbf{B}_{0I} + \mathbf{B}_{LI})^T \mathbf{S}^h \, d\Omega \\ \mathbf{P} &= \bigcup_{e=1}^{n_e} \sum_{I=1}^{n} \int_{\Omega_e} N_I \, \rho \mathbf{b}^h \, d\Omega + \int_{\partial\Omega_e} N_I \, \hat{\mathbf{t}}^h \, d(\partial\Omega) \end{aligned} \tag{2.10}$$

The linearization of (1.33) leads to the definition of the tangential stiffness matrix. With (1.22), (1.23), (1.29), (1.31), (2.1), (2.6) and (2.7) the following form of the tangent matrix is obtained

$$D\, G_e^h(\mathbf{v}, \boldsymbol{\eta}) \, \Delta\mathbf{v} = \sum_{I=1}^{n} \sum_{K=1}^{n} \boldsymbol{\eta}_I^T \int_{\Omega_e^h} [(\mathbf{B}_{0I} + \mathbf{B}_{LI})^T \mathbb{C} (\mathbf{B}_{0K} + \mathbf{B}_{LK}) + \mathbf{I} G_{IK}] \, d\Omega \, \Delta\mathbf{v}_K \tag{2.11}$$

where for two–dimensional problems G_{IK} is defined by

$$G_{IK} = [\, N_{I,1} \quad N_{I,2} \,] \begin{bmatrix} S_{11} & S_{12} \\ S_{21} & S_{22} \end{bmatrix} \begin{Bmatrix} N_{K,1} \\ N_{K,2} \end{Bmatrix} . \tag{2.12}$$

It should be noted that we have to use in (2.11) the constitutive tensor (1.31) which needs the computation of $(\mathbf{C}^{-1})^h$. This can be obtained using (2.5). The assembly procedure leads with $\mathbf{B} := \mathbf{B}_0 + \mathbf{B}_L$ to the global tangential stiffness matrix $\mathbf{K}_T$

$$\mathbf{K}_T = \bigcup_{e=1}^{n_e} \sum_{I=1}^{n} \sum_{K=1}^{n} \int_{\Omega_e^h} [\, \mathbf{B}_I^T \mathbf{C} \mathbf{B}_K + \mathbf{I} G_{IK} \,] \, d\Omega \tag{2.13}$$

In (2.10) and (2.11) the evaluation of the integrals is performed according to the Gaussian quadrature formulas, see e.g. *Zienkiewics, Taylor* (1989).

REMARK: We have assumed so far that the load vector $\mathbf{P}$ does not depend on the deformation. This is true for many engineering applications. However there are also technically important cases – like pressure loading – where the load indeed depends on the deformations. Then we obtain within the linearization process a contribution from the loading term $\mathbf{P}$ to the tangent matrix.

2.3 Spatial Finite–Element–Formulation

Isoparametric elements open the possibility to a very efficient and straightforward implementation of the spatial continuum formulation. This is due to the fact that within the isoparametric concept the metric is mapped to an unit square. Thus it does not make a difference whether the reference or the current configurations is mapped to this unit square. An advantage of the spatial formulation is the simple form of the constitutive tensor (1.38) which only contains the unit tensor of the current configuration.

Again we employ the same isoparametric interpolations with the one difference that now the current geometry $\mathbf{x}$ is approximated instead of the reference configuration $\mathbf{X}$

$$\mathbf{x}_e^h = \sum_{I=1}^{n} N_I \mathbf{x}_I \tag{2.14}$$

As before, $(..)^h$ characterizes the finite element approximation.

Since nothing else changes we can apply the same shape functions (2.2) as in section 2.2. Now the global coordinates are given with respect to the current configuration which means that we have to perform all differentiations with respect to $\mathbf{x}$

$$\frac{\partial \mathbf{u}_e^h}{\partial x_a} = \sum_{I=1}^{n} \frac{\partial N_I(\xi, \eta)}{\partial x_a} \mathbf{v}_I \,. \tag{2.15}$$

Again we apply the chain rule to compute the partial derivative of N_I with respect to x_i

$$\left\{ \begin{matrix} N_{I,1} \\ N_{I,2} \end{matrix} \right\} = \frac{1}{\det \mathbf{J}} \begin{bmatrix} x_{2,\eta} & -x_{2,\xi} \\ -x_{1,\eta} & x_{1,\xi} \end{bmatrix} \left\{ \begin{matrix} N_{I,\xi} \\ N_{I,\eta} \end{matrix} \right\} \quad with \quad \mathbf{J} = \begin{bmatrix} x_{1,\xi} & x_{2,\xi} \\ x_{1,\eta} & x_{2,\eta} \end{bmatrix} \tag{2.16}$$

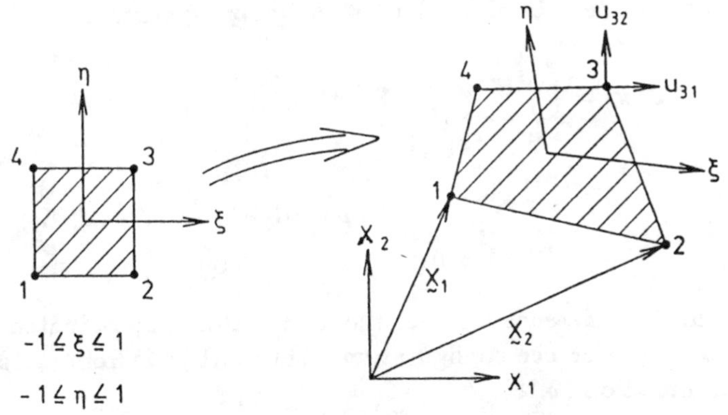

Fig. 3 Isoparametric concept for spatial formulation

where $det\,\mathbf{J}$ is the determinant of the Jacobi transformation between $d\xi$, $d\eta$ and dx_1, dx_2. Observe that the formalism is exactly the same as in sction 2.2. That means we can use the standard shape functions routines, with the one difference, that we have as an input the current position vector $\mathbf{x}^h$ instead of vector $\mathbf{X}^h$ associated with the reference configuration.

Within the finite element formulation the approximation of the stress divergence term in (1.36) has to be evaluated. For this purpose we need to compute the spatial gradient $grad\,\boldsymbol{\eta}^h$ which yields

$$(\text{grad}\,\boldsymbol{\eta})^h = \sum_{I=1}^{n} \mathbf{B}_I\,\boldsymbol{\eta}_I \quad \text{with} \quad \mathbf{B}_I = \begin{bmatrix} N_{I,1} & 0 \\ 0 & N_{I,2} \\ N_{I,2} & N_{I,1} \end{bmatrix}. \tag{2.17}$$

Inserting (2.17) in (1.36) leads to the discrete formulation of the weak form for one element Ω_e

$$g_e^h(\mathbf{x},\boldsymbol{\eta}) = \sum_{I=1}^{n} \boldsymbol{\eta}_I^{hT}\,[\int_{\Phi(\Omega_e)} (\mathbf{B}_I^T\,\boldsymbol{\sigma}^h - N_I\,\rho\,\bar{\mathbf{b}})\,d\Omega - \int_{\Phi(\partial\Omega_e)} N_I\,\bar{\mathbf{t}}\,d(\partial\Omega)\,] \tag{2.18}$$

where the Cauchy stress vector $(\boldsymbol{\sigma}^h)^T := \{\,\sigma_{11},\,\sigma_{22},\,\sigma_{12}\,\}$ has been introduced. The integrals in (2.18) are evaluated using Gaussian quadrature.

Since the integral in (1.36) is defined over the entire body we have to apply the standard assembly procedure to arrive at the discretized nonlinear equilibrium equations. Again we denote assembly by the operator $\cup$ which incorporates boundary and transition conditions for the displacements. With this notation the discretized form of the equilibrium equations is given by

$$\mathbf{G}(\,\mathbf{x}\,) = \mathbf{R}(\,\mathbf{x}\,) - \mathbf{P} = \mathbf{0} \qquad \text{with} \quad \mathbf{x} = \mathbf{X} + \mathbf{v};\quad \forall \mathbf{v} \in \mathbb{R}^N. \tag{2.19}$$

N is the number of degrees of freedom of the global system and

$$\mathbf{R}(\mathbf{x}) = \bigcup_{e=1}^{n_e} \sum_{I=1}^{n} \int_{\Phi(\Omega_e)} \mathbf{B}_I^T \boldsymbol{\sigma}^h \, d\Omega \,,$$

$$\mathbf{P} = \bigcup_{e=1}^{n_e} \sum_{I=1}^{n} \int_{\Phi(\Omega_e)} N_I \, \rho \, \bar{\mathbf{b}} \, d\Omega + \int_{\Phi(\partial\Omega_e)} N_I \, \bar{\mathbf{t}} \, d(\partial\Omega) \,. \tag{2.20}$$

It does not make a difference whether the load vector $\mathbf{P}$ is evaluated with respect to the current or the reference configuration. Thus either $\mathbf{P}$ from (2.10) or (2.20) can be used in formulation (2.18).

The stress $\boldsymbol{\sigma}^h$ in (2.18) is given by the hyperelastic constitutive law (1.22). For its evaluation within the finite element context we have to compute the approximations of the left Cauchy–Green tensor $\mathbf{b}^h = \mathbf{F}\,\mathbf{F}^T$ and of the Jacobi determinant J^h. In the spatial formulation we have only access to the current position vector $\mathbf{x}^h$ and not to $\mathbf{X}^h$. This means that we cannot compute the deformation gradient $\mathbf{F}^h$ which has to be used in $(1.3)_2$ from $\mathbf{F}^h = Grad\, \mathbf{X}^h$. However we can apply the following simple relation. Knowing that $\mathbf{x}^h = \mathbf{X}^h + \mathbf{u}^h$ we obtain

$$(\,\mathbf{F}^{-1}\,)^h = \mathbf{1} - \operatorname{grad} \mathbf{u}^h \,. \tag{2.21}$$

Thus we are able compute the inverse of $\mathbf{F}^h$ where $grad\,\mathbf{u}^h$ is the spatial displacement gradient. After some algebraic manipulation we can express the left Cauchy–Green tensor explicitly

$$\mathbf{b}^h = \begin{Bmatrix} b_{11} \\ b_{22} \\ b_{12} \end{Bmatrix}^h = J^h \begin{bmatrix} (1 - N_{I,2}\, v_{I2})^2 + (N_{I,1}\, v_{I2})^2 \\ (1 - N_{I,1}\, v_{I1})^2 + (N_{I,2}\, v_{I1})^2 \\ (1 - N_{I,2}\, v_{I2})(N_{I,1}\, v_{I2}) + (1 - N_{I,1}\, v_{I1})(N_{I,2}\, v_{I1}) \end{bmatrix} , \tag{2.22}$$

where

$$J^h = \det \mathbf{F}^h = [\,(1 - N_{I,1}\, v_{I1})(1 - N_{I,2}\, v_{I2}) - (N_{I,2}\, v_{I1})(N_{I,1}\, v_{I2})\,]^{-1} \,.$$

Now $\mathbf{b}^h$ is known in terms of the nodal displacements $\mathbf{v}_I$. The Cauchy stresses follow from (1.22). With the expression for the unit tensor $\mathbf{e}$ defined in $(2.6)_2$ we obtain

$$\boldsymbol{\sigma}^h = \frac{\lambda}{2\, J^h}\, [\,(J^h)^2 - 1\,]\, \mathbf{e} + \frac{\mu}{J^h}\, (\, \mathbf{b}^h - \mathbf{e}\,) \,. \tag{2.23}$$

The matrix form of the spatial constitutive tensor (1.38) takes the simple form

$$\mathbf{c}^h = \lambda\, (J^h)^2\, \mathbf{e}\,\mathbf{e}^T + 2\,\mu\, \{\, 1 - \frac{\lambda}{2\,\mu}\, [\,(J^h)^2 - 1\,]\,\}\, \mathbf{E} \,, \tag{2.24}$$

where **E** denotes the unit matrix.

When a Newton iteration is employed to solve the algebraic nonlinear system of equations (2.19) equation (2.18) has to be linearized. As in section 2.2 we obtain with the matrices

$$\hat{\boldsymbol{\sigma}}^h = \begin{bmatrix} \sigma_{11} & \sigma_{12} \\ \sigma_{21} & \sigma_{22} \end{bmatrix}^h, \quad \mathbf{G}_I = \begin{Bmatrix} N_{I,1} \\ N_{I,2} \end{Bmatrix}, \quad \text{und} \quad \mathbf{I} = \begin{bmatrix} 1 & 0 \\ 0 & 1 \end{bmatrix} \tag{2.25}$$

the tangent operator of an element Ω_e

$$Dg_e^h(\bar{\mathbf{x}}, \boldsymbol{\eta})\,\Delta\mathbf{v} = \sum_{I=1}^{4}\sum_{J=1}^{4} \boldsymbol{\eta}_I^{h^T} \int_{\Phi(\Omega_e)} \{\mathbf{B}_I^T\,\mathbf{c}^h\,\mathbf{B}_J + (\mathbf{G}_I^T\,\hat{\boldsymbol{\sigma}}^h\,\mathbf{G}_J)\mathbf{I}\}\; d\Omega\,\Delta\mathbf{v}_K\,. \tag{2.26}$$

The assembly procedure leads to the global tangent stiffness $\mathbf{K}_T$ of the spatial formulation

$$\mathbf{K}_T = \bigcup_{1}^{n_e}\sum_{I=1}^{4}\sum_{J=1}^{4} \int_{\Phi(\Omega_e)} \{\mathbf{B}_I^T\,\mathbf{c}^h\,\mathbf{B}_J + (\mathbf{G}_I^T\,\hat{\boldsymbol{\sigma}}^h\,\mathbf{G}_J)\mathbf{I}\}\; d\Omega\,. \tag{2.27}$$

In (2.27) appear only the matrices $\mathbf{B}_0$ defined in section 2.2 — compare (2.6) and (2.17) — which makes besides the simplicity of the constitutive tensor the spatial formulation especially attractive.

This completes the finite element discretization of hyperelastic bodies undergoing large deformations and strains. The solution of the nonlinear algebraic systems (2.9) or (2.19) will be discussed in the next section.

2.4 Solution of nonlinear finite element equations

Since Newton's method is very efficient for the solution of nonlinear systems like (2.9) or (2.19) the associated algorithm is stated below. Normally one cannot apply the entire load of a given nonlinear problem in one step. Thus an incremental approach has to be considered in combination with Newton's method. To open the possibility to apply the load in incremental steps we introduce a so called load parameter λ in (2.9) and (2.19) which leads to the general equation

$$\mathbf{G}(\mathbf{v}, \lambda) = \mathbf{R}(\mathbf{v}) - \lambda\mathbf{P} = \mathbf{0}\,. \tag{2.28}$$

Other incremental schemes like continuation procedures are described in section 3. With the linearization (2.13) or (2.26) of equation (2.28) we construct Newton's method from

$$D\,\mathbf{G}(\mathbf{v}, \lambda)\,\Delta\mathbf{v} + \mathbf{G}(\mathbf{v}, \lambda) = \mathbf{0}\,.$$

This leads to the algorithm stated in BOX 1 below

* Set initial values: $i = 0, \quad \mathbf{v}_0 = \mathbf{0}, \quad l = 0, \quad \lambda_0 = 0.$

1. Loop over load increments $\lambda_{l+1} = \lambda_l + \Delta\lambda_l$

   2. Iteration loop $i = 0, 1, \ldots$ *until convergence*

      2.1 IF spatial formulation: $\mathbf{y}_i = \mathbf{X} + \mathbf{v}_i$

      ELSE : $\mathbf{y}_i = \mathbf{v}_i$

      2.2 Convergence check

$$\| \mathbf{G}(\mathbf{y}_i, \lambda_{l+1}) \| \quad \begin{cases} \leq \text{TOL} \longrightarrow & \text{go to 3.)} \\ > \text{TOL} \longrightarrow & \text{go to 2.3)} \end{cases}$$

      2.3 Compute displacement increments: $\mathbf{K}_T(\mathbf{y}_i)\, \Delta\mathbf{v}_i = -\mathbf{G}(\mathbf{y}_i, \lambda_{l+1})$

      2.4) New displacement: $\mathbf{v}_{i+1} = \mathbf{v}_i + \Delta\mathbf{v}_i$ go to 2.)

3. IF $\lambda_{l+1} < \lambda_{max} \Longrightarrow l = l + 1$ go to 1.)

   ELSE $\Longrightarrow$ *STOP*

BOX 1 : Algorithm for Newton's method

In the described algorithm a Newton iteration is performed within every load step l. We observe that within the iteration loop a linear system of equations has to be solved which can be very costly for large systems when a direct solution method like Gaussian elimination is employed. Newton'method exhibits on the other hand a quadratic rate of convergence which leads to few iterations within each load cycle. Both characteristics of the algorithm are in contrast with respect to numerical efficiency.

Here we like to remark that a large amount of research work has been devoted to the topic of solving nonlinear finite element equations. We do not have the space to go into details. There are iterative solvers for the linear system in step 2.3 which can be very efficient if a good pre–conditioning scheme is known for the problem at hand. Quasi–Newton methods circumvent the need to solve the linear system in each iteration step by updating schemes which makes these methods attractive. However they exhibit only superlinear convergence and thus need more iterations than a pure Newton scheme. For further details see e.g. *Luenberger* (1984) or *Golub, Van Loan* (1989).

The following overview shows some of the methods which can be used to solve the system of equations (2.9 and (2.19)

* Linear algebraic systems of equations
  - Elimination techniques
    - Gaussian elimination,
    - Cholesky algorithm,
    - Frontal solution methods,
    - Block elimination methods,
  - Iterative methods
    - Conjugated gradient methods,
    - Overrelaxation methods,
  - Multigrid methods
  - Parallel solvers
* Nonlinear algebraic systems of equations
  - Fix point methods
  - Newton methods
  - Quasi Newton methods
  - Dynamic relaxation techniques
  - Continuation methods

Especially in nonlinear applications a good solution algorithm for (2.9) or (2.19) is a main factor for the efficiency of the method. This lies in the fact that finite element approximations lead to large systems with many degrees of freedom which have to be solved. Generally one can say that for small or medium sized problems elimination techniques are favorable whereas for large problems iterative solution techniques are preferable.

2.5 Example

As an example for a numerical simulation based on the theoretical framework given above we like to consider an ring undergoing large deformations and large strains in the plane strain case.

A is ring depicted in Figure 4 which is loaded by a single force at the top. The material constants are: $\mu = 6.000\ \mathrm{N/cm^2}$ and $\lambda = 24.000\ \mathrm{N/cm^2}$. 80 four–node finite elements were used to discretize the ring. The deformed configuration shown in Figure 4 is reached under a load of $P = 264$ N. The load was applied in only five load steps to arrive the final deformation. The following table shows the typical convergence behaviour of Newton's method within one step. Note that only 7 iterations (loop 2 in BOX 1) have been performed to converge to the solution up to computer precision.

Load factor λ	Residual $\|\|\mathbf{G}\|\|$
3.0	$1.32 \cdot 10^{2}$
3.0	$6.36 \cdot 10^{3}$
3.0	$1.42 \cdot 10^{2}$
3.0	$2.07 \cdot 10^{1}$
3.0	$2.42 \cdot 10^{-1}$
3.0	$6.80 \cdot 10^{-5}$
3.0	$4.27 \cdot 10^{-10}$

Newton convergence for load step 3

The next example is a plane strip which is stretched 100 %. Material data are as above. Here only one load step and 8 Newton iterations were sufficient to converge to the solution shown in Figure 5.

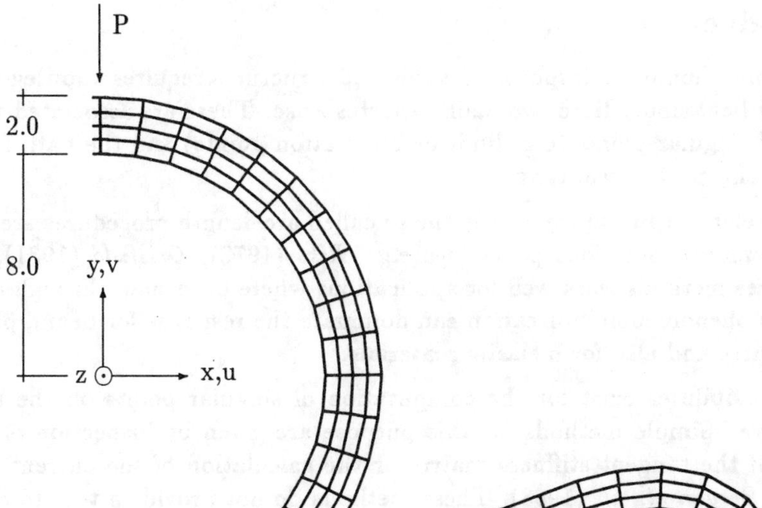

Fig. 4 System and deformation of a hyperelastic ring

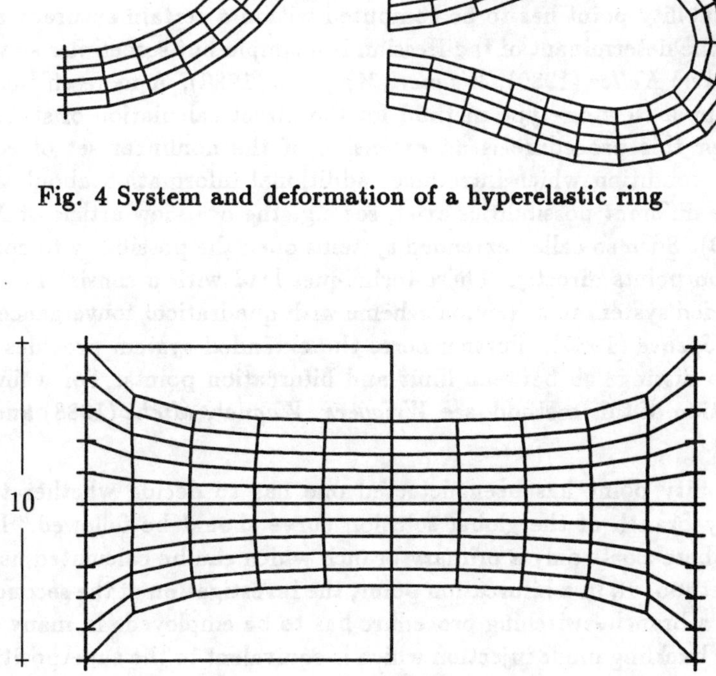

Fig. 5 System and deformation of a hyperelastic plane strip

3 Stability analysis and numerical algorithms

3.1 Introduction

Investigation of nonlinear response of solids and structures requires knowlegde about the stability behaviour. Here two main aspectcs arise. These are associated with the detection of singular points (e.g. limit or bifurcation points) and the path following in the pre– and postcritical range.

In the finite element literature mostly the so called arc–length procedures are applied to follow nonlinear solutions paths, see e.g. *Riks* (1972), *Crisfield* (1981), *Ramm* (1981). These methods work well for applications where pure snap-through appears. Besides this phenomenon bifurcation can dominate the response for beam, plate and shell structures and also for inelastic materials.

Different possibilities exist for the computation of singular points on the load deflection curve. Simple methods for this purpose are given by inspection of the determinant of the tangent stiffness matrix or the calculation of the current stiffness parameter, *Bergan et. al.* (1988). These methods do not provide a tool to calculate stability points accurately since the basis of path–following is an incremental procedure. If a stability point has to be computed within a certain accuracy a bi–section method for the determinant of the Hessian is a simple but sometimes slow approach, see e.g. *Decker, Keller* (1980), *Wagner, Wriggers* (1988). A more efficient approach is provided by a Newton–type method for the direct calculation of stability points. The basis for this procedure is an extension of the nonlinear set of equations by a constraint condition which introduces additional information about the stability points. Here different possibilities exist, see e.g. the overview article of *Mittelmann, Weber* (1980). Such so called extended systems open the possibility to compute limit or bifurcation points directly. These techniques lead with a consistent linearization of the extended system to a Newton scheme with quadratical convergence behaviour, see *Moore, Spence* (1980). Furthermore, the extended system provides enough information to distinguish between limit and bifurcation points. For a finite element implementation of this method, see *Wriggers, Wagner, Miehe* (1988) and *Wriggers, Simo* (1990).

Once a stability point has been detected one has to decide whether the primary or secondary branch of the global solution curve should be followed. In case of a limit point there exists only a primary branch which can be calculated using a path–following method. If, in a bifurcation point, the investigation of the secondary branch is necessary a branch–switching procedure has to be employed. In many engineering applications buckling mode injection which is equivalent to the superposition of buckling modes onto the displacement field at the bifurcation point is used to arrive at the secondary path. For this purpose an eigenvalue problem has to be solved which is expensive. However a sound mathematical treatment starts from the so called bifurcation equation, see *Decker, Keller* (1980), which need further information about the curvature of the parametrized solution curve.

The calculation of stability points via an extended system includes an additional advantage. First the buckling modes associated with simple bifurcation points are automatically obtained without extra costs and second the algorithms to compute the curvature of the parametrized solution curve are present.

At the end of this chapter an overall algorithm which is a combination of a path–following method and the use of the extended system for an effective calculation of stability points is given.

The algorithms based on the different approaches are illustrated by means of examples. We investigate the stability behaviour of a three–dimensional truss, a cylindrical shell and the diffuse necking of an elasto–plastic bar in detail.

3.2 Path–following methods

Standard finite element approximations lead to a system of algebraic equations, see (2.9) and (2.19) in section 2. This equation may be rewritten by introducing a scaling factor λ , also called load parameter,

$$\mathbf{G}(\mathbf{v},\lambda) = \mathbf{R}(\mathbf{v}) - \lambda\,\mathbf{P} = \mathbf{0}\,, \qquad \mathbf{v} \in \mathbb{R}^N \tag{3.1}$$

Algorithms for the solution of (3.1) are standard. However, if singular points occur standard methods like the Newton–Raphson procedure as discussed in section 2.4 fail. Then arc–length or continuation methods are advantageous. These methods are well established, see e.g. *Riks* (1972), *Ramm* (1981), *Schweizerhof, Wriggers* (1986). Since arc-length methods are essential to follow postbuckling branches, we will briefly discuss a general form of these procedures.

The main idea of continuation methods is to add a single constraint equation to the nonlinear set of equations (3.1). Thus a special form of an extended system is used which can be stated as

$$\tilde{\mathbf{G}}(\mathbf{v},\lambda) = \left\{ \begin{matrix} \mathbf{G}(\mathbf{v},\lambda) \\ f(\mathbf{v},\lambda) \end{matrix} \right\} = \mathbf{0}\,, \tag{3.2}$$

with a general form of the constraint equation $f(\mathbf{v},\lambda) = 0$. Based on this formulation special methods like load control, displacement control or arc–length procedures can be deduced. An important aspect for the iteration behaviour within the nonlinear calculations is the consistent linearization of the above given extended system at a known state $\mathbf{v}_i, \lambda_i$, which leads to

$$\begin{pmatrix} \nabla_v \mathbf{G} & -\nabla_\lambda \mathbf{G} \\ \nabla_v \mathbf{f}^T & \nabla_\lambda f \end{pmatrix}_i \left\{ \begin{matrix} \Delta \mathbf{v} \\ \Delta \lambda \end{matrix} \right\}_i = - \left\{ \begin{matrix} \mathbf{G} \\ f \end{matrix} \right\}_i \tag{3.3}$$

Here we have introduced the short hand notation for the directional derivatives

$$\nabla_v \mathbf{G}\,\Delta\mathbf{v} = \tfrac{d}{d\epsilon}[\mathbf{G}(\mathbf{v}_i + \epsilon\Delta\mathbf{v}, \lambda_i)]\,\big|_{\epsilon=0} \qquad \nabla_\lambda \mathbf{G}\,\Delta\lambda = \tfrac{d}{d\epsilon}[\mathbf{G}(\mathbf{v}_i, \lambda_i + \epsilon\Delta\lambda)]\,\big|_{\epsilon=0}$$
$$\nabla_v \mathbf{f}^T\,\Delta\mathbf{v} = \tfrac{d}{d\epsilon}[f(\mathbf{v}_i + \epsilon\Delta\mathbf{v}, \lambda_i)]\,\big|_{\epsilon=0} \qquad \nabla_\lambda f\,\Delta\lambda = \tfrac{d}{d\epsilon}[f(\mathbf{v}_i, \lambda_i + \epsilon\Delta\lambda)]\,\big|_{\epsilon=0} \tag{3.4}$$

$\nabla_v \mathbf{G}$ is in engineering literature known as tangent stiffness matrix $\mathbf{K}_T$ whereas $\nabla_\lambda \mathbf{G}$ is a given load pattern defined by $\mathbf{P} = -\nabla_\lambda \mathbf{G}$, see also equation (3.1).

The, in general, non-symmetric system of equations (3.3) is commonly solved by a partitioning method also called bordering algorithm leading to two equations for $\Delta\mathbf{v}_i$ and $\Delta\lambda_i$, From $(3.3)_1$ the two partial solutions

$$\Delta\mathbf{v}_{Pi} = (\mathbf{K}_{Ti})^{-1}\,\mathbf{P}\,, \quad \Delta\mathbf{v}_{Gi} = (\mathbf{K}_{Ti})^{-1}\,\mathbf{G}_i, \tag{3.5}$$

are available which can be combined to the total displacement increment

$$\Delta\mathbf{v}_i = \Delta\lambda_i \Delta\mathbf{v}_{Pi} + \Delta\mathbf{v}_{Gi} \tag{3.6}$$

The unknown increment of the load factor λ has to be calculated from the second eqation $(3.3)_2$

$$\Delta\lambda_i = -\frac{f_i + \mathbf{f}_{vi}^T \Delta\mathbf{v}_{Gi}}{f_{\lambda i} + \mathbf{f}_{vi}^T \Delta\mathbf{v}_{Pi}}\,. \tag{3.7}$$

Such methods are called arc–length procedures or in a more general context path–following or continuation methods. A large number of variants exists which only differ in the formulation of the constraint equation, see e.g. *Riks* (1972), *Keller* (1977), *Rheinboldt* (1981), *Crisfield* (1981), *Ramm* (1981). An overview may be found in e.g. *Riks* (1984). Necessary for quadratical convergence behaviour is the consistent linearization of the constraint equation., see e.g. *Schweizerhof, Wriggers* (1986).

The associated algorithm for continuation methods is given in BOX 2.

This algorithm differs only from the incremental Newton procedure stated in section 2.4 in BOX 1 by the fact that within the continuation procedure the constraint equation $f(\mathbf{v}, \lambda) = 0$ yields the load parameter for a chosen arc–length ds.

In the following BOX 3 some selected constraint equations are depicted.

A detailed description of these constraint equations may be found in the given references or in e.g. *Wagner, Wriggers* (1988).

1.	calculate predictor	$\mathbf{K}_T\,\Delta\,\mathbf{v}_{P0} = \mathbf{P}$
2.	calculate load increment	$\Delta\lambda_0 = \dfrac{ds}{\sqrt{(\Delta\mathbf{v}_{P0})^T\Delta\mathbf{v}_{P0}}}$
3.	loop over all iterations i	$\mathbf{K}_T\,\Delta\,\mathbf{v}_{Pi} = \mathbf{P}$ $\mathbf{K}_T\,\Delta\,\mathbf{v}_{Gi} = -\mathbf{G}(\mathbf{v}_i, \lambda_i)$
4.	calculate increments	$\Delta\lambda_i = -\dfrac{f_i + \mathbf{f}_{vi}^T\Delta\mathbf{v}_{Gi}}{f_{\lambda i} + \mathbf{f}_{vi}^T\Delta\mathbf{v}_{Pi}}$ $\Delta\mathbf{v}_i = \Delta\lambda_i\Delta\mathbf{v}_{Pi} + \Delta\mathbf{v}_{Gi}$
5.	Update	$\lambda_{i+1} = \lambda_i + \Delta\lambda_i\,,\quad \mathbf{v}_{i+1} = \mathbf{v}_i + \Delta\mathbf{v}_i$
6.	convergence criterion	if $\lVert\mathbf{G}(\mathbf{v}_{i+1}, \lambda_{i+1})\rVert \le$ TOL Stop else go to 3

BOX 2 : General algorithm for arc–length procedures

3.3 Calculation of Stability points

The continuation algorithm based on (3.3) can be used to compute arbitrary load deflections paths. However singular points cannot be detected using this incremental solution scheme. Also arc–length procedures do not provide information about the type of stability — limit or bifurcation points. Thus other algorithms are needed for the investigation of the global stability behaviour of a given structure.

Different methods for the computation of singular points (e.g. limit and bifurction points) in the nonlinear load–deflection path exist. These methods are based on so called extended systems. For this purpose a constraint which characterizes the singular point is added to the system of equations (3.1). As a result we obtain the load parameter, the displacement field and sometimes also the eigenvector at the singular point. Furthermore, Newton–type algorithms can be constructed.

Before proceeding further we summarize below standard characterizations of stability points, and review several proposed formulations of extended systems. The emphasis

name	Fig.	constraint equation
load control	3.1a	$f = \lambda - c$
displacement control *Batoz, Dhatt* (1979)	3.1b	$f = v_a - c$
arc–length method *Riks* (1972)	3.2	$f = (\mathbf{v}_m - \bar{\mathbf{v}})^T(\mathbf{v} - \mathbf{v}_m) + (\lambda_m - \bar{\lambda})\,(\lambda - \lambda_m)$
arc–length method *Crisfield* (1981)	3.3	$f = \sqrt{(\mathbf{v} - \bar{\mathbf{v}})^T(\mathbf{v} - \bar{\mathbf{v}}) + (\lambda - \bar{\lambda})^2} - ds$

BOX 3 : Examples for constraint equations

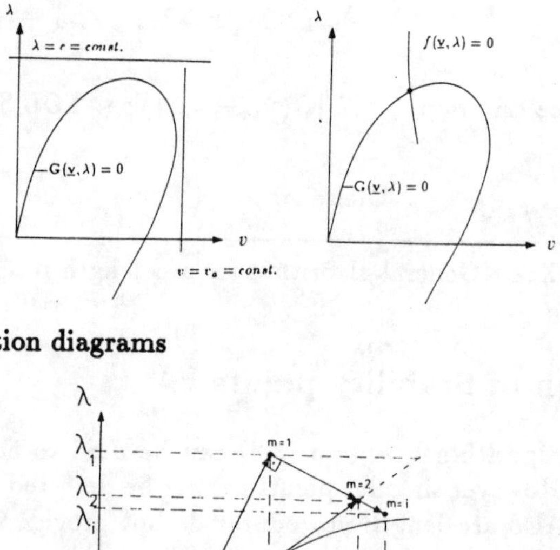

Fig. 3.1: Load-deflection diagrams

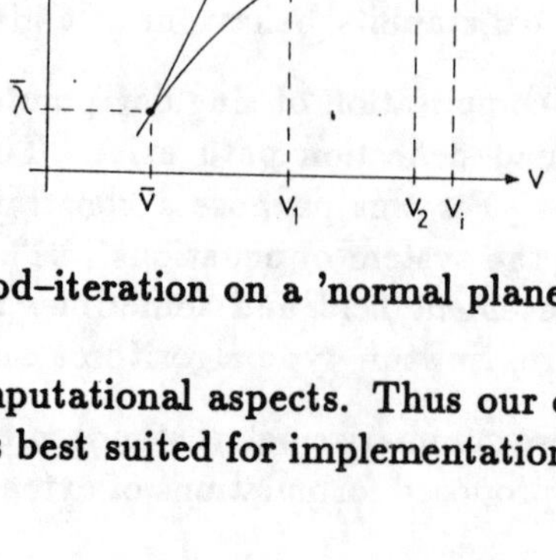

Fig. 3.2: Arc–length method–iteration on a 'normal plane'

in our discussion is on computational aspects. Thus our objective is to identify the class of algorithms which is best suited for implementation within the context of the finite element method.

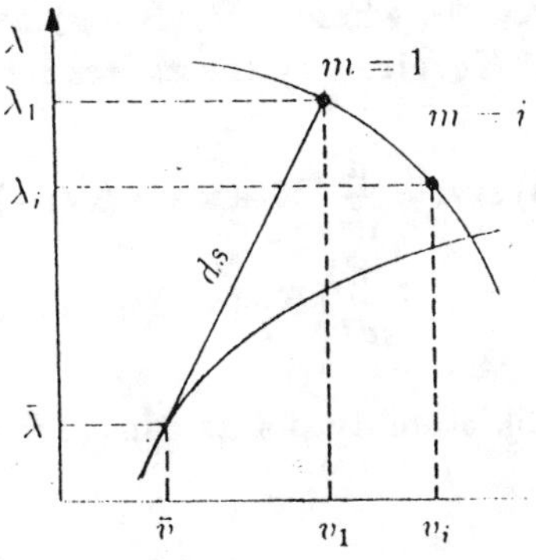

Fig. 3.3: Arc–length method–iteration on a 'sphere'

3.3.1 Criteria for singular points

A point $\mathbf{v} \in \mathbb{R}^N$ is a *stability point* (bifurcation or limit point) of the system (3.1) if the following conditions hold

$$\boxed{\begin{aligned}\mathbf{K}_T(\mathbf{v})\,\boldsymbol{\varphi} &= \mathbf{0},\\ \det \mathbf{K}_T(\mathbf{v}) &= 0\,.\end{aligned}} \tag{3.8}$$

If, on the other hand, $\det \mathbf{K}_T(\mathbf{v}) > 0$ the equilibrium path is stable at $\mathbf{v}$, whereas for $\det \mathbf{K}_T(\mathbf{v}) < 0$ the equilibrium path is is unstable.

Once a stability point is detected by means of conditions (3.8), it is relatively simple to differentiate limit from bifurcation points. Following *Spence, Jepson* [25], we define the standard criteria

Limit (turning) points:

$$\boldsymbol{\varphi}^T\,\mathbf{P} \neq 0\,, \tag{3.9$_1$}$$

Bifurcation points:

$$\boldsymbol{\varphi}^T\,\mathbf{P} = 0\,. \tag{3.9$_2$}$$

Further insight into the nature of the equilibrium path at the stability point can be gained by means of the following classification. Define the quantities

$$\begin{aligned} a &= \boldsymbol{\varphi}^T\,\nabla_v\,(\mathbf{K}_T\,\boldsymbol{\varphi})\,\boldsymbol{\varphi},\\ b &= \boldsymbol{\varphi}^T\,\nabla_\lambda\,(\mathbf{K}_T\,\boldsymbol{\varphi}) + \boldsymbol{\varphi}^T\,\nabla_v\,(\mathbf{K}_T\,\boldsymbol{\varphi})\,\bar{\mathbf{v}},\\ c &= \boldsymbol{\varphi}^T\,\nabla_\lambda\,\mathbf{P} + 2\boldsymbol{\varphi}^T\,\nabla_\lambda\,(\mathbf{K}_T\,\bar{\mathbf{v}}) + \boldsymbol{\varphi}^T\,\nabla_v\,(\mathbf{K}_T\,\bar{\mathbf{v}})\,\bar{\mathbf{v}},\\ d &= b^2 - a\,c. \end{aligned} \tag{3.10}$$

Here, $\bar{\mathbf{v}}$ is defined by $\bar{\mathbf{v}} = \mathbf{K}_T^{-1}\,\mathbf{P}$, whereas $\nabla_v(\mathbf{K}_T\,\boldsymbol{\varphi})\,\Delta\mathbf{v}$ and $\nabla_\lambda(\mathbf{K}_T\,\boldsymbol{\varphi})\,\Delta\lambda$ denote the directional derivatives of $\mathbf{K}_T$ given by the expressions

$$
\begin{aligned}
\nabla_v(\mathbf{K}_T\,\boldsymbol{\varphi})\,\Delta\mathbf{v} &:= \frac{d}{d\epsilon}\left[\mathbf{K}_T(\mathbf{v}+\epsilon\,\Delta\mathbf{v},\,\lambda)\,\boldsymbol{\varphi}\right]\Big|_{\epsilon=0}, \\
\nabla_\lambda(\mathbf{K}_T\,\boldsymbol{\varphi})\,\Delta\lambda &:= \frac{d}{d\epsilon}\left[\mathbf{K}_T(\mathbf{v},\,\lambda+\epsilon\,\Delta\lambda)\,\boldsymbol{\varphi}\right]\Big|_{\epsilon=0}.
\end{aligned}
\tag{3.11}
$$

Following *Spence, Jepson* [25], stability points can be further classified according to the criteria

Limit points:

$$
\begin{cases}
\boldsymbol{\varphi}^T\,\mathbf{P} \neq 0 \text{ and } a \neq 0: & \text{simple quadratic turning point,} \\
\boldsymbol{\varphi}^T\,\mathbf{P} \neq \text{ and } a = 0: & \text{simple cubic turning point.}
\end{cases}
\tag{3.12$_1$}
$$

Bifurcation points:

$$
\begin{cases}
\boldsymbol{\varphi}^T\,\mathbf{P} = 0 \text{ and } a \neq 0 \text{ and } d > 0: & \text{simple trans-critical bifurcation point,} \\
\boldsymbol{\varphi}^T\,\mathbf{P} = 0 \text{ and } a = 0 \text{ and } b \neq 0: & \text{simple pitch-fork bifurcation point,} \\
\boldsymbol{\varphi}^T\,\mathbf{P} = 0 \text{ and } d < 0: & \text{simple isola formation point.}
\end{cases}
\tag{3.12$_2$}
$$

These criteria can be readily implemented if information about the directional derivative of $\mathbf{K}_T$ according to equation (3.11) is available. As shown below, the numerical implementation of extended systems for the determination of stability points requires knowledge of these derivatives. Consequently, as a by product of the algorithmic implementation, one is able to determine the type of stability at a point $(\mathbf{v}_0, \lambda_0)$, with no additional effort.

3.3.2 Formulation of extended systems

The addition of the constraint $det\,\mathbf{K}_T$ to the nonlinear equlilbrium equation (3.1) leads to the following extended system

$$
\bar{\mathbf{G}}(\mathbf{v},\lambda) = \left\{ \begin{array}{c} \mathbf{G}(\mathbf{v},\lambda) \\ det\,\mathbf{K}_T(\mathbf{v},\lambda) \end{array} \right\} = 0\,. \tag{3.13}
$$

This system was considered by *Abott* (1978) for the indirect computation of nontrivial bifurcation points. Strategies based on equation (3.13) have been widely used in engineering applications, see e.g. *Brendel, Ramm* (1982). However no method for a direct computation of singular points has been developed so far in the engineering literature using equation (3.13). In most applications the constraint $det\,\mathbf{K}_T = 0$ is basis for an accompanying investigation to detect stability points or for extrapolation techinques. Often (3.13) is combined with arc–length methods which yields the system

$$\tilde{\mathbf{G}}(\mathbf{v},\lambda) = \left\{ \begin{array}{c} \mathbf{G}(\mathbf{v},\lambda) \\ f(\mathbf{v},\lambda) \\ det\, \mathbf{K}_T(\mathbf{v},\lambda) \end{array} \right\} = 0 \tag{3.14}$$

where $f(\mathbf{v},\lambda)$ is a constraint equation associated with path–following procedures. Here $det\,\mathbf{K}_T$ checks whether a singular point has been passed during path following. With this knowledge we can construct a bi–section method in which the direction and arc–length of the continuation method is changed according to the sign of $det\,\mathbf{K}_T$. The associated algorithm, see BOX 4, is very simple to implement, however, it only converges linearly to the singular point.

* Follow nonlinear solution path until $det\,\mathbf{K}_T$ changes sign:

1. Do $i = 1, 2, ...until\ convergence$
2. Change load direction and arc–length: $ds_i = ds_{i-1}/2$
3. Compute new solution with algorithm in BOX 2

   3.1 Convergence check: IF $det\,\mathbf{K}_T < EPS \Longrightarrow STOP$

   3.2 IF sign of $det\,\mathbf{K}_T$ changes again go to 2.)

   ELSE change arc–length: $ds_{i+1} = ds_i/2$ and go to 3.)

BOX 4 : Bi–section algorithm to compute singular points

The algorithm has been applied successfully to elasticity problems. For the stability analysis of specimen which exhibit inelastic material behavior we have to change the algorithm in BOX 4 such that unloading is excluded. This can be simply done by avoiding the change in load direction using a restart from the previous step once a sign change of $det\,\mathbf{K}_T$ has been detected, see also section 3.4.

In the following we like to construct an algorithm which exhibits faster convergence than the bi–section scheme. In section 2.4 we have seen that Newton's method is advantageous. However a full Newton scheme for equations (3.13) is not obvious. It can be formally obtained by linearization of the constraint equation $det\,\mathbf{K}_T = 0$ according to the well-known relation

$$\frac{d}{d\epsilon}\left[\det\,\mathbf{K}_T(\mathbf{v}+\epsilon\,\Delta\,\mathbf{v})\right]\Big|_{\epsilon=0} = \det\,\mathbf{K}_T\,\mathrm{tr}\,(\mathbf{K}_T^{-1}\,\nabla_v\mathbf{K}_T\,\Delta\,\mathbf{v}).$$

Note that evaluation of this expression requires the computation of the directional derivative of the tangent matrix $\nabla_v \mathbf{K}_T$. The Newton scheme for equation (3.13) is now given by

$$\begin{aligned} \mathbf{K}_{T\,i}\, \Delta \mathbf{v}_{i+1} - \mathbf{P}\, \Delta \lambda_{i+1} &= -\mathbf{G}\,(\mathbf{v}_i, \lambda_i), \\ \operatorname{tr}\,(\mathbf{K}_T^{-1}\, \nabla_v \mathbf{K}_T\, \Delta\, \mathbf{v}_{i+1}) &= -1, \\ \mathbf{v}_{i+1} &= \mathbf{v}_i + \Delta \mathbf{v}_{i+1}, \\ \lambda_{i+1} &= \lambda_i + \Delta \lambda_{i+1}. \end{aligned} \tag{3.15}$$

Unfortunately, this algorithm is not well suited for numerical implementation in a finite element method since the trace operation of two matrices can not be implemented in an element-by-element fashion. In fact, the formulation of equation $(3.15)_2$ requires the full assembly of the derivative $\nabla_v \mathbf{K}_T$ of the tangent matrix; a rather costly operation.

An alternative approach to construct an extended system for the computation of stability points bases on the eigenvalue problem

$$(\mathbf{K}_T - \omega\, \mathbf{1})\, \boldsymbol{\varphi} = \mathbf{0} \tag{3.16}$$

which is equivalent to the constraint $det\, \mathbf{K}_T = 0$. Since at singular points the eigenvalue ω of $\mathbf{K}_T$ is zero we obtain $\mathbf{K}_T \boldsymbol{\varphi} = \mathbf{0}$. Thus an extended system associated with this constraint may be stated as follows

$$\widehat{\mathbf{G}}(\mathbf{v}, \lambda, \boldsymbol{\varphi}) = \left\{ \begin{array}{c} \mathbf{G}(\mathbf{v}, \lambda) \\ \mathbf{K}_T(\mathbf{v}, \lambda)\, \boldsymbol{\varphi} \\ l\,(\boldsymbol{\varphi}) \end{array} \right\} = \mathbf{0}\,. \tag{3.17}$$

This extended system has been used e.g. in *Werner, Spence* (1984) and *Weinitschke* (1985) for the calculation of limit and simple bifurcation points, for a combination with finite element formulations, see *Wriggers, Wagner, Miehe* (1988).

In equation (3.17) $l\,(\boldsymbol{\varphi})$ denotes some normalizing functional which has to be included to ensure a non zero eigenvector $\boldsymbol{\varphi} \in \mathbb{R}^N$ at the singular point. The following expressions

$$\begin{aligned} l\,(\boldsymbol{\varphi}) &= \|\, \boldsymbol{\varphi} \,\| - 1 = 0\,, \\ l\,(\boldsymbol{\varphi}) &= \mathbf{e}_i^T\, \boldsymbol{\varphi} - \hat{\varphi}_0 = 0\,, \\ l\,(\boldsymbol{\varphi}) &= \mathbf{P}^T\, \boldsymbol{\varphi} - 1 = 0\,, \end{aligned} \tag{3.18}$$

where $\mathbf{e}_i$ is a vector containing zeros and a value 1 at the a chosen place associated with i. The constant scalar φ_0 is computed prior to the first iteration from the starting vector $\boldsymbol{\varphi}_0$ for the eigenvector: $\hat{\varphi}_0 = \frac{\mathbf{e}_i^T \boldsymbol{\varphi}_0}{\|\boldsymbol{\varphi}_0\|}$. Note, that constraint $(3.17)_3$ leads only to limit points which comes from $(3.9)_1$. In our computations we have used the first equation which seems to be more robust. Extended systems like (3.17) are associated with $2N+1$ unknowns. Thus the numerical effort to solve such systems is considerably increased. However this apparent disadvantage can be circumvented by using a bordering algorithm for the solution of the extended set of equations.

It should be noted that equation (3.17) automatically yields the eigenvector $\boldsymbol{\varphi}$ associated with the limit or bifurcation point which may be applied for branch–switching.

3.3.3 Newton scheme for the solution of the extended system

The incremental solution scheme using Newtons method leads to the formulation

$$\begin{aligned} \widehat{\mathbf{K}}_{T\,i}\,\Delta\mathbf{w}_i &= -\widehat{\mathbf{G}}(\mathbf{w}_i) \\ \mathbf{w}_{i+1} &= \mathbf{w}_i + \Delta\mathbf{w}_i \end{aligned} \tag{3.19}$$

where

$$\widehat{\mathbf{K}}_{T\,i}\Delta\mathbf{w} = \frac{d}{d\epsilon}[\widehat{\mathbf{G}}(\mathbf{w}_i + \epsilon\Delta\mathbf{w})]\bigg|_{\epsilon=0} \quad \text{with} \quad \mathbf{w} = \begin{Bmatrix} \mathbf{v} \\ \lambda \\ \boldsymbol{\varphi} \end{Bmatrix}.$$

Here the vector $\mathbf{w}$ contains $2N+1$ unknown variables as defined above.

In detail equation (3.19) looks as follows

$$\begin{bmatrix} \mathbf{K}_T & \mathbf{0} & -\mathbf{P} \\ \nabla_v(\mathbf{K}_T\boldsymbol{\varphi}) & \mathbf{K}_T & \nabla_\lambda(\mathbf{K}_T\boldsymbol{\varphi}) \\ \mathbf{0}^T & \frac{\boldsymbol{\varphi}^T}{\|\boldsymbol{\varphi}\|} & 0 \end{bmatrix} \begin{Bmatrix} \Delta\mathbf{v} \\ \Delta\boldsymbol{\varphi} \\ \Delta\lambda \end{Bmatrix} = -\begin{Bmatrix} \mathbf{G}(\mathbf{v},\lambda) \\ \mathbf{K}_T(\mathbf{v},\lambda)\boldsymbol{\varphi} \\ \|\boldsymbol{\varphi}\| - 1 \end{Bmatrix} \tag{3.20}$$

where $\nabla_v(\mathbf{K}_T\boldsymbol{\varphi})$ and $\nabla_\lambda(\mathbf{K}_T\boldsymbol{\varphi})$ have already be defined in (3.11). The computation of these derivatives within the finite element setting is discussed in the next section.

Using a block elimination technique a solution scheme of (3.20) can be developed which only needs one factorization of the tangent stifness $\mathbf{K}_T$. BOX 5 below shows the algorithm for one iteration step.

The derivation of the tangent matrix can be calculated using standard linearization procedures. To obtain the closed form expression of $\nabla_v(\mathbf{K}_T\boldsymbol{\varphi})$ we assume the simplest case of the St. Venant material (1.24) which has a constant constitutive tensor. Thus we do not have to consider the differentiation of the constitutive tensor as we would have for (1.31).

Before computing the directional derivative of $\mathbf{K}_T\boldsymbol{\varphi}$ we note that the resulting matrix is never needed in the algorithm described in BOX 5. Thus we compute only the vector

$$\begin{aligned} \mathbf{h} &= \nabla_v(\mathbf{K}_T\boldsymbol{\varphi})\Delta\mathbf{v} \\ &= \bigcup_{e=1}^{n_e}\int_{\Omega_e}\{\mathbf{B}_{nl}^T(\Delta\mathbf{v}_e)\mathbf{D}\mathbf{B}(\mathbf{v}_e) + \mathbf{B}^T(\mathbf{v}_e)\mathbf{D}\mathbf{B}_{nl}(\Delta\mathbf{v}_e) + \mathbf{G}^T\Delta\bar{\mathbf{S}}\mathbf{G}\}\boldsymbol{\varphi}_e\,d\Omega. \end{aligned} \tag{3.21}$$

1. **Solve**

$$\mathbf{K}_T\,\Delta\,\mathbf{v}_P = \mathbf{P}\,, \quad \mathbf{K}_T\,\Delta\,\mathbf{v}_G = -\mathbf{G}\,.$$

2. **Compute within a loop over all finite elements**

$$\begin{aligned} \mathbf{h}_1 &= \nabla_v\,(\mathbf{K}_T\,\boldsymbol{\varphi})\,\Delta\mathbf{v}_P + \nabla_\lambda\,(\mathbf{K}_T\,\boldsymbol{\varphi})\,, \\ \mathbf{h}_2 &= \nabla_v\,(\mathbf{K}_T\,\boldsymbol{\varphi})\,\Delta\mathbf{v}_G \end{aligned}\,.$$

3. **Solve**

$$\mathbf{K}_T\,\Delta\boldsymbol{\varphi}_1 = -\mathbf{h}_1\,, \quad \mathbf{K}_T\,\Delta\boldsymbol{\varphi}_2 = -\mathbf{h}_2\,.$$

4. **Compute new increments**

$$\Delta\lambda = \frac{-\boldsymbol{\varphi}^T\,\Delta\boldsymbol{\varphi}_2 + \|\boldsymbol{\varphi}\|}{\boldsymbol{\varphi}^T\,\Delta\boldsymbol{\varphi}_1}\,,$$

$$\Delta\mathbf{v} = \Delta\lambda\,\Delta\mathbf{v}_P + \Delta\mathbf{v}_G\,.$$

5. **Update displacements, eigenvector and load parameter**

$$\lambda = \lambda + \Delta\lambda\,, \quad \mathbf{v} = \mathbf{v} + \Delta\mathbf{v} \quad \text{and} \quad \boldsymbol{\varphi} = \Delta\lambda\,\Delta\boldsymbol{\varphi}_1 + \Delta\boldsymbol{\varphi}_2\,.$$

BOX 5: Algorithm for the direct computation of singular points

where $\Delta\bar{\mathbf{S}}$ contains the incremental stresses given by $\Delta\mathbf{S} = \mathbf{D}\,\mathbf{B}(\mathbf{v}_e)\,\Delta\mathbf{v}_e$. A detailed description can be found in *Wriggers, Wagner, Miehe* (1988).

For the solution of extended system (3.20) we need the vectors $\mathbf{h}_1$ and $\mathbf{h}_2$. Equation (3.21) is the basis for the calculation of these vectors. With $\Delta\mathbf{v}_e = \Delta\mathbf{v}_{Pe}$ we obtain the first part of $\mathbf{h}_1$ while setting $\Delta\mathbf{v}_e = \Delta\mathbf{v}_{Ge}$ is necessary for the calculation of $\mathbf{h}_2$. The term $\nabla_\lambda\,(\mathbf{K}_T\,\boldsymbol{\varphi})$ in $\mathbf{h}_1$ vanishes for dead loading. Thus the complete calculation of the vectors $\mathbf{h}_1$ and $\mathbf{h}_2$ for the extended system is given.

Within the derivation of (3.21) we have assumed St. Venant material which has the constant constitutive tensor $\mathbb{C}_L$. In case of nonlinear hyperelastic response functions like (1.31) the dependence of $\mathbb{C}$ on the deformation has to be considered when linearizing the tangent matrix (3.20). Analogous situations with very complex expressions of the tangent operator appear when a geometrically exact beam or shell model is investigated. Thus for a general purpose implementation a numerical differentiation scheme is advantageous to compute h. The following procedure is due to *Wriggers, Simo* (1990).

To construct a numerical approximation that preserves the quadratic rate of convergence of Newton's method, we first recall that the vector $\mathbf{K}_T\boldsymbol{\varphi}$ can be represented as the directional derivative of the residual $\mathbf{G}$ in the direction of $\boldsymbol{\varphi}$, according to the expression

$$\mathbf{K}_T\boldsymbol{\varphi} = \nabla_v \mathbf{G}(\mathbf{v},\lambda)\boldsymbol{\varphi} = \frac{d}{d\epsilon}\mathbf{G}(\mathbf{v}+\epsilon\boldsymbol{\varphi},\lambda)\Big|_{\epsilon=0}. \tag{3.22}$$

By exploiting the symmetry of the second derivative of $\mathbf{G}$, we can express the directional derivative of $\mathbf{K}_T\boldsymbol{\varphi}$ in the direction of $\Delta\mathbf{v}$ in the following equivalent form

$$\begin{aligned}\nabla_v[\mathbf{K}_T\boldsymbol{\varphi}]\Delta\mathbf{v} &= \nabla_v[\nabla_v\mathbf{G}(\mathbf{v},\lambda)\boldsymbol{\varphi}]\Delta\mathbf{v}\\ &= \nabla_v[\nabla_v\mathbf{G}(\mathbf{v},\lambda)\Delta\mathbf{v}]\boldsymbol{\varphi}.\end{aligned} \tag{3.23}$$

With this expression in hand, we show next that the vectors $\mathbf{h}_1$ and $\mathbf{h}_2$ in the algorithm described in BOX 5 can be computed with only one additional evaluation of the tangent stiffness. To this end, we use the definition of directional derivate to obtain the following expression which replaces (3.11):

$$\nabla_v[\mathbf{K}_T\boldsymbol{\varphi}]\Delta\mathbf{v} = \frac{d}{d\epsilon}[\mathbf{K}_T(\mathbf{v}+\epsilon\boldsymbol{\varphi})]\Delta\mathbf{v}\Big|_{\epsilon=0}. \tag{3.24}$$

This equation is now recast in an alternative format amenable to numerical approximation. We set

$$\nabla_v[\mathbf{K}_T\boldsymbol{\varphi}]\Delta\mathbf{v} = \lim_{\epsilon=0}\frac{1}{\epsilon}[\mathbf{K}_T(\mathbf{v}+\epsilon\boldsymbol{\varphi})\Delta\mathbf{v} - \mathbf{K}_T(\mathbf{v})\Delta\mathbf{v}]. \tag{3.25}$$

By suitable selecting the parameter ϵ we obtain

$$\boxed{\nabla_v[\mathbf{K}_T\boldsymbol{\varphi}]\Delta\mathbf{v} \approx \frac{1}{\epsilon}[\mathbf{K}_T(\mathbf{v}+\epsilon\boldsymbol{\varphi})\Delta\mathbf{v} - \mathbf{K}_T(\mathbf{v})\Delta\mathbf{v}].} \tag{3.26}$$

The application of this approximation to the directional derivative to the algorithm devised in section 3.33 leads to following algorithmic expressions for the vectors $\mathbf{h}_1$ and $\mathbf{h}_2$

$$\boxed{\begin{aligned}\mathbf{h}_1 &\approx -\frac{1}{\epsilon}[\mathbf{K}_T(\mathbf{v}+\epsilon\boldsymbol{\varphi})\Delta\mathbf{v}_P - \mathbf{P}],\\ \mathbf{h}_2 &\approx -\frac{1}{\epsilon}[\mathbf{K}_T(\mathbf{v}+\epsilon\boldsymbol{\varphi})\Delta\mathbf{v}_G + \mathbf{G}].\end{aligned}} \tag{3.27}$$

Observe that at this stage $\mathbf{P}$ and $\mathbf{G}$ are already computed. Consequently, the computation of $\mathbf{h}_i$, $(i = 1, 2)$, through the expressions in (3.27) involves only one additional evaluation of the stiffness matrix; i.e., $\mathbf{K}_T(\mathbf{v}+\epsilon\boldsymbol{\varphi})$.

REMARK:

i. The function evaluation of $\mathbf{K}_T$ has to performed only once. Furthermore the matrix multiplications in (3.27) can be carried out at the element level, and the matrix $\mathbf{K}_T(\mathbf{v} + \epsilon\boldsymbol{\varphi})$ need not be assembled at the global level. The standard assembly procedure has to be performed only on the vectors $\mathbf{h}_i$.

ii. A suitable selection of the parameter ϵ in (3.27) is crucial to the success of the method. This choice depends on the vectors $\mathbf{v}$ and $\boldsymbol{\varphi}$ and on the computer precision. An estimate for ϵ may be found in *Dennis, Schnabel* (1983), and leads to the following formula

$$\epsilon = \max_{1<k<n} v_k \, \eta_{TOL} \, . \tag{3.28}$$

Here v_k denotes the kth–component of $\mathbf{v} \in \mathbb{R}^N$, and η_{TOL} is a machine precision constant.

iii. It is clear that during the direct computation of the singular point the tangent matrix $\mathbf{K}_T$ will become singular. Thus the solution of the equations in BOX 5 is no longer possible and the algorithm can break down. To circumvent this problem several techniques exist. In *Wriggers, Simo* (1990) a penalty method which has also been described in a different context by *Felippa* (1987) was applied successfully. However also the deflation technique, well known in the mathematical literature, can be applied to overcome this problem.

3.4 Overall algorithm for nonlinear stability problems

Since extended systems provide a tool to calculate stability points directly it is advantageous to combine them with path–following procedures. Within such an algorithm an arc–length method is used to follow a solution branch. Near a stability point the extended system can be applied to compute this point in an efficient way. For this purpose an estimation is needed how to switch between the two solution procedures.

Basically two heuristic approaches can be applied. The first one is based on an inspection of the determinant of the tangent stiffness matrix $det\,\mathbf{K}_T$. Turning points of $det\,\mathbf{K}_T$ are used to switch from the arc–length method to the extended system. Thus large steps compared to an arc–length method are possible. Within the second approach the solution path is followed until the number of negative diagonal elements of $\mathbf{K}_T$ changes. This indicates that a singular point has been passed and the extended system is used to compute the stability point. For elasto–plastic problems only the latter approach has been applied together with a one sided bi–section procedure which is described in BOX 6 below.

The overall algorithm for the computation of nonlinear load deflection curves and their singular points is given in BOX 6 below.

Starting from a stability point the arc-length procedure can be used to follow the postcritical branch. In case of bifurcation points a branch-switching procedure is necessary, see e.g. *Wagner, Wriggers* (1988), if a secondary branch should be investigated. For this purpose the eigenvector $\boldsymbol{\varphi}$ associated with the buckling load is essential. This vector is computed automatically by the extended system. Thus in case of simple bifurcation points an additional solution of an eigenvalue problem is not required.

5 Bifurcation point, employ algorithm for branch-switching:

5.1 Save information for restart of problem

5.2 If calculation of primary branch is desired: Go to 1

5.3 Else, if extended system is not used solve eigenvalue problem

$$(\mathbf{K}_T - \omega\,\mathbf{1})\,\boldsymbol{\varphi} = \mathbf{0}$$

5.4 Add scaled eigenvector $\boldsymbol{\varphi}$ to displacement vector $\bar{\mathbf{v}}$:

$$\mathbf{v} = \bar{\mathbf{v}} + \xi\,\frac{\boldsymbol{\varphi}}{\|\boldsymbol{\varphi}\|}$$

5.5 Calculate an equlibrium state on secondary branch using arc-length method

$$\mathbf{G}(\mathbf{v},\lambda) = \mathbf{R}(\mathbf{v}) - \lambda\mathbf{P} = \mathbf{0}$$

5.6 If secondary branch is reached: Go to 6

5.7 Else: choose new scaling factor ξ: Go to 5.4

6 Calculation of equlibrium states on secondary branch using arc–length method

7 If another branch has to be calculated: Restart and Go to 5.2

BOX 7: Calculation of secondary branches

A mathematically sound branch-switching technique follows from the bifurcation

1 Calculation of equilibrium states with arc-length method along solution path, see algorithm in BOX 2

$$\mathbf{G}(\mathbf{v},\lambda) = \mathbf{R}(\mathbf{v}) - \lambda \mathbf{P} = \mathbf{0}$$

2 Observation of diagonal elements or determinant of $\mathbf{K}_T$:

2.1 If determinant passes a maximum or minimum: Go to 3

2.2 Ifthe number of negative diagonal elements changes: stability point has been passed: Go to 3

3 Compute singular point directly

3.1 In case of an elastic problem use extended system:

3.1.1 Calculate starting vector for φ via first step of an inverse iteration

$$\varphi_0 = \mathbf{K}_T^{-1}\,\mathbf{1}$$

3.1.2 Employ Newton's scheme to solve extended system using algorithm given in BOX 4

$$\widehat{\mathbf{G}}(\mathbf{v},\lambda,\varphi) = \mathbf{0}$$

3.2 In case of an elasto–plastic problem use one sided bi–section procedure:

3.2.1 restart at last converged solution

3.2.2 choose half arc–length

3.2.3 calculate next load step

3.2.4 if $det\,\mathbf{K}_T$ is less than TOL: Go to 4

3.2.5 if number of neg. diagonal elements is constant: Go to 3.2.2

3.2.5 else: Go to 3.2.1

4 Criterion for type of stability point

$$\varphi^T\mathbf{P} = \begin{cases} \neq 0 \ldots \text{ limit} & \text{point: Go to 1} \\ = 0 \ldots \text{ bifurcation} & \text{point: Go to 5} \end{cases}$$

BOX 6: Calculation of Stability Points

equation

$$a\alpha^2 + 2b\alpha\lambda + c\lambda^2 = 0 \tag{3.29}$$

which is derived in *Decker, Keller* (1980). The constants a, b, c are defined by (3.10). Thus with the knowledge of (3.21) or (3.27) this constants can be computed. Then the above quadratic equation can be solved for a given α which can be interpreted as the scaling factor ξ in BOX 7 since $\Delta \mathbf{v} = \Delta\lambda\,\bar{\mathbf{v}} + \alpha\,\boldsymbol{\varphi}$. For a further discussion we refer to *Decker, Keller* (1980). However, it should be remarked that for the practically important case of symmetric bifurcation the procedure in BOX 7 coincides with the use of the bifurcation equation (3.29).

The computation of limit and bifurcation points of elasto–plastic specimen is in case of associated flow rules often performed by using the linear comparsion solid introduced by *Hill* (1958). For associated J_2 plasticity the bifurcation criterion can be written as

$$I = \frac{1}{2}\int\limits_B \left[\tilde{\dot{\mathbf{S}}}\cdot\tilde{\dot{\mathbf{E}}} + \mathbf{S}\cdot\text{Grad}^T\,\tilde{\dot{\mathbf{u}}}\,\text{Grad}\,\tilde{\dot{\mathbf{u}}}\right] dv = 0 \tag{3.30}$$

with $\tilde{(..)} = (..)_2 - (..)_1$ where the indices 1 and 2 denote to neighbouring solutions. The linear comparison solid is introduced now by using a constant incremental constitutive tensor $\mathbb{D}_c$ which leads to the incremental constitutive equation

$$\tilde{\dot{\mathbf{S}}} = \mathbb{D}_c\,[\tilde{\dot{\mathbf{E}}}], \qquad \text{with}\mathbb{D}_c = \mathbb{C} - \alpha\,\mathbf{N}\otimes\mathbf{N}\,. \tag{3.31}$$

Thus unloading is explicitly prevented. The combination of (3.30) and (3.31) yields

$$F = \frac{1}{2}\int\limits_B \left[\tilde{\dot{\mathbf{E}}}\cdot\mathbb{D}_c\,[\tilde{\dot{\mathbf{E}}}] + \mathbf{S}\cdot\text{Grad}^T\,\tilde{\dot{\mathbf{u}}}\,\text{Grad}\,\tilde{\dot{\mathbf{u}}}\right] dv = 0\,. \tag{3.32}$$

This formulation provides a bound for the bifurcation load, see e.g. *Needleman* (1972).

A modern and efficient finite element method for large strain elasto–plasticity is given in *Simo* (1988). In contrast to most of the existing derivations of elasto–plasticity at finite strains this formulation provides an explicit expression for the exact algorithmic tangent operator $\mathbf{K}_T$. The resulting structure of $\mathbf{K}_T$ is equal to (2.26) when the algorithmic consistent constitutive matrix for elasto–plasticity is introduced. We observe a similar structure as in equation (3.32). It follows that in case we exclude unloading from our computation of singular points of elasto–plastic problems the tangent stiffness matrix indicates singularities like in the elastic range. The difference to the linear comparison solid is that here the true tangent stiffness at the singular point is used. Thus there is no need for the introduction of a linear comparison solid.

3.5 Numerical examples

In this section we present a selected number of numerical simulations that demonstrate the applicability of the algorithms and methods developed in the previous section to nonlinear engineering problems.

3.5.1 Necking bifurcation of an elasto–plastic bar

This example is concerned with the classical problem of necking bifurcation from a homogeneous state in the single tension test of an elastoplastic material at finite strains. The first numerical investigations of this problem using finite element methods go back to the work of *Neddleman* (1972) who makes use of the linear comparison solid of *Hill* (1958) to compute the bifurcation load. Here we use the global algorithm in its elasto-plastic extension to compute the limit and bifurcation points on the load-deflection curve, see BOX 7.

The material properties of the J_2–plasticity model, along with a summary of existing experimental results, are reported e.g. in *Hallquist* (1979). Due to symmetry considerations, only one quater of the axi-symmetric specimen is modeled with the finite element mesh shown in Figure 3.4

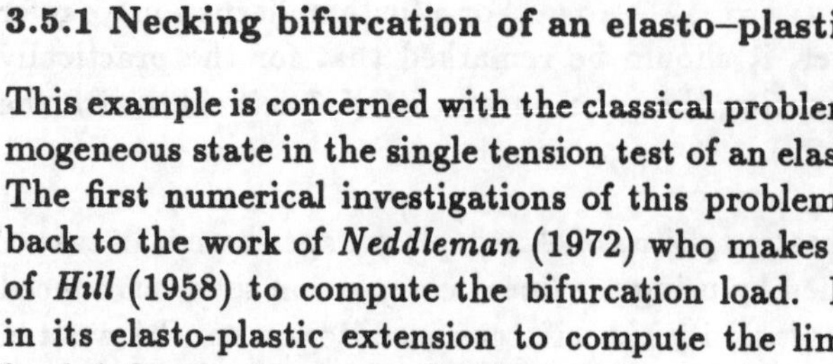

Data: Nonlinear hardening law

$$\kappa(e_p) = Y_0 + H' e_p + (Y_\infty - Y_0)(1 - e^{-\alpha e_p})$$

Constitutive constants	Value
Shear modulus μ	80.1938 GPa
Bulk modulus K	164.206 GPa
Initial flow stress Y_0	0.45 GPa
Residual flow stress Y_∞	0.715 GPa
Linear hardening coefficient H'	0.12924 GPa
Saturation exponent α	16.93
Isotropic hardening	0.0

$L = 53.334$ mm

$R = 6.413$ mm

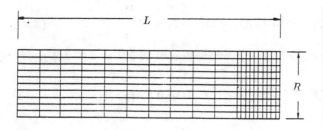

Figure 3.4 FE-Mesh and data of the specimen

The values of the material properties are also shown in Figure 3.4. Since the chosen boundary conditions lead to a homogeneous stress and strain field in the specimen necking can only occur as a result of a bifurcation analysis. The arc–length method was used to follow all solution branches in this problem.

The primary solution path is shown in Figure 3.5, and exhibits one limit point and two bifucation points. In this example, we compute the first bifurcation point via the method described in section 3.4. Then we calculated the eigenvector associated with the zero eigenvalue and performed branch–switching which lead to the secondary path in Figure 3.5. The other load deflection curves in Figure 3.5 show the solution for a specimen with an imperfection introduced via a slight reduction in the

radius of the original specimen at the middle cross section. Note the imperfection sensitiveness exhibited by the analysis. Necking develops earlier with a larger imperfection. However after necking is initiated, the behaviour of the three specimens with different geometries is essentially the same. Figure 3.6 shows the deformed shapes for different load levels.

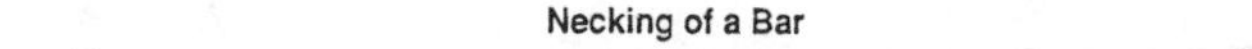

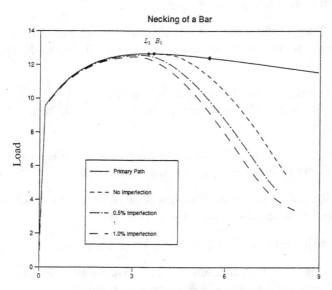

Figure 3.5. Load-deflection curve

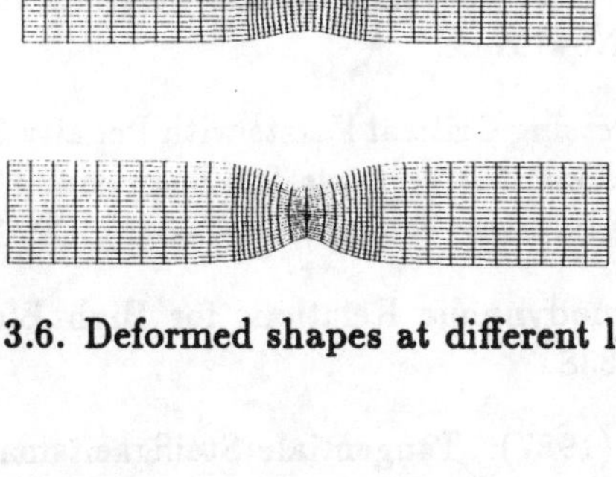

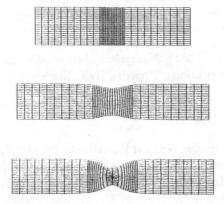

Figure 3.6. Deformed shapes at different load levels

References

Abbot, J. P. (1978): An Efficient Algorithm for the Determination of certain Bifurcation Points, J. Comp. Appl. Math., 4, 19–27.

Batoz, I.L., Dhatt, G. (1979): Incremental Displacement Algorithms for Non-linear Problems, Int. J. Num. Meth. Engng., 14, 1262–1267.

Belytschko, T.; Tsay, C. S. (1983): A Stabilization Procedure for the Quadrilateral Plate Element with One–Point Quadrature. Int. J. Num. Meth. Engng., 19, 405–419.

Bergan, P. G., Horrigmoe, G., Krakeland, B., Soreide, T. H. (1978): Solution Techniques for Non-linear Finite Element Problems, Int. J. Num. Meth. Engng., 12, 1677–1696.

Brendel, B., Ramm, E. (1982): Nichtlineare Stabilitätsuntersuchungen mit der Methode der Finiten Elemente, Ing. Archiv 51, 337–362.

Bushnell, D., (1985): Computerized Buckling Analysis of Shells, Mechanics of Elastic Stability, Vol. 9, Martinus Nijhoff Publishers, Boston.

Ciarlet, P.G. (1988): Mathematical Elasticity, Volume I, North-Holland, Amsterdam.

Crisfield, M. A. (1981): A Fast Incremental/Iterative Solution Procedure that Handles Snap Through, Computers & Structures, 13, 55–62.

Decker, D. W., Keller, H.B. (1980): Solution Branching — A Constructive Technique, in P.J. Holmes, ed., New Approaches to Nonlinear Problems in Dynamics (SIAM, Philadelphia) 53–69.

Dennis J.E., Schnabel R.B., (1983): Numerical Methods for Unconstrained Optimization and Nonlinear Equations, Prentice Hall, Englewood Cliffs, New Jersey.

Doyle, T.C., Ericksen, J.L., (1956): Nonlinear Elasticity, in Advances in Appl. Mech. IV, Academic Press, New York.

Felippa, C. A. (1987): Traversing Critical Points with Penalty Springs, in: Transient/Dynamic Analysis and Constitutive Laws for Engineering Materials, ed. Pande, Middleton (Nijhoff, Dordrecht, Boston, Lancaster), C2/1–C2/8.

Flory, P.J., (1961): Thermodynamic Relations for High Elastic Materials, Trans. Faraday Soc.,57, 829-838.

Gruttmann, F.; Stein, E. (1987): Tangentiale Steifigkeitsmatrizen bei Anwendung von Projektionsverfahren in der Elastoplastizitätstheorie. Ing. Archiv, 58, 15–24.

Hallquist, J. (1979): NIKE2D: An Implicit, Finite Deformation, Finite Element Code for Analysing the Static and Dynamic Response of Two–Dimensional Solids, University of California, LLNL, Rep. UCRL-52678.

Hill, R. (1958): A General Theory of Uniqueness and Stability in Elastic–Plastic Solids, J. Mech. Phys. Solids, 6, 236–249.

Hughes, T. R. J. (1987): The Finite Element Method, Prentice Hall, Englewood Cliffs, New Jersey.

Hughes, T. J. R., Pister, K. S. (1978): Consistent Linearization in Mechanics of Solids and Structures, Computers & Structures, 8, 391–397.

Johnson, C. (1987): Numerical Solution of Partial Differential Equations by the Finite Element Method, Cambridge, U.K.

Keller, H.B. (1977): Numerical Solution of Bifurcation and Nonlinear Eigenvalue Problems. In: Rabinowitz, P. (ed.): Application of Bifurcation Theory, New York: Academic Press 1977, 359–384.

Luenberger, D.G. (1984), Linear and Nonlinear Programming, Addision-Wesley, Masachusetts.

Malkus, D.S.; Hughes, T.R.J. (1983): Mixed Finite Element Methods - Reduced and Selective Integration Techniques: a Unification of Concepts. Comp. Meth. Appl. Mech. Engng., 15, 63–81.

Malvern, L.E. (1969): Introduction to the Mechanics of a Continuous Medium, Prentice-Hall, Englewood Cliffs.

Marsden, J.E.; Hughes, T.R.J. (1983): Mathematical Foundations of Elasticity, Englewood Cliffs: Prentice-Hall.

Mittelmann, H.-D.; Weber, H. (1980): Numerical Methods for Bifurcation Problems - a Survey and Classification. In: Mittelmann, Weber (ed.): Bifurcation Problems and their Numerical Solution, ISNM 54, (Basel, Boston, Stuttgart: Birkhäuser), 1–45.

Moore, G., Spence, A. (1980): The Calculation of Turning Points of Nonlinear Equations, SIAM J. Num. Anal., 17, 567–576.

Needleman, A. (1972): A Numerical Study of Necking in Circular Cylindrical Bars, J. Mech. Phys. Solids, 20, 111–127.

Ogden, R.W. (1972): Large Deformation Isotropic Elasticity on the Correlation of Theory and Experiment for Incompressible Rubberlike Solids, Proc. R. Soc., London, A(326), 565–584.

Ogden, R. W.(1984): Non-linear elastic deformations, Ellis Horwood, Chichester.

Ramm, E., (1981): Strategies for Tracing the Nonlinear Response Near Limit Points, in: Nonlinear Finite Element Analysis in Structural Mechanics, ed. Wunderlich, Stein Bathe, Springer, Berlin-Heidelberg-New-York.

Rheinboldt, W.C. (1981): Numerical Analysis of Continuation Methods for Nonlinear Structural Problems. Comp. & Struct., 13, 103–113.

Riks, E. (1972): The Application of Newtons Method to the Problem of Elastic Stability. J. Appl. Mech., 39, 1060–1066.

Riks, E., (1984), Some computational aspects of stability analysis of nonlinear structures, Comp. Meth. Appl. Mech. Engng., 47, 219–260.

Schweizerhof, K., Wriggers, P. (1986): Consistent Linearizations for Path Following Methods in Nonlinear FE Analysis, Computer Methods in Applied Mechanics and Engineering, 59, 261–279.

Seydel, R. (1979): Numerical Computation of Branch Points in Nonlinear Equations, Numer. Math., 33, 339–352.

Simo, J.C.; Taylor, R.L. (1985): Consistent Tangent Operators for Rate-independent Elastoplasticity. Comp. Meth. Appl. Mech. Engng., 48, 101–118.

Simo, J.C. (1988): A Framework for Finite Strain Elastoplasticity Based on the Multiplicative Decomposition and Hyperelastic Relations. Part I: Formulation, Comp. Meth. Appl. Mech. Engng., 66, 199–219.

Simo, J.C. (1988): A Framework for Finite Strain Elastoplasticity Based on the Multiplicative Decomposition and Hyperelastic Relations. Part II: Computational Aspects, Comp. Meth. Appl. Mech. Engng., 67, 1–31.

Simo, J. C., Wriggers, P., Schweizerhof, K.H., Taylor, R.L. (1986): Post-buckling Analysis involving Inelasticity and Unilateral Constraints, Int. J. Num. Meth. Engng., 23, 779–800.

Simo, J. C., K. D. Hjelmstad and R. L. Taylor (1984): Numerical Formulations for Finite Deformation Problems of Beams Accounting for the Effect of Transverse Shear, Comp. Meth. Appl. Mech. Engng., 42, 301–330.

Simo, J. C.; Marsden, J. E. (1984): On the Rotated Stress Tensor and the Material Version of the Doyle Erickson Formula, Arch. Rat. Mech. Anal., 86, 213–231.

Simo, J.C., & T.J.R. Hughes (1992): Plasticity and Viscoplasticity: Numerical Analysis and Computational Aspects, Springer–Verlag, Berlin.

Spence, A., Jepson A. D. (1985): Folds in the Solution of Two Parameter Systems and their Calculation. Part I, SIAM J. Numer. Anal., 22, 347–368.

Stein, E.; Wagner, W.; Wriggers, P. (1988): Concepts of Modeling and Discretization of Elastic Shells for Nonlinear Finite Element Analysis. In: Whiteman, J. (ed.): The Mathematics of Finite Elements and Applications VI, Proceedings of MAFELAP 87, London: Academic Press.

Taylor, R.L., (1985), Solution of linear equations by a profile solver, Engineering Computations, 2, 334-350.

Truesdell, C.; Noll, W. (1965): The Non-Linear Field Theories of Mechanics. In: Flügge, S. (ed.): Handbuch der Physik, Band III/3, Berlin: Springer.

Wagner, W., Wriggers, P. (1988): A Simple Method for the Calculation of Postcritical Branches, Engineering Computations, 5, 103–109.

Weber, H. (1981): On the Numerical Approximation of Secondary Bifurcation Problems, in: Numerical Solution of Nonlinear Equations, ed. Allgower, Glashoff, Peitgen, Lecture Notes in Mathematics 878, Springer, Berlin, 407–425.

Werner, B., Spence, A. (1984): The Computation of Symmetry-Breaking Bifurcation Points, SIAM J. Num. Anal., 21, 388–399.

Werner, B. (1984): Regular Systems for Bifurcation Points with Underlying Symmetries, in: Bifurcation Problems and their Numerical Solution , ISNM 70, ed. Mittelmann, Weber, Birkhäuser, Basel, Boston, Stuttgart, 562–574.

Weinitschke, H. J. (1985): On the Calculation of Limit and Bifurcation Points in stability Problems of Elastic Shells, Int. J. Solids Structures, 21, 79–95.

Wriggers, P. (1988): Konsistente Linearisierungen in der Kontinuumsmechanik und ihre Anwendung auf die Finite–Element–Methode, Forschungs– und Seminarberichte aus dem Bereich der Mechanik der Universität Hannover, Bericht F 88/4, Hannover.

Wriggers, P.; Wagner, W.; Miehe, C. (1988): A Quadratically Convergent Procedure for the Calculation of Stability Points in Finite Element Analysis. Comp. Meth. Appl. Mech. Engng., 70, 329–347.

Wriggers, P., Gruttmann, F. (1989): Large deformations of thin shells: Theory and Finite–Element–Discretization, in Analytical and Computational Models of Shells, ed. A. K. Noor, T. Belytschko, J.C. Simo, ASME, CED-Vol. 3, 135–159.

Wriggers, P., Simo, J. C. (1990): A General Procedure for the Direct Computation of Turning and Bifurcation Points, Int. J. Num. Meth. Engng., 30, 155–176.

Zienkiewicz, O. C., Taylor, R. L. (1989): The Finite Element Method, Vol. 1, Mc Graw–Hill,London.